11th Simulation Workshop (SW23)

Southampton, United Kingdom
27 – 29 March 2023

Editors:

Christine Currie
Luke Rhodes-Leader

ISBN: 978-1-7138-7095-1

Printed from e-media with permission by:

Curran Associates, Inc.
57 Morehouse Lane
Red Hook, NY 12571

CURRAN ASSOCIATES INC.
proceedings
.com

Some format issues inherent in the e-media version may also appear in this print version.

The Operational Research Society
Seymour House
12 Edward Street
Birmingham
B1 2RX
United Kingdom

Phone: +44 (0)121 233 9300

www.theorsociety.com

Additional copies of this publication are available from:

Curran Associates, Inc.
57 Morehouse Lane
Red Hook, NY 12571 USA
Phone: 845-758-0400
Fax: 845-758-2633
Email: curran@proceedings.com
Web: www.proceedings.com

WELCOME TO THE 11TH OPERATIONAL RESEARCH SOCIETY SIMULATION WORKSHOP SW23

This year we meet in person for the first time since the COVID-19 pandemic, and we are very excited at the prospect! It seems a long time since SW18, the last Simulation Workshop that took place 'in real life'. The location, the National Oceanography Centre in Southampton (NOCS), is a stunning waterside venue adjacent to the Ocean Village marina and the cruise liner terminal, yet within easy walking distance of the city centre.

As ever, the programme includes a wide range of exciting talks, bringing you up to date with recent developments in simulation research as well as providing introductory tutorials in a variety of topics. This year we have introduced a new 'case study' category, describing real-world applications of simulation; the idea was to encourage practitioners to present without the need to write an academic-style paper. There will be many opportunities for informal networking and a range of social activities, including a Gala Dinner, and a drinks reception kindly sponsored by Taylor & Francis, publishers of the Journal of Simulation (JoS). This will include a brief introduction to the journal from two of its editors, Christine Currie and Navonil Mustafee, who will be able to advise prospective authors on writing good papers. There will be a special issue of JoS dedicated to work presented at SW23 (further information to follow).

The scientific programme includes 20 contributed papers, 14 posters, 5 case studies, 2 keynote lectures and 5 tutorials. The posters will be on display throughout the workshop, and there will be a 'lightning' poster session in which the presenters will each be given three minutes to describe their poster. There will be awards for the best poster and the best paper.

The keynote speakers are Professor Raghu Pasupathy from Purdue University, Illinois, and Professor Susan Howick from the University of Strathclyde. Raghu will present a new method for quantifying the error in simulation output based on the idea of batching, traditionally regarded as a variance reduction technique. Susan will discuss mixing OR methods to tackle complex problems, and will reflect on over 25 years of combining system dynamics with other methods. Susan's talk exemplifies how the different simulation communities, formerly separated along methodological lines, are growing ever closer. Both JoS and the Simulation Workshops have included system dynamics and agent-based simulation for several years, and the 2023 UK System Dynamics conference (which is also taking place at NOCS) was purposely timed for the two days immediately after SW23 so that people could attend both meetings.

On the final day of the workshop, Tillal Eldabi will chair a panel session in honour of Ray Paul, who sadly died last year after a long battle with Parkinson's Disease. Ray, a world leader in simulation theory and practice, was the driving force behind the creation of the Simulation Workshops and JoS. The panellists will each take a field in which Ray worked and will argue that his ideas were decades ahead of their time. This year's tutorials will cover conceptual modelling, hybrid simulation, business process modelling, metamodeling, and system dynamics for supply chains. Another of our sponsors, Simul8 Corp, are running a workshop session related to the cost of living crisis.

In conclusion, we would like to thank everyone who contributed to SW23; authors, speakers, sponsors, session chairs and reviewers. We are particularly grateful to those people whose hard work over the past year has made the workshop possible: Caitlin Griffin and her colleagues in the OR Society Events Team, the staff at NOCS, Christine Currie and Luke Rhodes-Leader (Programme Chairs), Laura Boyle (Poster Chair) and Martin Kunc (Publicity and Sponsorship Chair). We also thank Stewart Robinson and Simon Taylor for their guidance and advice. We are really looking forward to seeing old friends again and making new ones. Enjoy the conference!

Sally Brailsford and Siôn Cave

Conference Chairs

CONTENTS

Contents

Simulation Methodology

Simulation Modelling of Patient Flow

Simulation and Sustainability

Applications of Simulation Optimisation

Simulation Modelling of COVID-19

Contents

Proceedings of the Operational Research Society Simulation Workshop 2023 (SW23)
C. Currie and L. Rhodes-Leader, eds.
https://doi.org/10.36819/SW23.001

SIMULATION-ESTIMATOR INFERENCE

Raghu Pasupathy

Department of Statistics
Purdue University
West Lafayette, IN 47906, USA

ABSTRACT

Simulation, and now digital twins, excel in estimating *parameters* associated with complex and time-dependent stochastic processes, e.g., large airport, highway, and warehouse operations. Within such contexts, we consider *statistical inference*, whereby one seeks to quantify the error in the obtained simulation estimate. Historically, statistical inference in simulation has been considered challenging because parameters needing estimation are often complicated, and simulation output often autocorrelated and non-normal. However, we argue that the remarkably simple idea of *batching* can be used as an "omnibus" inference device in simulation settings. Batching for inference works in three simple steps: (i) divide simulation output data into overlapping batches; (ii) construct parameter estimates from each batch; and (iii) use the batch estimates, after accounting for dependence, to perform statistical inference. As we describe in this paper, the resulting procedures are usually trivial to implement in software, and are provably correct and efficient. Batching ideas originated in the 1950s, and have enjoyed steady development in the simulation community since the 1970s mostly within the problem of variance estimation. Our thesis is that batching ideas have much wider utility, and that batching should be considered alongside other resampling ideas such as *the bootstrap* or *the jackknife* in modern statistics.

1 INTRODUCTION

We describe batching methods for statistical inference in the context of simulation, whereby a simulationist, having estimated a desired unknown parameter θ associated with a simulated experiment, seeks some statistical measure of the extent of the error in the obtained estimator. We start with two concrete examples aimed at ensuring the reader's firm grasp of the notion of a simulation parameter θ, the simulation output dataset $(Y_1, Y_2, \ldots, Y_n)$, and the estimator $\hat{\theta}_n$.

1.1 Example I: Time-Dependent Inventory Levels in a Supply Chain

As a first example, consider the global supply chain introduced in the tutorial by (Ingalls 2014), where the simulationist wishes to analyze the delivery of computing servers produced in Europe to the Asia-Pacific region, with the specific intention of evaluating whether it may be wise to move production to Singapore. Due to the complexity and scale of such a supply chain, it is easy to see why a simulation model would be helpful in answering many narrow questions, e.g., effect on inventory, effect on on-time delivery, effect on costs and revenue, which together will be pertinent to the broader question of whether a move to Singapore is warranted.

Let's consider one such narrow question, that of *time-dependent inventory level*, that is, inventory as a function of time, at a specified location and observed over a horizon $[0,T]$ of interest. The simulationist executes n runs of the simulation, producing time-dependent inventory level $Y_j(t), t \in [0,T]$ during the j-th run. Importantly, notice that the j-th "observation" denoted $Y_j := Y_j(t), t \in [0,T]$ is a function of time, or a *random function*. Suppose the simulationist is especially interested in analyzing low inventory levels and so chooses the parameter θ to be the 20-th percentile inventory level as a function of time, that is, $\theta := \theta(t), t \in [0,T]$, where $\theta(t)$ is the 20-th percentile inventory at time t. Recognize again that the parameter θ is a fixed, unknown function in time. Recalling the "dataset" $(Y_1, Y_2, \ldots, Y_n)$ generated

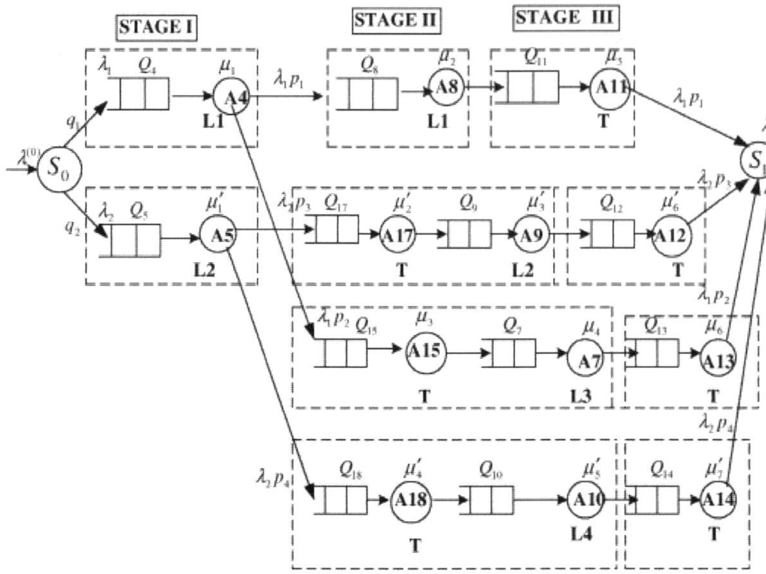

Figure 1: A queueing network model of a supply chain.

by n runs of the simulation, an estimator $\hat{\theta}_n := \hat{\theta}_n(t), t \in [0,T]$ of θ can then be constructed as:

$$\hat{\theta}_n(t) := \min\left\{ y : \frac{1}{n}\sum_{j=1}^{n} \mathbb{I}(Y_j(t) \leq y) \geq 0.2) \right\}, \quad t \in [0,T], \tag{1}$$

where the notation $\mathbb{I}(A) = 1$ if A is true and 0 otherwise.

The simulationist may have chosen a different parameter of interest, e.g., the mean vector of inventory levels at d specific locations $\ell_1, \ell_2, \ldots, \ell_d$ in the supply chain, at the fixed time instant T. In this case, denoting $\pi_{j,T}, j = 1,2,\ldots,d$ as the inventory level distribution at time T in location j, and denoting $Y_{i,j}(T), j = 1,2,\ldots,d; i = 1,2,\ldots,n$ as the i-th observed inventory level in location j at time T, we can write:

$$\theta := \left(\int_\ell \ell\, \pi_{1,T}(\mathrm{d}\ell), \int_\ell \ell\, \pi_{2,T}(\mathrm{d}\ell), \ldots, \int_\ell \ell\, \pi_{d,T}(\mathrm{d}\ell) \right).$$

In such a case, the estimator $\hat{\theta}_n$ becomes

$$\hat{\theta}_n := \left(\frac{1}{n}\sum_{i=1}^{n} Y_{i,1}(T), \frac{1}{n}\sum_{i=1}^{n} Y_{i,2}(T), \ldots, \frac{1}{n}\sum_{i=1}^{n} Y_{i,d}(T) \right).$$

Statistical inference on $\hat{\theta}_n$ refers to the collective activity of constructing various statistical measures on the nature of the error $\hat{\theta}_n - \theta$. The premise of statistical inference is that decision-making can be more informed if some measure of statistical error accompanies the estimator $\hat{\theta}_n$. ■

1.2 Example II: Steady State Congestion Pricing Using Variable Tolls

Variable toll pricing has become a popular method to manage traffic on highways, by shifting purely discretionary traffic to off-peak hours or other roadways. Accordingly, a question of immense interest involves identifying the relationship between the toll price and the resulting congestion levels at steady state, toward better congestion pricing policies.

Let's introduce notation to make this question more precise. Suppose $p = (p_1, p_2, \ldots, p_d), p_i \in [0,M]$ represents the prevailing toll price for d vehicle classes, and $\theta = \{(\theta_1(p), \theta_2(p), \ldots, \theta_d(p)), p \in [0,M]^d\}$ the corresponding expected steady state waiting time at the tolls for each of the d classes. Given the complicated relationship between the expected wait time and the toll price, a simulation (whose mechanics are not relevant for our purposes) is used to estimate the parameter θ. Suppose the simulation yields the

Figure 2: Variable toll pricing is a popular way of managing travel demand on congested roads. A manager might simulate numerous variable toll scenarios before deciding on a pricing policy.

output $(Y_1, Y_2, \ldots, Y_n)$, where $Y_i = (Y_{i1}(p), Y_{i2}(p), \ldots, Y_{id}(p)), p \in [0, M]^d$ represents the i-th realization of the wait time vector, that is, the vector wait times corresponding to the i-th vechicle in each of the d classes, with p held fixed. It is important to observe that each output observation Y_i in this example is a *random function or surface* of the toll price. A useful thought experiment that clarifies the nature of Y_i: fix and hold all "random elements" of the simulation while varying the toll price p to form a time series of observations, each of which is a function of the price p.

Since the parameter $\theta := \{(\theta_1(p), \theta_2(p), \ldots, \theta_d(p)), p \in [0, M]^d\}$ is the expected steady state wait time, we have

$$\theta(p) = \left(\int_{\mathcal{X}} x_i \pi_p(\mathrm{d}x), \int_{\mathcal{X}} x_2 \pi_p(\mathrm{d}x), \ldots, \int_{\mathcal{X}} x_d \pi_p(\mathrm{d}x) \right); \quad x = (x_1, x_2, \ldots, x_d), p \in [0, M]^d \tag{2}$$

where π_p is the (unknown) steady state distribution of the wait times of the d vehicle classes when the toll price is fixed at p. The expression in (2) suggests

$$\hat{\theta}_n(p) = \left(\frac{1}{n} \sum_{i=1}^{n} Y_{i1}(p), \frac{1}{n} \sum_{i=1}^{n} Y_{i2}(p), \ldots, \frac{1}{n} \sum_{i=1}^{n} Y_{id}(p) \right), \quad p \in [0, M]^d.$$

Alternatively, suppose the simulationist is interested in setting the tolls $p = (p_1, p_2, \ldots, p_d)$ so that the expected wait times for the d classes matches target wait times $\gamma_1, \gamma_2, \ldots, \gamma_d$, respectively. Then the parameter θ is the solution (in p) to the following nonlinear system of equations:

$$\begin{aligned} \int_{\mathcal{X}} x_1 \pi_p(\mathrm{d}x) &= \gamma_1; \\ \int_{\mathcal{X}} x_2 \pi_p(\mathrm{d}x) &= \gamma_2; \\ &\vdots \\ \int_{\mathcal{X}} x_d \pi_p(\mathrm{d}x) &= \gamma_d; \end{aligned} \tag{3}$$

Of course the solution θ to (3) is unknown, but can be estimated as $\hat{\theta}_n$ by solving the corresponding system constructed using the data generated by simulation, that is, by solving the system:

$$\begin{aligned} \frac{1}{n} \sum_{i=1}^{n} Y_{i1}(p) &= \gamma_1; \\ \frac{1}{n} \sum_{i=1}^{n} Y_{i2}(p) &= \gamma_2; \\ &\vdots \\ \frac{1}{n} \sum_{i=1}^{n} Y_{id}(p) &= \gamma_d; \end{aligned} \tag{4}$$

(There are existence and uniqueness issues pertaining to the solution of (4) but we omit discussion about such details here.) As in Example I, the simulation-estimator inference question here is whether anything can be inferred about the nature of the error $\hat{\theta}_n - \theta$. ■

2 WHAT IS SIMULATION-ESTIMATOR INFERENCE?

Suppose that in the service of estimating a desired parameter θ as in Example I or Example II, a simulation produces "output data" $(Y_1, Y_2, \ldots, Y_n)$ which are "processed" in an appropriate manner to yield an estimator $\hat{\theta}_n$. We are agnostic to whether $(Y_1, Y_2, \ldots, Y_n)$ are obtained using independent replications of the simulation or from a single long run. And, we have deliberately left the nature of each Y_j vague; for instance, in both examples discussed in the previous section, $Y_j, j = 1, 2, \ldots, n$ are "function-valued" random variables. By *simulation-estimator inference* we broadly mean inferring properties about the distribution of the estimator error $\hat{\theta}_n - \theta$, for example, by estimating summary measures, constructing confidence sets, or by statistically testing interesting hypotheses.

To make the notion of inference more concrete, let's discuss five of the most common tasks categorized as statistical inference, assuming $\theta, \hat{\theta}_n \in \mathbb{R}$. A standard reference on statistical inference is (Casella and Berger 2002).

(a) (Bias Estimation) Estimate the bias, denoted $\text{bias}(\hat{\theta}_n, \theta)$, of $\hat{\theta}_n$ with respect to θ:

$$\text{bias}(\hat{\theta}_n, \theta) := \mathbb{E}\left[\hat{\theta}_n\right] - \theta. \tag{5}$$

The bias measures the accuracy of the estimator $\hat{\theta}_n$ as the mean deviation of $\hat{\theta}_n$ from the true parameter θ. A useful but informal interpretation is that $\text{bias}(\hat{\theta}_n, \theta)$ represents the error in the estimator averaged over a large number of practitioners performing the same experiment. Thus, the estimator $\hat{\theta}_n$ is unbiased (or has zero bias) if $\hat{\theta}_n$ is "correct on average," with the average taken over an ensemble of identical experiments. Since parameters that are not population means are rarely unbiased, estimates of bias can be quite useful. However, it is not immediately clear how one might estimate $\text{bias}(\hat{\theta}_n, \theta)$, since the expression in (5) contains the unknown parameter θ. As we shall see, batching provides a possible remedy.

(b) (Standard Error Estimation) Estimate the *standard error* $\text{se}(\hat{\theta}_n)$ of the estimator $\hat{\theta}_n$, defined as

$$\text{se}(\hat{\theta}_n) := \sqrt{\text{Var}(\hat{\theta}_n)} = \sqrt{\mathbb{E}\left[\left(\hat{\theta}_n - \mathbb{E}\left[\hat{\theta}_n\right]\right)^2\right]}. \tag{6}$$

The standard error is a measure of estimator precision calculated as the square root of the mean (over repeated experiments) fluctuations of the estimator. Importantly, the standard error is generally considered of limited value because $\text{se}(\hat{\theta}_n)$ measures deviations of the estimator $\hat{\theta}_n$ about *its own mean* (and not about the true parameter θ). The standard error and bias together provide a fuller view of the quality of $\hat{\theta}_n$ as an estimator of θ. In particular, the *mean squared error*

$$\text{mse}(\hat{\theta}_n, \theta) := \mathbb{E}[(\hat{\theta}_n - \theta)^2] = (\text{se}(\hat{\theta}_n))^2 + (\text{bias}(\hat{\theta}_n, \theta))^2.$$

Like bias, it is not immediately clear how $\text{se}(\hat{\theta}_n)$ can be estimated since its expression contains the unknown quantity $\mathbb{E}\left[\hat{\theta}_n\right]$. As we shall see, batching again provides a possible remedy.

(c) (Quantile Estimation) Estimate the u-quantile, $u \in (0,1)$ of the squared error $\varepsilon_n^2 := \left(\hat{\theta}_n - \theta\right)^2$ defined as

$$Q_{\varepsilon_n^2}(u) := F_{\varepsilon_n^2}^{-1}(u) := \inf\{x : F_{\varepsilon_n^2}(x) \geq u\}, \tag{7}$$

where $F_{\varepsilon_n^2}$ is the cumulative distribution function of ε_n^2. Compared to the mean squared error, the quantiles of ε_n^2 provide a more complete picture of the distribution of the squared error. (The 0.5-quantile is called the median.) As with the bias and the standard error, we will see that batching facilitates estimating the quantiles of the squared error.

(d) (Confidence Set Construction) Construct an asymptotically valid $(1-\alpha), \alpha \in (0,1)$ *confidence set* $\mathscr{C}_n$, that is, form a set $\mathscr{C}_n$ from available simulation output data $(Y_1, Y_2, \ldots, Y_n)$ such that

$$\lim_{n \to \infty} P(\theta \in \mathscr{C}_n) = 1 - \alpha. \tag{8}$$

The statement in (8) guarantees, loosely speaking, that when n is large, $(1-\alpha)$-fraction of constructed sets $\mathscr{C}_n$ over repeated experiments will contain the true parameter θ.

(e) (Hypothesis Testing) In the simplest setting, suppose $\mathscr{P}_0$ and $\mathscr{P}_1$ are two competing "policies" with associated effects $\theta(\mathscr{P}_0)$, $\theta(\mathscr{P}_1)$. In example (ii) on tolls, $\mathscr{P}_0$ might represent the existing toll policy and $\mathscr{P}_1$ a proposed toll policy, with $\theta(\mathscr{P}_0), \theta(\mathscr{P}_1)$ the resulting unknown 90-th percentile delay. Then the simulationist would like to test the hypothesis $H_0 : \theta(\mathscr{P}_0) - \theta(\mathscr{P}_1) \leq 0$ against the alternate hypothesis $H_A : \theta(\mathscr{P}_0) - \theta(\mathscr{P}_1) > 0$. Two difficulties arise when performing such a test: first, the unknown parameter $\theta(\cdot)$ is not a population mean; and second, the simulation output data that will be used in constructing estimators $\hat{\theta}_n(\mathscr{P}_0)$ and $\hat{\theta}_n(\mathscr{P}_1)$ will be heavily autocorrelated.

2.1 Terminology

Simulation-estimator inference in this paper pertains to simulation output uncertainty *conditional on the given simulation model*. This is in contrast to another modern popular topic called simulation *input uncertainty* (Henderson 2003, Chick 2001, Cheng and Holland 1997, Barton 2012, Lam 2016), which quantifies the effect of errors in the input distributions that form the primitives to the simulation. In effect, simulation-estimator inference that we consider here provides a sense of how decision-making might be affected due to performing too few simulation runs, whereas input uncertainty deals with the corresponding effects due to a lack of adequate real-world data used when estimating the distributional input to the simulation.

Both output uncertainty and input uncertainty in simulation are subsumed by the recently phrased "umbrella" topic *uncertainty quantification* (Abdar et al. 2021, Najm 2009, Soize 2017) which should be understood loosely as the effort to quantify the effect of all sources of error, e.g., input parameters, structure, logic, and solution, within models that include, but not limited to, simulation. Some examples of models other than simulation are stochastic differential equations (Hoel et al. 1986), neural networks (Bottou et al. 2018), and regression (Wasserman 2004). A number of other terms such as *error estimation* and *reliability estimation* have also come into use recently (Barth 2011) and should be carefully distinguished from what we see as the narrow topic of simulation-estimator inference considered in this paper.

3 BATCHING FOR SIMULATION-ESTIMATOR INFERENCE

Batching is a simple idea that can serve as an omnibus inference device in contexts such as those described in Section 1. The key idea in batching is the ability to construct many *batch estimates* of the parameter θ by partitioning the existing data into individual batches. The batch estimates, along with the "grand estimator" $\hat{\theta}_n$, are then used together with what has been called the *plug-in* principle (Efron and Tibshirani 1994, pp. 35) to accomplish different inference tasks.

Let's introduce notation to make the batching idea precise. Partition the available "data" $Y_1, Y_2, \ldots, Y_n$ into b possibly overlapping batches each of size m as shown in Figure 3. The first of these batches consists of observations $Y_1, Y_2, \ldots, Y_m$, the second consists of observations $Y_{d+1}, Y_{d+2}, \ldots, Y_{d+m}$, and so on, and the last batch consists of observations $Y_{(b-1)d+1}, Y_{(b-1)d+2}, \ldots, Y_n$. It is important that the reader recognize $Y_j, j = 1, 2, \ldots, n$ to be output data obtained from executing the simulation.

The quantity $d \geq 1$ represents the offset between batches, with the choice $d = 1$ corresponding to "fully-overlapping" batches and any choice $d \geq m$ corresponding to "non-overlapping" batches. Notice then that the offset d and the number of batches b are related as $d = \frac{n-m}{b-1}$. Now use the data in batch 1 to construct the parameter estimate $\hat{\theta}_1$, data in batch 2 to construct the parameter estimate $\hat{\theta}_2$, and so on, giving the batch estimates $\hat{\theta}_i, i = 1, 2, \ldots, b$. Also recall the grand estimator $\hat{\theta}_n$ constructed using the entire dataset.

As we express next, each of the inferential tasks in (a)–(e) can now be accomplished rather simply using the batch estimators $\hat{\theta}_i, i = 1, 2, \ldots, b$ and the grand estimator $\hat{\theta}_n$.

(a') Mimicking the expression in (5), we obtain

$$\hat{\text{bias}}(\hat{\theta}_n, \theta) = \left(\frac{1}{b} \sum_{i=1}^{b} \hat{\theta}_i \right) - \hat{\theta}_n \tag{9}$$

as an estimator of $\text{bias}(\hat{\theta}_n, \theta)$.

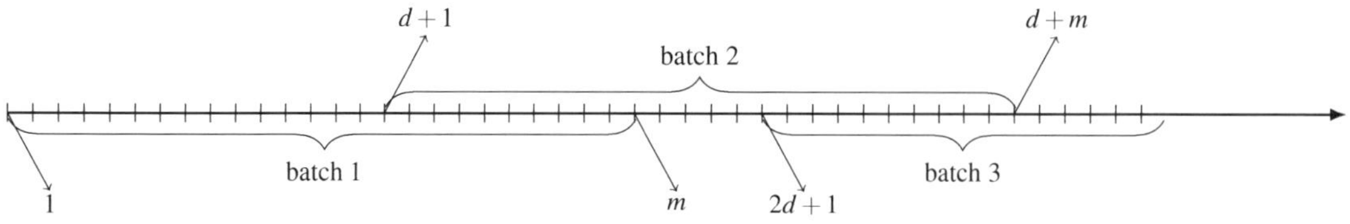

Figure 3: The figure depicts partially overlapping batches. Batch 1 consists of observations $X_j, j = 1, 2, \ldots, m$; batch 2 consists of observations $X_j, j = d+1, d+2, \ldots, d+m$, and so on, with batch i consisting $X_j, j = (i-1)d+1, (i-1)d+2, \ldots, (i-1)d+m$. There are thus $b := d^{-1}(n-m)+1$ batches in total, where n is the size of the dataset.

(b') Mimicking the expression in (6), we obtain

$$\hat{\text{se}}(\hat{\theta}_n) = \sqrt{\frac{1}{1-(m/n)} \times \frac{m}{b}\sum_{i=1}^{b}(\hat{\theta}_i - \hat{\theta}_n)^2} \tag{10}$$

as an estimator of $\text{se}(\hat{\theta}_n)$. The factor $1/(1-m/n)$ that appears inside the radical sign in (10) is a "bias correction" that, under some regularity conditions, ensures that $\hat{\text{se}}(\hat{\theta}_n)^2$ is asymptotically unbiased, that is,

$$\lim_{n\to\infty} \mathbb{E}\left[\left(\hat{\text{se}}(\hat{\theta}_n)\right)^2\right] = \left(\text{se}(\hat{\theta}_n)\right)^2.$$

(c') The u-quantile $Q_{\varepsilon_n^2}(u)$ can be estimated as the inverse of the empirical cdf of the squared error estimates, $(\hat{\theta}_{i,n} - \theta_n)^2, i = 1, 2, \ldots, b$. Defining the empirical cdf

$$F_n(x) = \frac{1}{b}\sum_{i=1}^{b} \mathbb{I}\{\left(\hat{\theta}_{i,n} - \hat{\theta}_n\right)^2 \leq x\},$$

the u-quantile estimate

$$\hat{Q}_{\varepsilon_n^2}(u) := F_n^{-1}(u) := \min\{u : F_n(x) \geq u\}.$$

(d') As in classical statistics, a confidence set $\mathscr{C}_n$ can be constructed as follows:

$$\mathscr{C}_n := \hat{\theta}_n \pm t_{\text{OB-S},\alpha/2}\, \hat{\text{se}}(\hat{\theta}_n). \tag{11}$$

Notice that the expression in (11) uses the "critical value" denoted $t_{\text{OB-S},\alpha/2}$ in lieu of the usual Student's t critical value from classical statistics. Loosely, $t_{\text{OB-S}}$ accounts and corrects for the potential dependence in the simulation output data $(Y_1, Y_2, \ldots, Y_n)$. (Tables and code for the critical value $t_{\text{OB-S},\alpha/2}$ are now available (Su et al. 2023).)

3.1 Why is this Important?

The simplicity of (a')–(d') belie their utility and effectiveness. The first key observation is that even though $\theta, \hat{\theta}_n$ have been assumed to be real-valued in writing (a)–(e) and (a')–(e'), analogous expressions can be written when $\theta, \hat{\theta}_n$ are vector-valued or function-valued as in both examples of Section 1. This breadth of applicability is due to batching facilitating the construction of many batch estimates from the same output data.

The second key observation relates to the correctness of the estimators and procedures in (a')–(d'). Specifically, if the number of batches b is chosen appropriately, and under certain regularity conditions (on the stochastic process generating the output data) that appear to hold widely (Su et al. 2023), the estimators $\hat{\text{bias}}(\hat{\theta}_n, \theta), \hat{\text{se}}(\hat{\theta}_n)^2, \hat{Q}_{\varepsilon_n^2}(u)$ appearing in (a')–(c') converge (in a certain precise sense) to their respective true counterparts $\text{bias}(\hat{\theta}_n, \theta), \text{se}(\hat{\theta}_n)^2, Q_{\varepsilon_n^2}(u)$ as the amount of data $n \to \infty$. Indeed, such convergence seems to occur as rapidly as statistically possible in many contexts. Similarly, the confidence set $\mathscr{C}_n$ constructed in (d') can be shown to be asymptotically valid (Su et al. 2023) as in (8).

Third, it is noteworthy that the estimators and procedures outlined in (a')–(c') are trivial to implement in software. The batching procedure we have outlined works on existing simulation output data and requires no additional simulation execution.

3.2 Modification for Independent Output Data.

The treatment thus far has assumed that the available simulation output $(Y_1, Y_2, \ldots, Y_n)$ might be an initial segment of a time series with the implication that the data might exhibit heavy serial correlation. This is the reason why batches were formed using contiguous observations.

When it is known that the simulation output data $(Y_1, Y_2, \ldots, Y_n)$ are independent and identically distributed, the estimators and procedures in (a')–(d') can be improved by constructing even more batch estimates. Specifically, notice that we can partition the n observations into $\binom{n}{m}$ batches of size m. (This is in contrast to the $(n-m)/d+1$ batches that were constructed in the non-iid context.) Each of these batches can then be used to construct a batch estimate, thus giving $\hat{\theta}_{i,n}, i = 1,2,\ldots,b = \binom{n}{m}$ along with the grand estimator $\hat{\theta}_n$. Inference can then proceed exactly as before.

4 TEN FURTHER POINTS OF DISCUSSION

The following notes are salient and will be discussed further during the oral presentation of this paper.

(a) Batching ideas seem to have originated in the 1950s and later developed within the simulation community in the late 70's as a method to estimate the variance parameter. See (Aktaran-Kalaycı et al. 2009, Su et al. 2023) for an overview and for key references. Batching ideas have steadily matured over the ensuing four decades, finding use in confidence region construction on statistical functionals.

(b) The parameter θ is routinely not real-valued, that is, they can be vector-valued or function valued as in the examples we described. In such cases, the interpretation of simulation output $Y_j, j = 1,2,\ldots,n$, and the ensuing inference, needs to be performed carefully even though the fundamental insights do not change. Theory associated with (a')–(d') can be found in (Su et al. 2023) for the statistical functional context.

(c) Our treatment assumes that the simulation output data $(Y_1, Y_2, \ldots, Y_n)$ are in steady state. This is usually not the case and leads to what has been called the *initial transient problem*. See (Pasupathy and Schmeiser 2010) for an annotated bibliography on this problem. For appropriate inference, ideas from removing the initial transient need to be used in concert with batching, constituting what is an interesting research question.

(d) There exist methods of standard error estimation different from that outlined in (b'). See (Aktaran-Kalaycı et al. 2009, Su et al. 2023).

(e) Batching ideas are closely related to resampling ideas in statistics, e.g., bootstrapping (Efron 1979, Hall 1992, Efron and Tibshirani 1994, Davison and Hinkley 1997, Cheng 2017), sub-sampling (Politis et al. 1999) and jackknifing (Shao and Tu 2012). A tremendous amount has been written on the topic of resampling and how the various existing methods compare against each other, but batching has been overlooked somewhat.

(f) The question of batch sizing and spacing, that is, selecting m and d, is a subject of ongoing research (Su et al. 2023).

(g) Virtually all discussion in this paper applies to estimators constructed in the context of *digital twins* (Biller et al. 2022).

(h) Parametric batching, analogous to parametric bootstrap (Cheng 2017), has not been sufficiently explored and should form a topic of future research.

(i) Bias estimation in (a') tends to be tricky and delicate, and should be performed with care. This issue is not specific to batching and similar caution has been issued even in the context of the bootstrap and the jackknife (Efron and Tibshirani 1994).

(j) There is a deep and interesting connection between variance estimation and certain types of input model uncertainty, as explained through semi-parametric estimation (Kosorok 2008).

5 THE "TAKE HOME" MESSAGE FOR THE SIMULATIONIST

Simulation excels at estimating parameters that result from complicated mathematical and logical operations involving random objects. However, having a measure of the *error in the parameter estimate*, in addition to the parameter estimate itself, can be very useful to a practitioner. For example, in addition to an estimate on the tolls that achieve the target expected wait times in Example II, a

95 percent confidence interval on the toll estimate will provide the practitioner a sense of how much the estimate might fluctuate if the experiment was repeated. Batching methods outlined in this paper afford this through statistical inference procedures, whereby properties about the error in the obtained estimator can be inferred through confidence regions, standard error estimation, and hypothesis testing. Importantly, batching-based statistical inference procedures are trivial to implement in software, and often work without having to perform additional simulation runs.

REFERENCES

Abdar, M. et al. 2021. "A review of uncertainty quantification in deep learning: Techniques, applications and challenges". *Information Fusion* 76:243–297.

Aktaran-Kalaycı, T., C. Alexopoulos, D. Goldsman, and J. R. Wilson. 2009. "Optimal Linear Combinations of Overlapping Variance Estimators for Steady-State Simulation". In *Advancing the Frontiers of Simulation*, edited by C. Alexopoulos, D. Goldsman, and J. R. Wilson. Springer, NY.

Barth, T. 2011. "A brief overview of uncertainty quantification and error estimation in numerical simulation". *NASA Ames Research Center, NASA Report*.

Barton, R. R. 2012. "Tutorial: Input uncertainty in output analysis". In *Proceedings of the 2012 Winter Simulation Conference*, edited by R. P. O. R. C. Laroque, J. Himmelspach, and A. Uhrmacher, 1–12. Piscataway, New Jersey: Institute of Electrical and Electronics Engineers, Inc.

Biller, B., J. Xi, J. Yi, and P. Venditti. 2022. "Simulation: The Critical Technology in Digital Twin Development". In *Proceedings of the 2022 Winter Simulation Conference*, edited by B. Feng, G. Pedrielli, Y. Peng, S. Shashaani, E. Song, C. G. Corlu, L. H. Lee, E. P. Chew, T. Roeder, and P. Lendermann, 1340–1355. Piscataway, New Jersey: Institute of Electrical and Electronics Engineers, Inc.

Bottou, L., F. Curtis, and J. Nocedal. 2018. "Optimization Methods for Large-Scale Machine Learning". *SIAM Review* 60 (2).

Casella, G., and R. L. Berger. 2002. *Statistical inference*. 2nd ed. ed. Duxbury advanced series. Duxbury/Thomson Learning.

Cheng, R. 2017. *Non-standard parametric statistical inference*. Oxford University Press.

Cheng, R. C., and W. Holland. 1997. "Sensitivity of computer simulation experiments to errors in input data". *Journal of Statistical Computation and Simulation* 1-4 (57): 219–241.

Chick, S. E. 2001. "Input distribution selection for simulation experiments: Accounting for input uncertainty". *Operations Research* 49 (5): 744–758.

Davison, A. C., and D. V. Hinkley. 1997. *Bootstrap Methods and their Application*. Cambridge University Press.

Efron, B. 1979. "Bootstrap Methods: Another Look at the Jackknife". In *Breakthroughs in Statistics*, Springer Series in Statistics, 569–593. New York, NY: Springer New York.

Efron, B., and R. Tibshirani. 1994. *An Introduction to the Bootstrap*. Monographs on statistics and applied probability ; 57. New York: Chapman & Hall.

Hall, P. 1992. *Principles of Edgeworth Expansion*, 39–81. New York, NY: Springer New York.

Henderson, S. G. 2003. "Input model uncertainty: Why do we care and what should we do about it?". In *Proceedings of the 2003 Winter Simulation Conference*, edited by S. Chick, P. J. Sánchez, D. Ferrin, and D. J. Morrice, 1–12. Piscataway, New Jersey: Institute of Electrical and Electronics Engineers, Inc.

Hoel, P. G., S. C. Port, and C. J. Stone. 1986. *Introduction to stochastic processes*. Waveland Press.

Ingalls, R. 2014. "INTRODUCTION TO SUPPLY CHAIN SIMULATION". In *Proceedings of the 2014 Winter Simulation Conference*, 36–50: ACM.

Kosorok, M. R. 2008. *Introduction to empirical processes and semiparametric inference*. Springer.

Lam, H. 2016. "Advanced Tutorial: Input Uncertainty and Robust Analysis in Stochastic Simulation". In *Proceedings of the 2016 Winter Simulation Conference*, edited by T. M. K. Roeder, P. I. Frazier, R. Szechtman, E. Zhou, T. Huschka, and S. E. Chick, 1–12. Piscataway, New Jersey: Institute of Electrical and Electronics Engineers, Inc.

Najm, H. N. 2009. "Uncertainty quantification and polynomial chaos techniques in computational fluid dynamics". *Annual review of fluid mechanics* 41:35–52.

Pasupathy, R., and B. W. Schmeiser. 2010. "The Static Single-Replication Initial-Transient Problem: Using the MSER Statistic". In *Proceedings of the 2010 Winter Simulation Conference*, edited by B. Johansson, S. Jain, J. Montoya-Torres, J. Hugan, and E. Yücesan: Institute of Electrical and Electronics Engineers: Piscataway, New Jersey.

Politis, D. N., J. P. Romano, and M. Wolf. 1999. *Subsampling*. 1st ed. 1999. ed. Springer Series in Statistics.

Shao, J., and D. Tu. 2012. *The jackknife and bootstrap*. Springer Science & Business Media.

Soize, C. 2017. *Uncertainty quantification*. Springer.

Su, Z., R. Pasupathy, Y. Yeh, and P. Glynn. 2023. "Overlapping Batch Confidence Intervals on Statistical Functionals Constructed from Time Series: Application to Quantiles, Optimization, and Estimation". *ACM Transactions on Modeling and Computer Simulation (TOMACS)*.

Wasserman, L. 2004. *All of statistics: a concise course in statistical inference*, Volume 26. Springer.

AUTHOR BIOGRAPHIES

RAGHU PASUPATHY is Professor of Statistics at Purdue University. His current research interests lie broadly in stochastic optimization and statistical inference. He has been actively involved with the Winter Simulation Conference for the past 20 years. Raghu Pasupathy's email address is pasupath@purdue.edu, and his web page https://web.ics.purdue.edu/~pasupath contains links to papers, software codes, and other material.

Proceedings of the Operational Research Society Simulation Workshop 2023 (SW23)
C. Currie and L. Rhodes-Leader, eds.
https://doi.org/10.36819/SW23.002

MIXING METHODS: REFLECTIONS FOR SIMULATION

Professor Susan Howick

University of Strathclyde
Glasgow, UK
susan.howick@strath.ac.uk

ABSTRACT

Complex problems often benefit from being modelled by multiple Operational Research (OR) methods and for a number of years there has been increasing interest in the literature in how such methods can be effectively mixed. This includes simulation methods where (i) multiple simulation methods have been effectively combined and (ii) individual simulation methods have been combined with other types of OR methods. Although a key focus of the presenter's research and work with industry has included using system dynamics, much of her work has also focused on combining system dynamics with other methods, including other simulation methods. This presentation will reflect on over 25 years of mixing system dynamics with other OR methods and will consider lessons for simulation. This will include topics such as the skills required and client value.

AUTHOR BIOGRAPHIES

SUSAN HOWICK is a Professor of Management Science and Vice-Dean (Academic) for Strathclyde Business School. Her main research interest lies in taking a systems perspective when using models to support decision-makers, including the use of system dynamics. She has developed approaches to systemic risk evaluation and modelled highly disrupted projects to understand the causes of project failure. She is also interested in exploring approaches to integrating models that promote client value from the modelling process. Susan has been Vice-President of the Operational Research Society and President of the Policy Council for the UK Chapter of the System Dynamics Society. Susan is on the Editorial Board of the European Journal of Operational Research and is an Associate Editor of the System Dynamics Review. Research funding has included grants from H2020, EPSRC and NERC. Her consultancy experience includes projects for Bombardier, Strathclyde Police, Reed Elsevier, Pricewaterhouse Coopers, White & Case, Scottish and Southern Energy.

Proceedings of the Operational Research Society Simulation Workshop 2023 (SW23)
C. Currie and L. Rhodes-Leader, eds.

PANEL IN MEMORY OF PROFESSOR RAY PAUL

Prof. Tillal Eldabi

School of Management,
University of Bradford,
Richmond Road, Bradford,
West Yorkshire BD7 1DP, UK

t.a.eldabi@bradford.ac.uk

Prof. Sally Brailsford

Southampton Business School,
University of Southampton,
Southampton SO17 1BJ, UK

S.C.Brailsford@soton.ac.uk

Prof. Navonil Mustafee

Centre for Simulation, Analytics and Modelling,
University of Exeter Business School,
Streatham Court, Rennes Drive,
Exeter, EX4 4ST, UK

n.mustafee@exeter.ac.uk

Prof. Stewart Robinson

Newcastle University Business School
5 Barrack Road
Newcastle upon Tyne
NE1 4SE, UK

stewart.robinson@ncl.ac.uk

Prof. Simon Taylor

Department of Computer Science
Brunel University London
Uxbridge, UB8 3PH, UK

simon.taylor@brunel.ac.uk

ABSTRACT

Professor Ray Paul founded the Centre for Applied Simulation Modelling (CASM) at the LSE in the mid-1980s and later on at Brunel University London in the early 1990s, spanning over a period of 30 years. Throughout his academic life, he was an inspirational and thought leader in the field of Simulation. Ray's vision for CASM was to build a comprehensive simulation environment that is accessible to modellers as well as, more importantly, to actual decision makers and users. His philosophy was that modellers should be engaged in solving problems rather than coding and developing software. Throughout the 80s and 90s and 00s he developed various ideas in collaboration with many colleagues and doctoral students, researching into the various aspects of the simulation lifecycle, depending on their field of experience. Most of his papers tended to be visionary with futuristic gaols. To commemorate the first anniversary of his passing, this panel will look back at some the important works of Professor Ray Paul and attempt map them on the current state of Simulation, given the shift in technology, and expand on them by laying roadmaps and visions that will help in improving the use and application of Simulation.

Proceedings of the Operational Research Society Simulation Workshop 2023 (SW23)
C. Currie and L. Rhodes-Leader, eds.
https://doi.org/10.36819/SW23.003

A TUTORIAL ON CONCEPTUAL MODELLING FOR DISCRETE-EVENT SIMULATION

Prof. Stewart Robinson

Newcastle University Business School
5 Barrack Road, Newcastle upon Tyne, NE1 4SE, United Kingdom
stewart.robinson@newcastle.ac.uk

ABSTRACT

Conceptual modelling is the abstraction of a simulation model from the part of the real world it is representing; in other words, choosing what to model, and what not to model. This is generally agreed to be the most difficult, least understood and most important task to be carried out in a simulation study. In this tutorial we define the term 'conceptual model' and go on to identify the artefacts of conceptual modelling and hence the role of conceptual modelling in the simulation project life-cycle. The discussion then focuses on the requirements of a conceptual model, the benefits and approaches for documenting a conceptual model, and frameworks for guiding the conceptual modelling activity. One specific framework is described and illustrated in more detail. The tutorial concludes with a discussion on the level of abstraction.

Keywords: Conceptual Modelling, Simulation

1 INTRODUCTION

One of the most difficult issues in simulation modelling is determining the content of the simulation model. The job of the modeller is to understand the real system that is the subject of the simulation study and to turn this into an appropriate simulation model. The chosen model could range from a very simple single server and queue, through to a model that tries to encapsulate every aspect of the system. In effect, there are an infinite number of models that could be selected within this range, each with a slightly, or even very, different content. The question is: which model should we choose? We explore the answer to this question in this tutorial.

On the surface we might suggest the answer is to build the model that contains as much detail as possible. After all, this model will be the closest to the real system and so surely the most accurate. This might be true if we had complete knowledge of the real system and a very large amount of time available to develop and run the model. But what if we only have limited knowledge of the real system and limited time? Indeed, we rarely have the luxury of vast quantities of either knowledge or time, not least because the real system rarely exists at the time of modelling (it is a proposed world) and a decision needs to be made according to a tight time schedule. Further, a simpler model is often sufficient to address the problem at hand and so there is no need for a more complex model.

So, if we need to develop a simpler model, we need to determine the level of abstraction at which to work. This process of abstracting a model from the real world is known as conceptual modelling. In this tutorial we define conceptual modelling, its requirements and the process for forming a conceptual model.

2 WHAT IS CONCEPTUAL MODELLING?

Conceptual modelling is the abstraction of a simulation model from the part of the real world it is representing ('the real system'). The real system may, or may not, currently exist. Abstraction implies the need for a simplified representation of the real system in the simulation model. The secret to good conceptual modelling is to get the level of simplification correct, that is, to abstract at the right level.

Because all models are simplifications of the real world, all simulation modelling involves conceptual modelling. Even the most complex and detailed simulation still makes various assumptions about the real world and chooses to ignore certain details (simplifications).

2.1 Definition of a Conceptual Model

More formally we define a conceptual model as follows: '... a non-software specific description of the computer simulation model (that will be, is or has been developed), describing the objectives, inputs, outputs, content, assumptions and simplifications of the model.' (Robinson, 2008a)

Let us explore this definition in some more detail. First, this definition highlights the separation of the conceptual model from the computer model. The latter is software specific, that is, it represents the conceptual model in a specific computer code. The conceptual model is not specific to the software in which it is developed. It forms the foundation for developing the computer code.

Second, the conceptual model describes the computer simulation model. It describes how we conceive the model; it does not describe the real system. In other words, the conceptual model describes how we have abstracted the model away from our understanding of the real world. This distinction is important because of the need for model abstraction in simulation. Consider a model such as Schelling's model of segregation (Schelling, 1971): a simple checkerboard representation of a population consisting of two types of resident, represented by the colour of the checkers, making location choices based on the similarity of their neighbours. We will not describe the model in any more detail here, but importantly for our discussion Schelling's model contains no real world data and only the simplest possible representation of the phenomena under study. In this respect, Schelling's conceptual model is very distinct (and 'far') from our description of the real world.

Third, it is stated that the description is of a computer simulation model that 'that will be, is or has been developed.' This serves to highlight the persistent nature of the conceptual model. It is not an artefact that gets created and is then dispensed with once the computer code has been written. It describes the concept of the computer model prior to development, during development and after development. Indeed, the conceptual model persists long beyond the end of the simulation study, since we cannot dispose of the model concept. Of course, because the modelling process is iterative in nature (Balci, 1994; Willemain, 1995; Robinson, 2014), the conceptual model is continually subject to change throughout the life-cycle of a simulation study.

Finally, the definition is completed by a list of what a conceptual model describes. It is vital that the *objectives* of the model are known in forming the conceptual model. The model is designed for a specific purpose and without knowing this purpose it is impossible to create an appropriate simplification. Consider what would happen if we tried to form a model without properly understanding its purpose. We would almost certainly be driven to a model which closely corresponded to the real world and so, by nature, is a much more complex model. This would provide assurance that we could answer a wide range of questions, whatever they might be. Poorly understood modelling objectives can lead to an overly complex model. Clearly understanding the purpose of a model is the basis on which we can identify the appropriate level of simplification.

It is useful to know the model *inputs* and *outputs* prior to thinking about the content of the model. The *inputs* are the experimental factors that are altered in order to try and achieve the modelling objectives. The *outputs* are the reports that inform us as to whether the modelling objectives are being achieved and if not, why they are not being achieved.

Knowing the objectives, inputs and outputs of the model helps to inform the *content* of the model. In particular, the model must be able to receive the inputs and it must provide the outputs. The model content can be thought of in terms of two dimensions:

- *The scope of the model*: the model boundary or the breadth of the real system that is to be included in the model.
- *The level of detail*: the detail to be included for each component in the model's scope.

The final two items in the list of what a conceptual model describes are the *assumptions* and *simplifications* of the model. These are quite distinct concepts:

- *Assumptions:* in the presence of inaccurate, incomplete or absent information, an assumption is the best possible information available that is considered acceptable to enable a model to be completed. That information may later be investigated through the model or changed in the light of new information.

- *Simplifications:* are reductions in model complexity that are incorporated into a model to enable easier and more rapid model building, testing, use and maintenance; and/or to improve the transparency of the model.

So, assumptions are a facet of limited knowledge, while simplifications are a facet of the desire to create simple models.

2.2 Artefacts of Conceptual Modelling

To understand conceptual modelling further it is useful to set it within the wider context of the modelling process for simulation. Figure 1 shows the key artefacts of conceptual modelling. The 'cloud' represents the real world (current or future) within which the problem situation resides; this is the problem that is the basis for the simulation study. The four rectangles represent specific artefacts of the (conceptual) modelling process. These are as follows:

- *System description*: a description of the problem situation and those elements of the real world that relate to the problem.
- *Conceptual model*: as defined in Section 3.1
- *Model design*: the design of the constructs for the computer model (data, components, model execution, etc.) (Fishwick, 1995).
- *Computer model*: a software specific representation of the conceptual model.

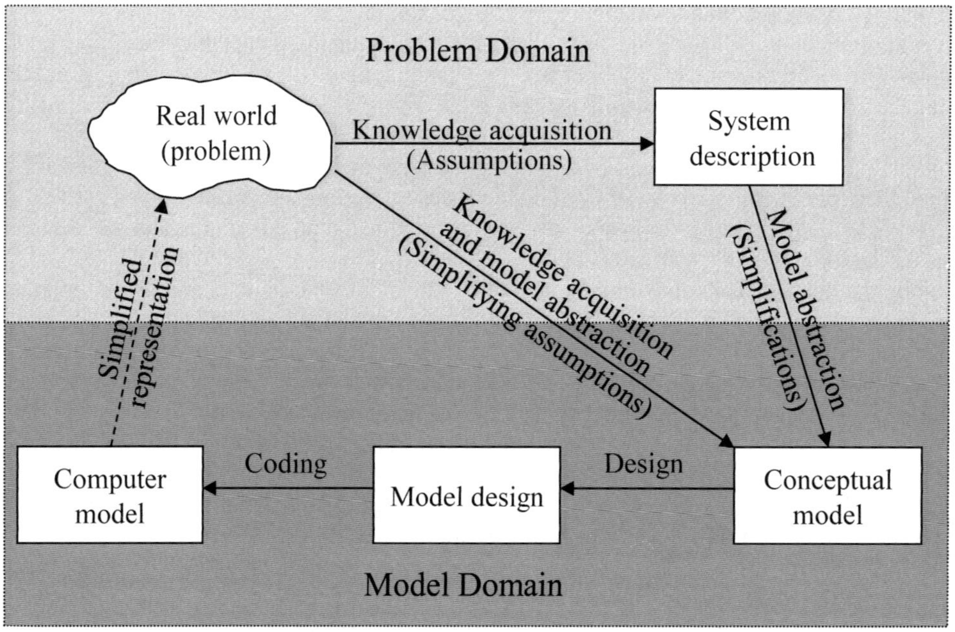

Figure 1 *The Artefacts of Conceptual Modelling (Adapted from Robinson, 2011)*

These artefacts are quite separate. This is not to say that they are always explicitly expressed, with the exception of the computer model. For instance, the system description, conceptual model and model design may not be (fully) documented and can remain within the minds of the modeller and the problem owners. It is, of course, good modelling practice to document each of these artefacts and to use this as a means of communicating their content with the simulation project clients.

The model design and computer model are not strictly part of conceptual modelling, but they do embody the conceptual model within the design and code of the model. These artefacts are included in Figure 1 for completeness. Our main interest here is in the system description and conceptual model which make up the process of conceptual modelling; as represented by the shape with a dashed outline in Figure 1. Unlike the model design and computer model, these two artefacts are independent of the software that will ultimately be used for developing the simulation model.

It is important to recognize the distinction between the system description and the conceptual model. The system description relates to the problem domain, that is, it describes the problem and those elements of the real world that relate to the problem. The conceptual model belongs to the model domain in that it describes those parts of the system description that are included in the simulation model and at what level of detail. The author's experience is that these two artefacts are often confused and seen as indistinct. Indeed, a major failure in any simulation project is to try and model the system description (i.e. everything that is known about the real system) and to not attempt any form of model abstraction; this leads to overly complex models.

The arrows in Figure 1 represent the flow of information, for instance, information about the real world feeds into the system description. The processes that drive the flow of information are described as knowledge acquisition, model abstraction, design and coding. The arrows are not specifically representative of the ordering of the steps within the modelling process, which we know are highly iterative (Balci, 1994; Willemain, 1995; Robinson, 2014). In other words, a modeller may return to any of the four processes at any point in a simulation study, although there is some sense of ordering in that information from one artefact is required to feed the next artefact.

The specific and different roles of assumptions and simplifications are highlighted in Figure 1. Assumptions relate to knowledge acquisition, that is, they fill in the gaps in the knowledge that can be acquired about the real world. Meanwhile, simplifications relate to model abstraction, since they are deliberate choices to model the world more simply. Figure 1 also highlights the idea of a simplifying assumption, that is a simplification that is a direct consequence of making an assumption. For a detailed discussion on simplifications in simulation modelling see van der Zee (2019).

The dashed arrow in Figure 1 shows that there is a correspondence between the computer model and the real world. The degree of correspondence depends on the degree to which the model contains assumptions that are correct, the simplifications maintain the accuracy of the model, and the model design and computer code are free of errors. Because the model is developed for a specific purpose, the correspondence with the real world only relates to that specific purpose. In other words, the model is not a general model of the real world, but a simplified representation developed for a specific purpose. The issue of whether the level of correspondence between the model and the real world is sufficient is an issue of validation (Landry, Malouin, and Oral 1983; Balci, 1994; Robinson, 1999; Sargent, 2013). Both conceptual modelling and validation are concerned with developing a simulation of sufficient accuracy for the purpose of the problem being addressed. As a result, there is a strong relationship between the two topics, conceptual modelling being concerned with developing an appropriate model and validation being concerned with whether the developed model is appropriate.

The artefacts described in this section are similar to Zeigler's concepts of the real system, the experimental frame, the base model, the lumped model, and the computer. The interested reader is referred to Zeigler (1976).

3 REQUIREMENTS OF A CONCEPTUAL MODEL

Before discussing how to perform conceptual modelling, let us consider what makes for a good conceptual model. The key requirements are that the model should be valid, credible, feasible and have utility (Robinson, 2008a). By these we mean the model should:

- Produce sufficiently accurate results for the purpose (*validity*);
- Be believed by the clients (*credibility*);
- Be *feasible* to build within the constraints of the available data and time;
- Have *utility*, that is, sufficiently easy to use, flexible, visual and quick to run.

Overarching all of this is the requirement to build the simplest model possible to meet the objectives of the simulation study. There are several discussions on the benefits of simplification: Innis and Rexstad (1983), Fishwick (1988), Ward (1989), Sevinc (1991), Salt (1993), Brooks and Tobias (2000), Chwif et al. (2000), Lucas and McGunnigle (2003), Urenda Moris et al. (2008) and Tako et al. (2020). Table 1 provides a summary of the main advantages identified by these authors.

Table 1 *Summary of Benefits of Simpler Simulation Models Identified in the Literature*

Model Building and Testing • Less time consuming to develop and maintain • Require less input data • Easier to test, verify and validate
Model Use • Less time to run • Easier to perform sensitivity analysis • Easier to learn to use
Model Maintenance • More flexible: easier to change • Easier to throw away and start again • Easier to combine with another model
Model Understanding • Easier to make the model transparent • Assumptions are more apparent and accessible • Easier to interpret and understand why a result has happened

As such, the need to abstract a conceptual model from the system description becomes even more pertinent. This does not, of course, mean that we should never develop more complex models, but that we should only develop them if they are required to meet the modelling objectives. For further discussion on the topic of model complexity, with respect to how a model is used, see Pidd (2010).

Figure 2 illustrates the relationship between model accuracy and model complexity (scope and level of detail). It shows that with increasing levels of complexity we obtain diminishing returns in terms of accuracy, never reaching 100% accuracy. Eventually we may even find that the accuracy of the model reduces. This is because we do not have the knowledge or data to support the complexity that is being included in the model and we start to make assumptions that are incorrect.

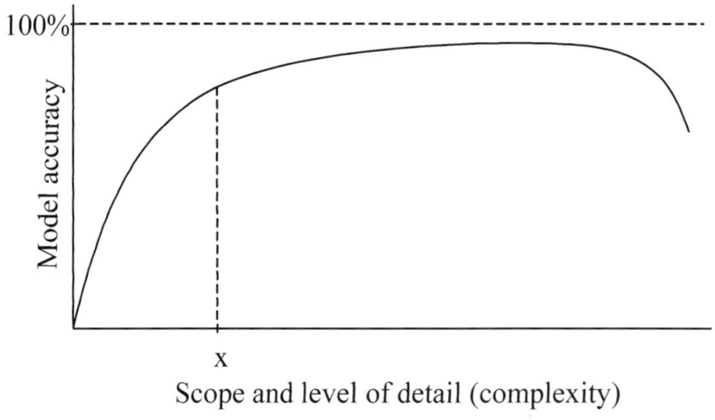

Figure 2 *How Simulation Model Accuracy Changes with the Complexity of the Model (Robinson, 2008a)*

A recent study by Robinson (2022) explores the veracity of the relationship proposed in Figure 2. He finds some evidence of diminishing returns to accuracy from increased complexity, but also identifies instances where this does not hold, for example, there can be increasing returns from increased complexity. He concludes by stating that while the relationship between accuracy and complexity in Figure 2 does not strictly hold, it remains "a useful heuristic for guiding modellers when considering the scope and level of detail at which to model a system."

So which conceptual model should we choose? We might argue that the model at point x in Figure 2 is the best. At this point we have gained a high level of accuracy for a low level of complexity. Moving beyond x will only marginally increase accuracy and adding further complexity generally requires ever increasing effort. Of course, if we have a specific need for an accuracy level greater than that provided by x, we will need to increase the complexity of the model. Indeed, Sargent (2013) suggests that a model's 'acceptable range of accuracy' should be determined early in the modelling process so it can guide both model development and model validation.

The difficulty is in finding point x. Conceptual modelling frameworks, such as the ones listed below, aim to help us in that quest, but conceptual modelling is more of an art than a science (we might prefer to use the word 'craft'). As a result, we can only really hope to get close to x. In other words, there may be a 'best' model, but we are extremely unlikely to find it among an infinite set of models. What we should hope to do is identify the best model we can. As such, our quest is for better models, not necessarily the best.

4 DOCUMENTING THE CONCEPTUAL MODEL

As stated above, the conceptual model is not always explicitly expressed, but can remain within the mind of the modeller. That said, it is good practice to document the conceptual model and in so doing to provide a means of communication between all parties in a simulation study (e.g. the modeller, code developers, domain experts, end users and clients). In so doing it helps to build a consensus, or least an accommodation, about the nature of the model and its use. A documented conceptual model:

- Minimizes the likelihood of incomplete, unclear, inconsistent and wrong requirements
- Helps build the credibility of the model
- Guides the development of the computer model
- Forms the basis for model verification and guides model validation
- Guides experimentation by expressing the modelling objectives, and model inputs and outputs
- Provides the basis of the model documentation
- Can act as an aid to independent verification and validation when it is required
- Helps determine the appropriateness of the model or its parts for model reuse and distributed simulation

There are no set standards for documenting discrete-event simulation conceptual models, but a range of approaches have been proposed, including (references to examples are provided in brackets):

- Component list
- Process flow diagram (Robinson, 2014)
- Activity cycle diagram (Robinson, 2014)
- Logic flow diagram (Robinson, 2014)
- List of assumptions and simplifications
- Unified modelling language (UML) (Richter and März, 2000)
- Petri nets (Torn, 1981)
- Condition specification (Overstreet and Nance, 1985)

The documentation for a conceptual model should be kept simple and focus on identifying *what* is to be modelled and *what* is not to be modelled. There is no need for elaborate documentation because detailed decisions about *how* to model something are not being taken during conceptual modelling. These decisions are taken while creating the model design from the conceptual model. For example, in conceptual modelling a decision is taken to model a flight schedule; in model design the way in which that flight schedule is to be modelled is determined. Hence, the conceptual model documentation only needs to state that the flight schedule is to be modelled, while the model design documentation needs to provide the detail of how to model the flight schedule.

5 FRAMEWORKS FOR CONCEPTUAL MODELLING

A framework for conceptual modelling provides a set of steps and tools that guide a modeller through the development of a conceptual model. It is also useful for teaching conceptual modelling, especially to novice modellers. The simulation literature, however, provides very few such frameworks. Some examples, that the reader may wish to explore further are:

- Conceptual modelling framework for manufacturing (van der Zee, 2007)
- The ABCmod conceptual modelling framework (Arbez and Birta, 2010)
- Karagöz and Demirörs (2010) present a number of conceptual modelling frameworks: Conceptual Model Development Tool (KAMA), Federation Development and Execution Process (FEDEP), Conceptual Models of the Mission Space (CMMS), Defense Conceptual Modelling Framework (DCMF), and Base Object Model (BOM)
- Conceptual modelling with Onto-UML (Guizzardi and Wagner, 2012)
- Conceptual modelling using the Structured Analysis and Design Technique (Ahmed, Robinson, and Tako, 2014)
- The PartiSim framework (Tako and Kotiadis, 2015)

In general these frameworks have been developed independently of one another, although the final two in the list are built with at least some reference to Robinson's framework which is described below. For a more detailed discussion on conceptual modelling frameworks see Robinson et al. (2010).

Here, a very brief outline and illustration of Robinson's framework for conceptual modelling is given. For a more detailed account, and an illustration of the framework in use, see Robinson (2008b).

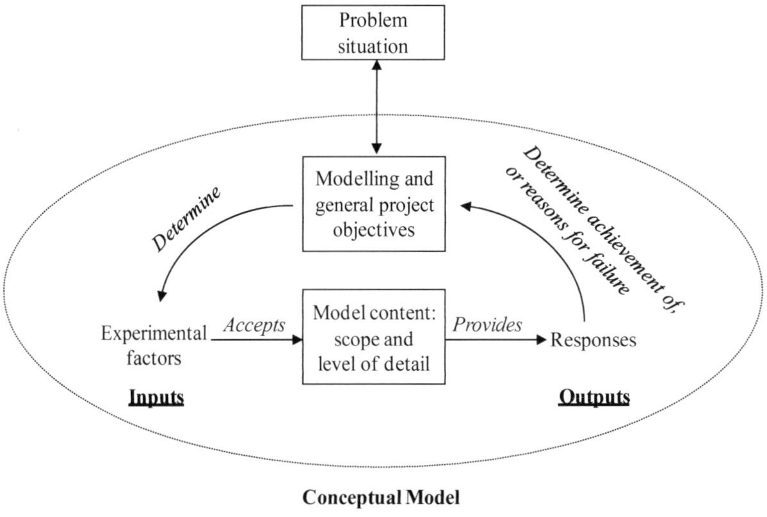

Figure 3 *A Framework for Conceptual Modelling (Robinson, 2008b)*

Figure 3 outlines Robinson's conceptual modelling framework. In this framework, conceptual modelling involves five activities that are performed roughly in this order:

- Understanding the problem situation
- Determining the modelling and general project objectives
- Identifying the model outputs (responses)
- Identify the model inputs (experimental factors)
- Determining the model content (scope and level of detail), identifying any assumptions and simplifications

Starting with an understanding of the problem situation, a set of modelling and general project objectives are determined. These objectives then drive the derivation of the conceptual model, first by defining the outputs (responses) of the model, then the inputs (experimental factors), and finally the model content in terms of its scope and level of detail. Assumptions and simplifications are identified throughout this process.

The ordering of the activities described above is not strict. Indeed, we would expect much iteration between these activities and with the other activities involved in a simulation study: data collection and analysis, model coding, verification and validation, experimentation and implementation.

The framework is supported by a conceptual model template which provides a set of tables that describe each element of the conceptual model. These tables describe:

- Modelling and general project objectives (organisational aim, modelling objectives, general project objectives)
- Model outputs/responses (outputs to determine achievement of objectives, outputs to determine reasons for failure to meet objectives)
- Experimental factors
- Model scope
- Model level of detail
- Modelling assumptions
- Model simplifications

Beyond completing these tables, it is also useful to provide a diagram of the model. For instance, process flow diagrams (Robinson, 2014) are useful for communicating the conceptual model.

The modeller works through these tables with the support of the stakeholders and domain experts, iteratively improving them to the point that the modeller and stakeholders are satisfied that the conceptual model meets the requirements for validity, credibility, feasibility and utility. This provides a structured framework for making the conceptual modelling decisions explicit (documentation) and for debating ways of improving the conceptual model.

6 LEVELS OF ABSTRACTION

In Section 2.1 we briefly mention Schelling's model of segregation (Schelling 1971) as an example of a 'far' abstraction. By this we mean that the conceptual model involves many simplifications and so it is removed a long way from the system description. The implication of this is that the computer model is a highly simplified representation of the real world. At the extreme, such as in the case of Schelling's model, a far abstraction can lead to a (conceptual) model that bears little resemblance to the real world. Despite being a far abstraction, Schelling's model has certainly attracted a lot of attention.

However, we would not want to leave the impression that conceptual models have to be so far abstracted. Indeed it is not always desirable to abstract to this degree and for some simulation studies it is appropriate to model much of the scope and detail in the problem domain. We refer to this as 'near' abstraction. For an example, see the Ford engine plant model described in Robinson (2008a, 2008b). These papers describe a simulation that was designed to determine the throughput of a new engine assembly plant. The model contained much detail about the real system and took a considerable time to develop.

The level of abstraction should be determined by the requirement for the model to be valid, credible, feasible and have utility. One danger with far abstraction is that whilst the model may be valid, it may lack credibility. Hence, we may need to reduce the level of abstractness, making the model nearer to the system description, to increase the credibility of the model. Although there are many benefits gained from simplification (as described in Section 3), we are not arguing that all models should be simple. The level of simplification, or abstraction, should be appropriate to the problem at hand.

7 CONCLUSION

Conceptual modelling is the abstraction of a simulation model from the part of the real world it is representing. It is probably the most important aspect of any simulation study. Get the conceptual model right and the rest of the simulation work will be more straightforward, providing the right information in the right time-scale.

The discussion in this tutorial defines conceptual modelling, its artefacts and its requirements; it also discusses the benefits and approaches to documenting the conceptual model. From this base, some frameworks for conceptual modelling are listed and Robinson's framework is outlined in more detail. The framework aims to guide a modeller through the process of creating and documenting a conceptual model. We also discuss levels of abstraction, from near to far.

Conceptual modelling is not a science, but a craft or even an art. As with any craft, it can be learned and it can be improved upon with experience. Frameworks provide a good way of learning about conceptual modelling and for helping to do it better. At present, however, there are very few examples of conceptual modelling frameworks and this is an area where more research needs to be undertaken.

ACKNOWLEDGEMENTS

This paper is based on Robinson, S. (2015). A Tutorial on Conceptual Modeling for Simulation. *Proceedings of the 2014 Winter Simulation Conference*; and Robinson, S. (2016). Conceptual Modelling for Simulation: A Tutorial. *Proceedings of the 8th Simulation Workshop (SW16) (A. Anagnostou, K. Hoad, M.Kunc, eds.)*, Operational Research Society, Birmingham, pp. 13-24.

REFERENCES

Ahmed F, Robinson S and Tako A A (2014). Using the structured analysis and design technique (SADT) in simulation conceptual modeling. In Tolk A, Diallo S D, Ryzhov I O, Yilmaz L, Buckley S and Miller J A (eds). *Proceedings of the 2014 Winter Simulation Conference.* Institute of Electrical and Electronics Engineers: Piscataway, New Jersey, pp 1038-1049.

Arbez G and Birta L G (2010). The ABCmod conceptual modeling framework. In Robinson S, Brooks R J, Kotiadis K and van der Zee D-J (eds). *Conceptual Modeling for Discrete-Event Simulation.* Chapman and Hall/CRC: Boca Raton, FL, pp 133-178.

Balci O (1994). Validation, verification, and testing techniques throughout the life cycle of a simulation study. *Annals of Operations Research* **53**: 121-173.

Brooks R J and Tobias A M (2000). Simplification in the simulation of manufacturing systems. *International Journal of Production Research* **38 (5)**: 1009–1027.

Chwif L, Barretto M R P and Paul R J (2000). On simulation model complexity. In: Joines J A, Barton R R, Kang K and Fishwick P A (eds). *Proceedings of the 2000 Winter Simulation Conference.* Institute of Electrical and Electronics Engineers: Piscataway, New Jersey, pp 449–455.

Fishwick P A (1988). The role of process abstraction in simulation. *IEEE Transactions on Systems, Man, and Cybernetics* **18 (1)**: 18–39.

Fishwick P A (1995). *Simulation Model Design and Execution: Building Digital Worlds*. Prentice-Hall, Inc.: Upper Saddle River, New Jersey.

Guizzardi G and Wagner G (2012). Tutorial: conceptual simulation modeling with Onto-UML. In Laroque C, Himmelspach J, Pasupathy R, Rose O and Uhrmacher A M (eds). *Proceedings of the 2012 Winter Simulation Conference.* Institute of Electrical and Electronics Engineers: Piscataway, New Jersey, pp 52-66.

Innis G and Rexstad E (1983). Simulation model simplification techniques. *Simulation* **41 (1)**: 7-15.

Karagöz N A and Demirörs O (2010). Conceptual modeling notations and techniques. In Robinson S, Brooks R J, Kotiadis K and van der Zee D-J (eds). *Conceptual Modeling for Discrete-Event Simulation.* Chapman and Hall/CRC: Boca Raton, FL, pp 179-209.

Landry M, Malouin J L and Oral M (1983). Model validation in operations research. *European Journal of Operational Research* **14 (3)**: 207-220.

Lucas T W and McGunnigle J E (2003). When is model complexity too much? Illustrating the benefits of simple models with Hughes' salvo equations. *Naval Research Logistics* **50**: 197-217.

Overstreet M C and Nance R E (1985). A specification language to assist in analysis of discrete event simulation models. *Communications of the ACM* **28 (2)**: 190-201.

Pidd M (2010). Why modelling and model use matter. *Journal of the Operational Research Society* **61 (1)**: 14-24.

Richter H and März L (2000). Toward a standard process: the use of uml for designing simulation models. In: Joines J A, Barton R R, Kang K and Fishwick P A (eds). *Proceedings of the 2000 Winter Simulation Conference.* Institute of Electrical and Electronics Engineers: Piscataway, New Jersey, pp 394-398.

Robinson S (1999). Simulation verification, validation and confidence: a tutorial. *Transactions of the Society for Computer Simulation International* **16 (2)**: 63-69.

Robinson S (2008a). Conceptual modelling for simulation part I: definition and requirements. *Journal of the Operational Research Society* **59 (3)**: 278-290.

Robinson S (2008b). Conceptual modelling for simulation part II: a framework for conceptual modelling. *Journal of the Operational Research Society* **59 (3)**: 291-304.
Robinson S (2011). Conceptual modeling for simulation. In Cochran J J (ed). *Encyclopedia of Operations Research and Management Science*. Wiley: New York.
Robinson S (2014). *Simulation: The Practice of Model Development and Use, 2nd ed.* Palgrave: London, UK.
Robinson S (2022). Exploring the relationship between simulation model accuracy and complexity. *Journal of the Operational Research Society, forthcoming.*
Robinson, S, Brooks R J, Kotiadis K and van der Zee D J (2010). *Conceptual Modelling for Discrete-Event Simulation*. Taylor and Francis: FL.
Salt J (1993). Simulation Should be Easy and Fun. In Evans G W, Mollaghasemi M, Russell E C and Biles W E (eds). *Proceedings of the 1993 Winter Simulation Conference*. Institute of Electrical and Electronics Engineers: Piscataway, New Jersey, pp 1-5.
Sargent R G (2013). Verification and validation of simulation models. *Journal of Simulation* **7 (1):** 12–24.
Schelling T C (1971). Dynamic models of segregation. *Journal of Mathematical Sociology* **1**: 143-186.
Sevinc S (1991). Theories of discrete event model abstraction. In Nelson B L, Kelton W D and Clark G M (eds.). *Proceedings of the 1991 Winter Simulation Conference.* Institute of Electrical and Electronics Engineers: Piscataway, New Jersey, pp 1115-1119.
Tako A A and Kotiadis K (2015). partisim: a multi-methodology framework to support facilitated simulation modelling in healthcare. *European Journal of Operational Research* **244 (2)**: 555-564.
Tako A A, Tsioptsias N and Robinson S (2020). Can we learn from simple simulation models? An experimental study on user learning. *Journal of Simulation* **14 (2)**: 130-144.
Torn A A (1981). Simulation graphs: a general tool for modeling simulation designs. *Simulation* **37 (6)**: 187-194.
Urenda Moris M, Ng A H C and Svensson J (2008). Simplification and aggregation strategies applied for factory analysis in conceptual phase using simulation. In Mason S J, Hill R R, Mönch L, Rose O, Jefferson T and Fowler J W (eds.). *Proceedings of the 2008 Winter Simulation Conference.* Institute of Electrical and Electronics Engineers: Piscataway, New Jersey, pp 1913–1921.
van der Zee D-J (2007). Developing participative simulation models: framing decomposition principles for joint understanding. *Journal of Simulation* **1 (3)**: 187-202.
van der Zee D-J (2019). Model Simplification in manufacturing simulation – review and framework. *Computers & Industrial Engineering* **127**: 1056–1067.
Ward S C (1989). Arguments for constructively simple models. *Journal of the Operational Research Society* **40 (2)**: 141-153.
Willemain T R (1995). Model formulation: what experts think about and when. *Operations Research* **43 (6)**: 916-932.
Zeigler B P (1976). *Theory of Modeling and Simulation*. Wiley: New York.

AUTHOR BIOGRAPHY

STEWART ROBINSON is Dean and Professor of Operational Research at Newcastle University Business School, UK. His research focuses on the practice of simulation model development and use. Key areas of interest are conceptual modelling, model validation, output analysis and alternative simulation methods (discrete-event, system dynamics and agent based). Professor Robinson is author/co-author of six books on simulation, co-founder of the Journal of Simulation and co-founder of the UK Simulation Workshop conference series. He was President of the Operational Research Society (2014-2015). www.stewartrobinson.co.uk

Proceedings of the Operational Research Society Simulation Workshop 2023 (SW23)
C. Currie and L. Rhodes-Leader, eds.
https://doi.org/10.36819/SW23.004

BUSINESS PROCESS MODELLING AND SIMULATION WITH DPMN, ANYLOGIC AND SIMIO – A TUTORIAL

Dr. Gerd Wagner

Brandenburg University of Technology
Konrad-Wachsmann-Allee 5
03046 Cottbus, GERMANY
g.wagner@b-tu.de

ABSTRACT

Business process modelling and simulation has been a core research topic both in the area of Discrete Event Simulation and in the area of Business Process Management for a long time. However, both areas have largely ignored each other's research results. In particular, Discrete Event Simulation research has not managed to establish a general conceptual foundation and modelling language for Processing Networks, so vendors are using their own proprietary terminology and diagram language in their "process modelling" tools, ignoring the process modelling language BPMN, which has been established as a widely adopted standard in Business Process Management.

In this tutorial, we present the conceptual foundations of, and a modelling language for, Activity Networks and Processing Networks. The Discrete Event Process Modelling Notation (DPMN) proposed by Wagner (2018) combines the visual syntax of BPMN with the rigorous semantics of Event Graphs (Schruben 1983). DPMN allows making platform-independent visual simulation models that can be implemented with Discrete Event Simulation platforms such as AnyLogic or Simio.

Keywords: Business Process Modelling, BPMN, DPMN, AnyLogic, Simio

1 INTRODUCTION

Business processes have been modelled and simulated both in the area of *Discrete Event Simulation (DES)* and in the area of *Business Process Management (BPM)* for a long time. However, research in both areas has not managed to develop a unified conceptual framework for business process (BP) modelling and simulation.

While Schruben (1983) has proposed *Event Graphs* as a visual modelling language for the DES paradigm of *Event-Based Simulation (ES)*, no common modelling language has been established for the DES paradigm of *Processing Networks* (often called "process modelling"). Each "process modelling" tool (such as Arena, Simio, AnyLogic, etc.) is using their own proprietary terminology and diagram language. In BPM, the BP modelling notation BPMN, officially called 'Business Process Model and Notation' (OMG 2014), has been established as a widely adopted standard. BPMN has been largely ignored in DES, while Event Graphs have been ignored in BPM.

The term *Discrete Event Simulation* has been established as an umbrella term subsuming various kinds of computer simulation approaches, all based on the general idea of modelling the dynamics of a discrete system as a series of (explicit or implicit) events that change the system's state over time.

In the DES literature, it is often stated that DES is based on "entities flowing through a system". While this narrative applies to the DES paradigm of Processing Networks, called "process modelling" by Pegden (2010), it characterizes a special (yet important) kind of DES only, and it does not apply to all discrete event systems. The "process modelling" paradigm should be better called *Processing Network (PN)* paradigm, since it is not about process modelling in general, but only about modelling a particular kind of discrete processes that happen in Processing Networks (which generalize *Queuing*

Networks). It has been pioneered by GPSS (Gordon 1961) and SIMAN/Arena (Pegden and Davis 1992) and is implemented in various forms by all modern off-the-shelf simulation tools, including Simio and AnyLogic.

1.1 Some Remarks on the History of DES

Pegden (2010) explains that the history of DES has been shaped by three fundamental paradigms: Markowitz, Hausner, and Karr (1962) pioneered *Event-Based Simulation (ES)* with *SIMSCRIPT*, Gordon (1961) pioneered *Processing Network Simulation (PNS)* with *GPSS*, and Dahl and Nygaard (1967) pioneered *Object-Orientation (OO)* and the computational concept of co-routines for asynchronous programming with their simulation language *Simula*.

Notice, however, that OO does not represent a DES paradigm, but rather an information/data modelling paradigm for conceptual modelling and software design modelling, as well as a programming paradigm. In fact, the Simula paradigm is characterized by a combination of OO modelling and using co-routines for implementing the dynamics of a discrete system without an explicit computational concept of events in a way that is sometimes called 'process interaction' approach.

In Pegden (2010), there is no mention of the pioneering work of Tocher on the first general-purpose simulator, the *General Simulation Program (GSP)*, for which he conceived simulation models in the form of activity flow diagrams (called 'wheel diagrams') where he considered activity start events as 'conditional events' since they depend on the availability of required resources (Tocher and Owen 1960). In Tocher's *Three-Phase Method* for an activity-based simulator, first the simulation time is advanced to that of the next scheduled activity end event(s), then these events are processed, and finally the simulator checks for which activities a start event can be processed due to available resources.

According to Pegden (2010), ES has been widely used during the first 20 years of simulation, due to its great flexibility allowing to efficiently model a wide range of complex systems. Later, however, the PNS paradigm, implemented by tools like Arena, Simul8, FlexSim, Simio and AnyLogic, became the dominant approach in practical applications of simulation because it is based on the higher-level concept of processing activities and allows no-code (or low-code) simulation engineering with graphical user interfaces and appealing visualizations.

After OO had been established as the predominant paradigm in software engineering in the 1990s, it was also adopted by many simulation tools, which weaved it into their PNS approach. Pegden (2010) remarks that "Many process and object based simulation tools maintain an event capability as a 'backdoor' for flexibility". Consequently, modern DES tools allow combining PN models with OO modelling and event scheduling. As argued by Pegden (2010), ES is the most fundamental DES paradigm since the other paradigms also use events, at least implicitly. However, Pegden does not explain in which way the other paradigms are built upon ES.

1.2 Fragmentation and Conceptual Confusion in DES

Today, after a history of more than 50 years, the field of DES is fragmented into many different paradigms and formalisms, based on different concepts and terminologies. Unlike other scientific fields, it didn't achieve much conceptual unity regarding its foundations, which would be crucial for facilitating scientific progress.

There is a lot of conceptual confusion in the field of DES. For instance, Banks et al (2010) define that DES is "the modelling of systems in which the state variable changes only at a discrete set of points in time". Remarkably, this definition does not even mention the concept of events, possibly for accommodating approaches that do not explicitly refer to events, such as Petri Nets, Activity Cycle Diagrams and DEVS.

While some authors, such as Pegden (2010), distinguish between three fundamental DES approaches: ES, PNS and OO, others, such as Banks (1998), distinguish between four approaches: ES, PNS (called "process-interaction"), Activity Scanning and the Three-Phase Method. This disagreement about fundamental concepts, and the different terms used by different authors for the same concepts (e.g., the PNS paradigm has the following names in the DES literature: "process-oriented", "process-based", "process-interaction", "process-centric"), must be very confusing for students of and beginners in DES.

We follow the view of Pegden (2010) that ES is the most fundamental DES paradigm, although in many DES textbooks, e.g., in (Banks et al 2010), this is not explained. Adopting ES as the most fundamental DES paradigm implies that other paradigms should extend it in a conservative manner such that its basic concepts (and their semantics) are preserved.

The lack of a scientifically established conceptual foundation of DES and an accompanying standard terminology is witnessed in the diversity of terminologies and diagram languages used in DES software packages. Typically, DES practitioners are locked into the terminology (and implementation idiosyncrasies) of the simulation platform they use, and are often not aware of the general, platform-independent and implementation-agnostic concepts.

The use of proprietary terminologies and diagram languages makes it hard for simulation beginners to learn how to use a tool and for expert users of a tool to switch, or interchange models, from their tool to another one. Notice especially the strange term "Agent" used by AnyLogic, instead of the Arena term "Entity", for processing objects like manufacturing parts in production systems or patients in hospitals. It is confusing to call a manufacturing part, such as a wheel in the production of a car, an "agent".

1.3 Object Event Modelling and Simulation

In Wagner (2018; 2020; 2021), we show how to extend ES by adding concepts like objects and activities resulting in *Object Event Modelling and Simulation (OEM&S)*, a new general DES paradigm based on the two most important ontological categories: objects and events (Guizzardi and Wagner 2010).

OEM&S combines OO modelling with the event scheduling approach of ES. *Object Event Simulation (OES)* is a conservative extension of ES. While ES defines the system state structure in the form of a set of global variables, OES defines it in the form of a set of objects (or object states), such that their attributes take the role of state variables.

In Wagner (2018), we have introduced a variant of BPMN, called *Discrete Event Process Modelling Notation (DPMN)*, and have shown how an OEM approach based on UML Class Diagrams and DPMN Process Diagrams allows defining a set of object types OT, a set of event types ET, and a set of event rules R. In Wagner (2017), we have shown that (a) these three sets define a state transition system, where the state space is defined by OT and ET, and the transitions are defined by R, and (b) such a transition system represents an *Abstract State Machine* (Gurevich 1985). This fundamental characterization of an OES model provides a formal semantics for OES.

Since OEM&S accommodates the concepts of resource-constrained activities and processing activities, it integrates important DES concepts and supports modelling general forms of (BPMN-style) *Activity Networks* and (GPSS-style) *Processing Networks*, which extend Activity Networks by adding processing objects flowing through the network, as summarized in Table 1.

As a visual modeling language, DPMN's purpose is primarily to allow a simulation engineer specifying a platform-independent simulation design model with all relevant computational details. It is not a means for communicating the model to a client and may have limited value in helping to validate a model with subject matter experts.

1.4 Discrete Processes and Business Processes

A **discrete event process**, or simply *discrete process (DP)*, consists of a partially ordered set of **events** such that each of them causes zero or more discrete state changes of affected **objects**. When two or more events within such a process have the same order rank, this means that they occur simultaneously. A discrete process may be an instance of a *discrete process type* defined by a **discrete process model**.

A **business process** (BP) is a discrete process that involves **activities** performed by *organizational agents* qua one of their *organizational roles* defined by their *organizational position*. Typically, a business process is an instance of a business process type defined by an organization or organizational unit (as the owner of the business process type) in the form of a **business process model**.

While there are DPs that do not have an organizational context (like, for instance, message exchange processes in digital communication networks or private conversations among human agents), a BP always happens in the context of an organization. The performance of a resource-dependent activity is constrained by the availability of the required **resources**, which may include *human resources* or other resource objects (such as rooms or devices). There are two kinds of business process models:

1. BPMN-style **Activity Networks** (ANs) consisting of *event nodes* and *activity nodes* (with task queues) connected by means of **event scheduling arrows** and **resource-dependent activity scheduling (RDAS) arrows**, such that event and activity nodes may be associated with objects representing their participants. In the case of an activity node, these participating objects include the resources required for performing an activity. Typically, an activity node is associated with a particular resource object representing the activity **performer**.
2. GPSS/Arena-style **Processing Networks** (PNs) consisting of *entry nodes*, *processing nodes* (with task queues and input buffers) and *exit nodes* connected by means of **processing flow arrows**, which overlay an RDAS arrow with an **object flow arrow**. The PN concept is a conservative extension of the AN concept, that is, a PN is a special type of AN.

Table 1 *The layers of DPMN*

	Layer	Elements/Concepts	Diagrams
Event-Based Simul.	**Event Graphs** (Schruben 1983)	Event Circles, Event Scheduling Arrows, Variable Value Assignments	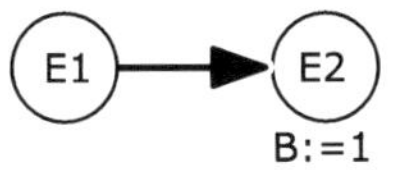
Object Event Modelling and Simulation (OEM&S)	**Object Event Graphs** (Basic DPMN)	+ Objects w/ State Changes	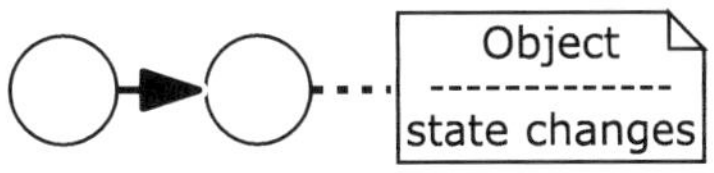
	Activity Networks (DPMN-A)	+ Activities + Resource Roles + Resource Cardinality Constraints + Resource Pools + Resource-Dependent Activity Scheduling Arrows	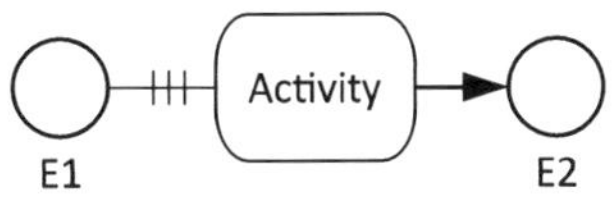
	Processing Networks (DPMN-PN)	+ Processing Activities + Entry/Processing/Exit Nodes + Processing Flow Arrows	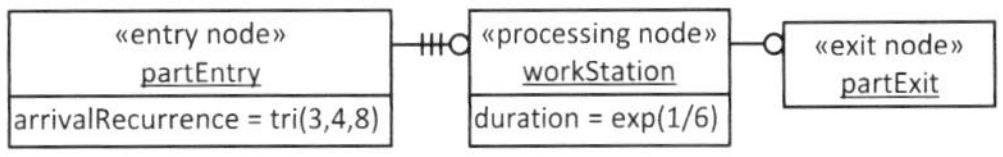

In an AN, all activity nodes have a task queue filled with tasks (or planned activities) waiting for the availability of the required resources. An RDAS arrow from an AN node to a successor activity node expresses the fact that a corresponding activity end event (or plain event) triggers the deferred scheduling of a successor activity start event, corresponding to the creation of a new task in the task queue of the successor activity node.

BP models focus on describing the possible sequences of events and activities, based on conditional and parallel branching, but they also describe the dependencies of activities on resource objects, either *declaratively*, as in BPMN and DPMN, by defining resource roles for activities, or *procedurally*, as in DES tools, by preceding and succeeding resource allocation and de-allocation steps.

1.5 Processing Networks Generalize Queuing Networks

A **PN process** is a business process that involves one or more *processing objects* and includes *arrival events*, *processing activities* and *departure events*. An arrival event for one or more *processing objects* happens at an *entry station*, from where they are routed to a *processing station* where *processing activities* are performed on them, before they are routed to another processing station or to an *exit station* where they leave the system via a departure event.

A **PN process model** defines a PN where each node represents a combination of a spatial object and an event or activity variable:

1. Defining an **entry node** means defining both an *entry station* object (e.g., a reception area or a factory entrance) and a variable representing *arrival* events for arriving *processing objects* (such as people or manufacturing parts).
2. Defining a **processing node** means defining both a *processing station* object (often used as a resource object, such as a workstation or a room) and a variable representing *processing activities*.
3. Defining an **exit node** means defining both an *exit station* object and a variable representing *departure* events.

In a PN, all processing nodes have a task queue and an input buffer filled with processing objects that wait to be processed. A PN where all processing activities have exactly one abstract resource (often called a "server") is also known as a **Queuing Network** in *Operations Research* where processing nodes are called "servers" and processing objects are called "entities" or "jobs".

2 FROM EVENT GRAPHS VIA ACTIVITY TO PROCESSING NETWORKS

In this section, summarizing Wagner (2018; 2020; 2021), we show how DPMN is constructed by incrementally extending Event Graphs by adding increasingly high-level modelling concepts: in the first step, we add the concept of objects, resulting in *Object Event Graphs*; in the second step, we add the concept of (resource-constrained) activities, resulting in *Activity Networks*; and in the third step we add the concept of processing activities, resulting in *Processing Networks*. Object Event Graphs, Activity Networks and Processing Networks are special forms of DPMN process models.

2.1 From Event Graphs to Object Event Graphs

Event Graphs define graphically, which state changes and follow-up events are triggered by an event. The Event Graph shown in Figure 1 models a manufacturing workstation, which is part of a manufacturing business system (viewed as a basic queuing system in *Operations Research*). It defines (a) two state variables: *L* for the length of an arrival queue and *B* for a performer being busy or not, as well as (b) three event variables representing *Arrival*, *ProcessingStart* and *ProcessingEnd* events, in the form of circles. In addition, it defines the sequencing of events of those types with the help of *Event Scheduling* arrows, together with caused state changes in the form of (possibly conditional) variable assignments (underneath event circle names).

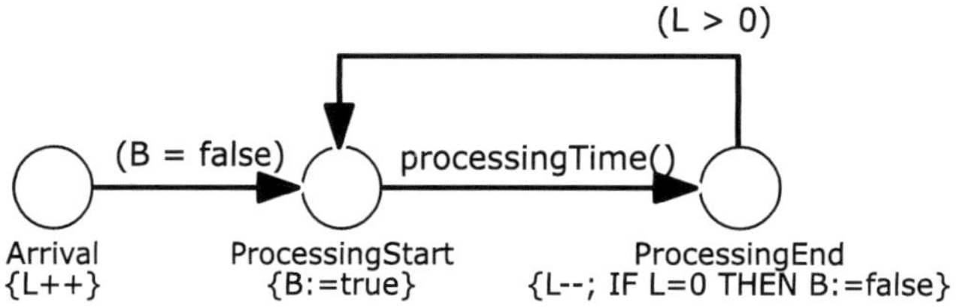

Figure 1 *An Event Graph defining an ES model with two state variables and three event types*

Event Graphs provide a visual modelling language with a precise semantics that captures the fundamental ES paradigm. However, Event Graphs are a rather low-level DES modelling language: they lack a visual notation for (conditional and parallel) branching, do not support OO state structure modelling (with attributes of objects taking the role of state variables) and do not support the concept of activities.

In OEM&S, object types and event types are modeled as special categories of classes («object type» and «event type») in a special kind of UML Class Diagram/Model, called *OE Class Diagram/Model*. *Random variables* are modeled as a special category of class-level operations (designated with «rv») constrained to comply with a specific probability distribution such that they can be implemented as static methods. Finally, *event rules* are modeled in DPMN process diagrams (and possibly also in pseudo-code), such that they can be implemented in the form of special *onEvent* methods of event classes.

In a simulation model, certain types of events are characterized as *exogenous*, while the others are endogenous (or *caused* by previous events). In an OE class model, we therefore categorize event types as either «exogenous event type» or just «event type». The exogenous events of a DES model correspond to the *Start* events of a BPMN process model, whereas the caused events correspond to BPMN *Intermediate* or *End* events.

In the OE class model shown in Figure 2, *PartArrival*, *ProcessingStart* and *ProcessingEnd* events are associated with a *WorkStation* object (as their only participant). This is the workstation where these events happen. As a class for exogenous events, the *PartArrival* class defines a random variable *recurrence*, which generally determines the recurrence frequency of exogenous events (the elapsed time between two consecutive events of the given type, also called *inter-occurrence time*). The *ProcessingStart* event class defines a random variable *processingTime*.

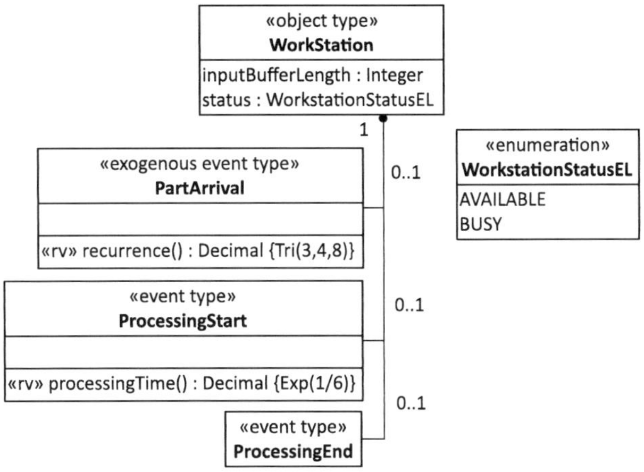

Figure 2 *An OE class model defining an object type and three event types*

Object Event (OE) Graphs, as a basic type of DPMN process diagrams, extend the Event Graph diagram language by adding object rectangles containing declarations of typed object variables and state change statements, as well as gateway diamonds for expressing conditional and parallel branching.

A DPMN process model, such as an OE Graph, is based on an underlying information model defining the types of its objects and events. The process model shown in Figure 3 is an OE Graph that is based on the OE class model shown in Figure 2.

Notice that in the OE Graph of Figure 3, the state variables of the Event Graph of Figure 1, *L* and *B*, have been replaced by the attributes *inputBufferLength* and *status* defined in the OE class model of Figure 2.

OE Graphs are a conservative extension of Event Graphs. This means that an OE Graph can be transformed to an Event Graph preserving its dynamics by replacing its objects with corresponding sets of state variables.

Like Petri Nets, OE Graphs have a formal semantics. But while Petri Nets are an abstract computational ("token flow") formalism without an ontological foundation, OE Graphs are based on the ontological categories of objects, events and causal regularities such that an OE Graph can be decomposed into a set of *event rules*, representing causal regularities, which define the transitions of an *Abstract State Machine* [15].

An OE Graph specifies a set of chained event rules, one rule for each event circle of the model. The OE Graph shown in Figure 3 specifies the following three event rules:

1. On each *PartArrival* event, the *inputBufferLength* attribute of the associated *WorkStation* object is incremented and if the workstation's *status* attribute has the value AVAILABLE, then a new *ProcessingStart* event is scheduled to occur immediately.

2. When a *ProcessingStart* event occurs, the associated *WorkStation* object's *status* attribute is changed to BUSY and a *ProcessingEnd* event is scheduled with a delay provided by invoking the *processingTime* function defined in the *ProcessingStart* event class.
3. When a *ProcessingEnd* event occurs, the *inputBufferLength* attribute of the associated *WorkStation* object is decremented and if the *inputBufferLength* attribute has the value 0, the associated *WorkStation* object's status attribute is changed to AVAILABLE. If the *inputBufferLength* attribute has a value greater than 0, a new *ProcessingStart* event is scheduled to occur immediately.

These event rules can be implemented as *onEvent* methods of their triggering event classes. The resulting model can be run at https://sim4edu.com/oesjs/core1/workstation-1/ as a web-based simulation.

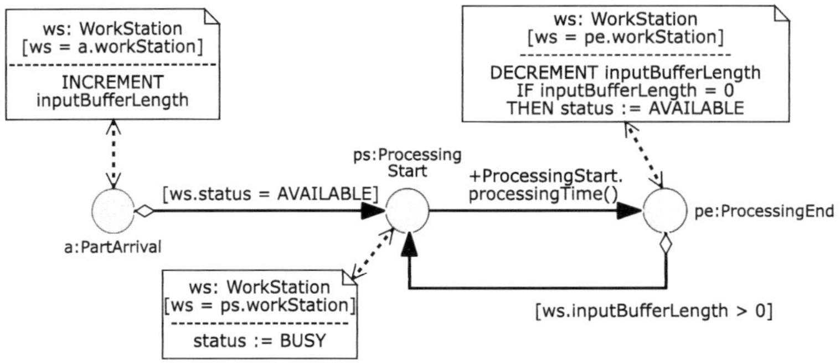

Figure 3 *An Object Event Graph based on the OE class model of Figure 2*

DPMN consists of three layers. The first layer, for modelling OE Graphs, corresponds to an extension of Event Graphs by adding the concept of *Objects*. The second layer (DPMN-A), for modelling Activity Networks, adds the concepts of *Resource-Constrained Activities* and *Resource-Dependent Activity Scheduling* based on resource roles and resource pools, while the third layer (DPMN-PN), for modelling Processing Networks, adds the concepts of *Processing Objects*, *Processing Activities* and *Processing Flows*.

2.2 From Object Event Graphs to Activity Networks

In Wagner (2020), we have shown how to extend OE Graphs by adding support for resource-constrained activities, resulting in DPMN-A, comprised of three new information modelling elements (*Activity Type*, *Resource Role*, *Resource Pool*) and two new process modelling elements (*Activity* and *Resource-Dependent Activity Scheduling Arrow*). DPMN-A diagrams allow modelling *Activity Networks*.

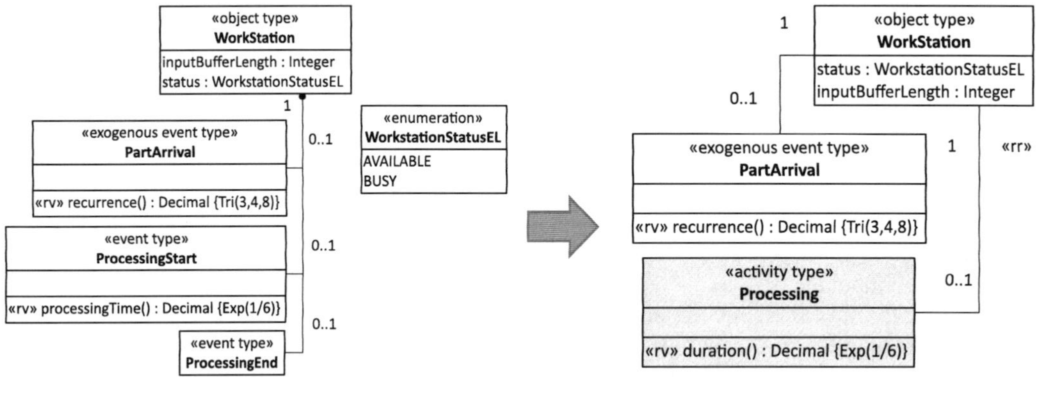

Figure 4 *Rewriting an OE class model with a pair of activity start and end event types to a model with a corresponding activity type (Processing)*

Conceptually, an activity is a composite event with a non-zero duration that is composed of, and temporally framed by, a pair of instantaneous start and end events. In the transformation shown in Figure 4 below, the pair of *ProcessingStart* and *ProcessingEnd* event types of the OE class model of Figure 2 is replaced with a corresponding *Processing* activity type. This replacement pattern is an essential part of the semantics of activities in DPMN: by reduction to a pair of corresponding start and end events.

The *Activity-Start-End Rewrite Pattern* exemplified in Figures 4 and 5 can also be applied in the inverse direction, replacing an Activity rectangle with a pair of Event circles. It allows reducing an Activity Network model with Activity rectangles to an OE Graph as a basic DPMN diagram without Activity rectangles. The target model of Figure 5 specifies two event rules:

1. On each *PartArrival* event: (a) if the workstation's *status* is AVAILABLE, then a local rule variable *wsAllocated* is set to *true* and the workstation's *status* is set to BUSY, else the workstation's *inputBufferlength* is incremented; (b) if the *wsAllocated* variable has the value *true*, then a new *Processing* activity is scheduled to start immediately (with a duration provided by invoking the *duration* function defined in the *Processing* activity class).
2. When a *Processing* activity ends: (a) if the workstation's *inputBufferlength* is equal to 0, then the workstation's *status* is set to AVAILABLE, else the local rule variable *wsAllocated* is set to *true* and the *inputBufferlength* is decremented; (b) if the *wsAllocated* variable has the value *true*, a new *Processing* activity is scheduled to start immediately (with a duration provided by invoking the *duration* function defined in the *Processing* activity class).

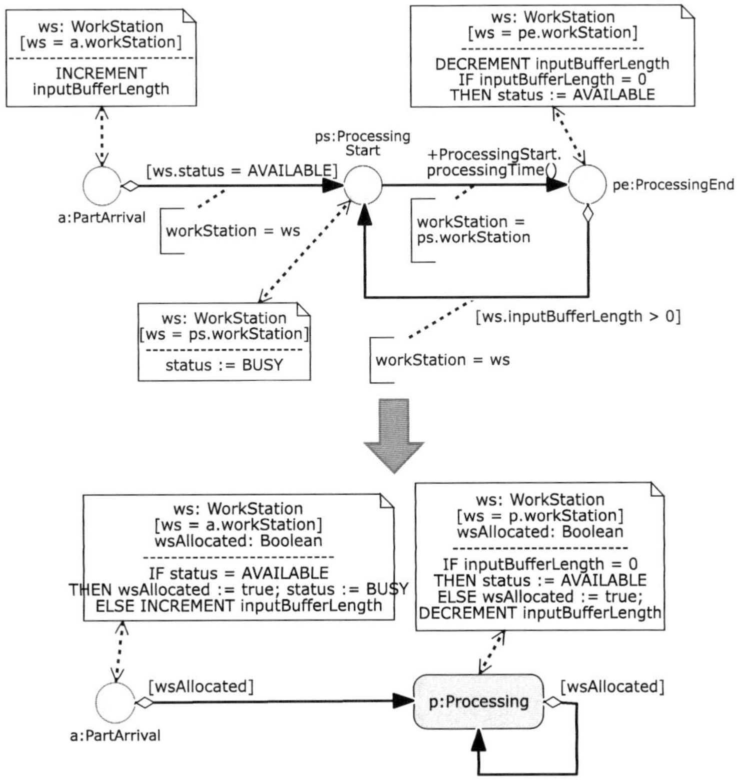

Figure 5 *Rewriting an OE Graph to a corresponding Activity Network model based on the target OE class model of Figure 4*

2.3 From Activity Networks to Processing Networks

The concept of discrete *Processing Networks* is a generalization of the Operations Research concept of *Queueing Networks*. A *Processing Object* enters a processing network via an *Arrival* event at an *Entry*

Station, is subsequently routed along a chain of *Processing Stations* where it is subject to *Processing Activities*, and finally exits the network via a *Departure* event at an *Exit Station*. In typical definitions of a queueing network, the processing station is the only required resource of the processing activities performed at that station, while in a processing network, processing activities can have many required (and optional) resources of various types being allocated with various methods.

PNs have been investigated in *operations management* (Loch 1998) and in the mathematical theory of queuing (Williams 2016), and have been the application focus of most industrial simulation software products, historically starting with GPSS (Gordon 1961) and SIMAN/Arena (Pegden and Davis 1992). They allow modelling many forms of *discrete processing processes* as can be found, for instance, in the manufacturing and services industries.

In the field of DES, PNs have often been characterized by the narrative of "entities flowing through a system". In fact, while in Activity Networks (DPMN-A), there is only a flow of events (including activities), in Processing Networks (DPMN-PN), this flow of events is over-laid with a flow of (processing) objects.

It is remarkable that the PN paradigm has dominated the DES software market since the 1990s and still flourishes today, often with object-oriented and "agent-based" extensions. Its dominance has led many simulation experts to view it as a synonym of DES, which is a conceptual flaw because the concept of DES, even if not precisely defined, is clearly more general than the PN paradigm.

A *Processing Activity* is a resource-constrained activity that is performed at a *processing station* and takes one or more objects as inputs and processes them in some way (possibly transforming them from one type of object to another type), creating one or more objects as outputs. The processed objects have been called "transactions" in GPSS and "entities" in SIMAN/Arena, while they are called *Processing Objects* in DPMN.

Each node in a PN model represents both an object and a typed event variable. An *Entry Node* represents both an *Entry Station* (e.g., a reception area or an entrance to an inventory) and an *Arrival* event variable. A *Processing Node* represents both a *Processing Station* (e.g., a workstation or a room) and a processing activity variable. An *Exit Node* represents both an *Exit Station* and a *Departure* event variable. A *Processing Flow* arrow connecting two processing nodes represents both an event flow and an object flow. Thus, the node types and the flow arrows of a PN are high-level modelling concepts that are overloaded with two meanings.

The (entry, processing and exit) stations of a PN define locations in a network space, which may be abstract or based on a two- or three-dimensional Euclidean geometry. Consequently, PN models are spatial simulation models, while Object Event Graphs and Activity Networks allow to abstract away from space. When a processing object is routed to a follow-up processing station, it moves to the location of that station. The underlying space model allows visualizing a PN simulation in a natural way with processing objects as moving objects.

A PN modelling language should have elements for modelling each of the three types of nodes. Consequently, DPMN-A has to be extended by adding new visual modelling elements for entry, processing and exit nodes, and for connecting them with processing flow arrows.

The simulation modelling concepts of the PN paradigm have been adopted by most DES software products, including Arena, Simio and AnyLogic. However, each of these products uses its own variants of the PN concepts, together with their own proprietary terminology (and proprietary diagram language), as illustrated by Table 2.

Table 2 *Different terminologies used for the same PN modelling concepts*

OEM/DPMN	Arena	Simio	AnyLogic
Processing Object	Entity	Token	Agent
Entry Node	Create	Source	Source
Processing Node	Process	Server	Seize+Delay+Release (or Service)
Exit Node	Dispose	Sink	Sink

For accommodating PN modelling, the Activity Network modelling language OEM-A is extended by adding pre-defined types for processing objects, entry nodes, arrival events, processing nodes, processing activities, exit nodes and departure events, resulting in *OEM-PN*.

As an illustrating example, we re-model the workstation system (considered above) as a Processing Network. Part arrivals are modeled with an «entry node» element (with name "partEntry"), the workstation is modeled with a «processing node» element, and the departure of parts is modeled with an «exit node» element (with name "partExit").

DPMN is extended by adding the new modelling elements of PN Node rectangles, representing node objects with associated event types, and Processing Flow arrows, representing combined object-event flows. PN Node rectangles take the form of stereotyped UML object rectangles, while PN Flow arrows have a special arrow head, as shown in Figure 6.

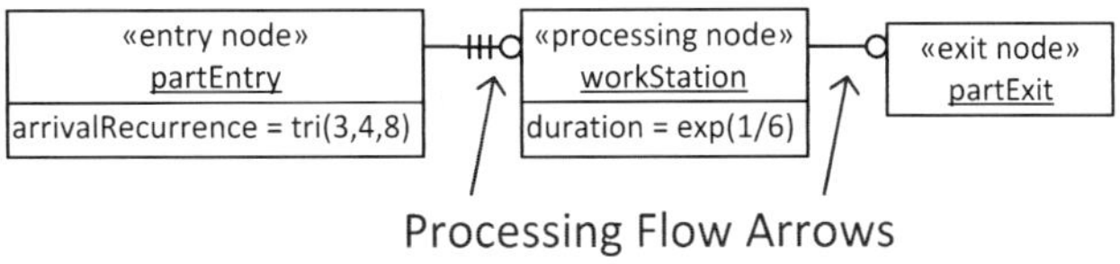

Figure 6 *A PN model of a workstation system using node rectangles and Processing Flow arrows*

The PN model implicitly defines a process model where the workstation process node stands both for a processing activity and a processing station resource object, as shown in the upper part of Figure 7, while using the built-in types shown in the lower part of Figure 7. Since the PN model is completely based on built-in types, no separate OE class model is needed. Notice how the entry node's *arrivalRecurrence* function defines the *recurrence* function of the corresponding *Arrival* event class, and the processing node's *duration* function defines the *duration* function of the corresponding *ProcessingActivity* event class.

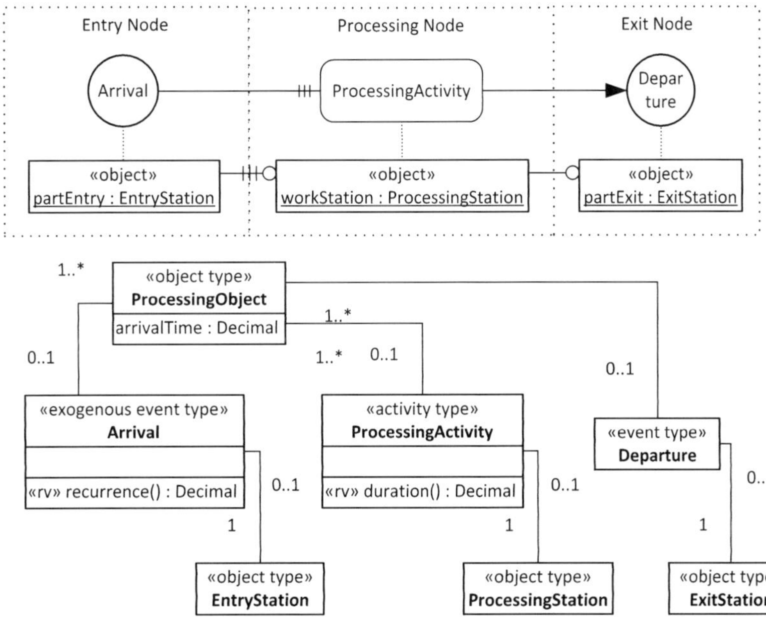

Figure 7 *The elements implicitly defined (or used) by the workstation PN model of Figure 6*

3 CASE STUDY: MAKE AND DELIVER PIZZA

We consider a simple model of a Pizza Service with three consecutive activities: (1) take order, (2) make pizza, and (3) deliver pizza, in a company with 2 order takers, 6 pizza makers, 3 ovens and 10 delivery scooter drivers and scooters. For getting a quick impression, you can run this model from the sim4edu.com website.

For lack of space, we omit discussing the conceptual modelling of this problem and jump directly to the simulation design with DPMN. For a conceptual model, see https://sim4edu.com/reading/des-engineering/make-and-deliver-pizza.

In our simulation design, we make the following simplifications. We consider only one particular pizza service company, which does not have to be modeled as an explicit object. Also, we abstract away from individual customers, orders and pizzas. And we merge the resource roles *delivery scooter driver* and *scooter*, keeping only *scooter*s as resources of *deliver pizza* activities.

We consider a scenario with the following resource pools: 2 order takers, 6 pizza makers, 3 ovens and 10 scooters.

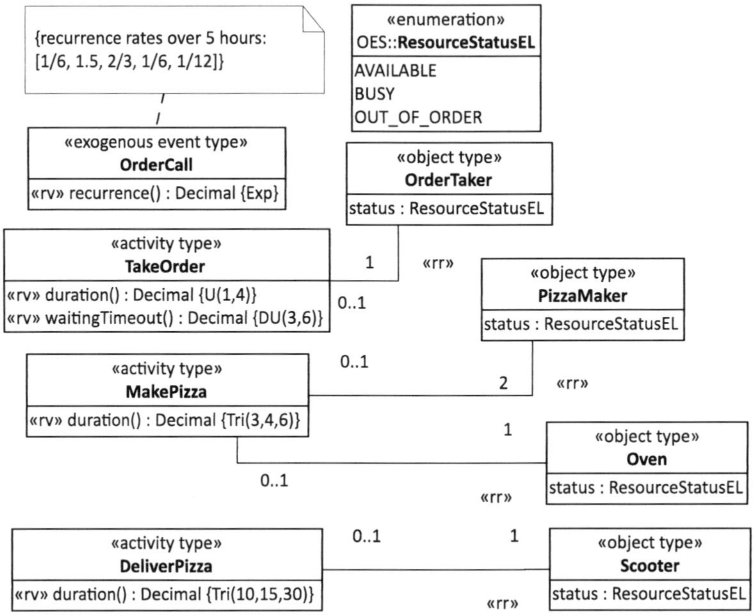

Figure 8 *An OE class model defining object, event and activity types for the Make-and-Deliver-Pizza process model*

Notice that the association end stereotype «rr» stands for "resource role". A resource role assigns resource objects to activities. The OE class model of Figure 8 specifies, for instance, that a *MakePizza* activity requires an oven and two pizza makers as resources.

Notice how functions representing random variables, like the duration function of all activity types, are marked with the keyword (or UML 'stereotype') «rv» standing for "random variable". These random variable functions sample from a probability distribution function (PDF), which is symbolically indicated with expressions like *Tri(30,40,50)* standing for the *triangular* PDF with lower and upper bounds 30 and 50 and a median of 40, or *DU(1,4)* standing for the *discrete uniform* PDF with lower and upper bounds 3 and 6.

In the case of the event type *OrderCall*, the random variable function *recurrence* samples from an *exponential* PDF with five different event rates given for the five consecutive hours during which the pizza service operates.

The activity type *TakeOrder* is associated with the object type *OrderTaker* via the implicit resource role *orderTaker* (with a resource cardinality constraint of "exactly 1"), indicated by the association end stereotype «rr».

Likewise, *MakePizza* is associated with *PizzaMaker* and *Oven* via the (implicitly named) resource roles *pizzaMakers*, having a resource cardinality constraint of "exactly 2", and *oven*, having a resource cardinality constraint of "exactly 1".

An OE class design diagram (like the one of Figure 8) defines resource roles (like *pizzaMakers*), resource role types (like *PizzaMaker*) and resource cardinality constraints (like "exactly 2") for all types of activities. Normally, in an OE simulation there is a one-to-one correspondence between resource role types and resource pools. By convention, a resource pool has the same name as the corresponding resource role type, yet pluralized and starting with a lowercase character. For instance, the name of the resource pool for the resource role type *PizzaMaker* is *pizzaMakers*.

In a simulation model, a resource pool can be implemented in one of two ways:

1. As a **count pool**, which abstracts away from individual resource objects and only counts their numbers.
2. As an **individual pool**, which represents a collection of individual resource objects such that their state changes can be tracked.

Notice that *OrderCall* events are exogenous, having a recurrence function defined case-wise for each of the five hours per day operation of the pizza service company (in the attached invariant box).

For implementing the waiting timeout event defined in the process model, the activity type *TakeOrder* has a class-level *waitingTimeout* function implementing a random variable with PDF *U(3,6)*.

A DPMN process design model (like the one shown in Figure 9) essentially defines the admissible sequences of events and activities (together with their dependencies and effects on participating objects), while its underlying OE class model (like the one shown in Figure 8) defines the types of objects, events and activities, together with the participation of objects in events and activities, including the resource roles of activities, as well as resource cardinality constraints.

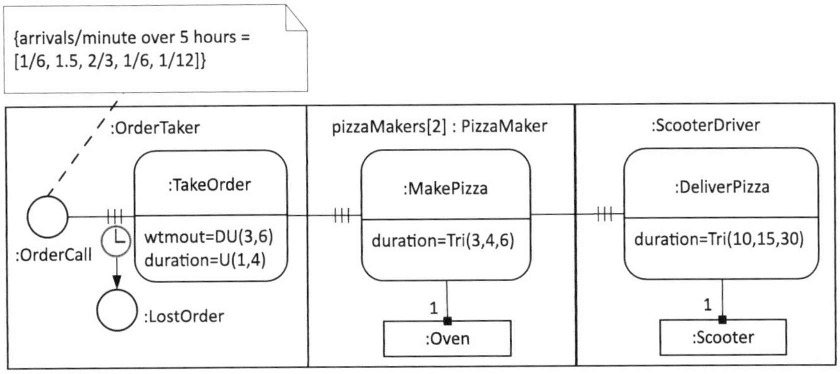

Figure 9 *A process design model for the Make-and-Deliver-Pizza business process*

The process model shown in Figure 9 is enriched by definitions of items from its underlying OE class model, such as activity duration functions or resource dependencies. Such an enriched DPMN process design model includes all computational details needed for an implementation without a separate explicit OE class design model, which is defined implicitly. For instance, the enriched DPMN model of Figure 9 implicitly defines the OE class model of Figure 8 above.

Notice that in the model of Figure 9, performer roles are defined in the form of *Lanes* consisting of a name (such as *pizzaMakers*) and a performer type name (such as *PizzaMaker*) denoting its range, separated by a colon. When the performer role name is appended by a multiplicity expression in brackets, as in *pizzaMakers[2]*, this denotes a resource cardinality constraint (stating that exactly 2 pizzaMakers are required). When only a performer type prefixed with a colon (such as *:OrderTaker*) is provided, this means that the implicit performer role name is obtained by lowercasing the performer type name (as in *orderTaker:OrderTaker*).

The model of Figure 9 does not include any element representing a resource pool. It is assumed that for any organizational position described in the underlying OE class model, the organization under consideration has a corresponding resource pool. By default, each resource role of an activity type is associated with a resource pool having the same (yet pluralized) name, such that its resource objects are instances of a corresponding resource role type, which is an organizational position in the case of human resources.

In the following subsections, we show how to implement the Make-and-Deliver-Pizza business process model shown in Figure 9 with AnyLogic and Simio. In addition to the *simulation model* expressed by this DPMN process diagram, we also need the data of a *simulation scenario*, like the initial states of variables and objects, including the resource pools, for being able to run a simulation. We use a baseline scenario with 2 order takers, 6 pizza makers, 3 ovens, and 10 scooters.

Since Simio and AnyLogic do not support Activity Networks, but only Processing Networks (PN) with "entities flowing through the system", we need to impose a PN view on the Make-and-Deliver-Pizza business process. This requires to figure out what could be used as "entities" for being able to make a PN model.

Since in the real pizza service system there are no entities that flow through the system, we need to assume an artificial abstract entity like "the order", which arrives at the order taker and then takes the form of the ordered pizza being delivered to the customer.

Notice that both the AnyLogic process diagram of Figure 10 and the Simio process diagram of Figure 11 are visually harder to read than the corresponding DPMN Activity Network diagram of Figure 9. They also miss displaying certain information, such as the association of resource pools to processing nodes.

3.1 Implementation with AnyLogic

An enriched DPMN process design model, like the Make-and-Deliver-Pizza process model above, can be implemented with AnyLogic's *Process Modelling Library* by taking the following steps:

1. An exogenous event circle, like *OrderCall*, is turned into an entry node *OrderCall* (an AnyLogic "Source" element). Since the *OrderCall* event recurrence is defined case-wise for each of the five hours of the pizza service's operation, we set the field *Arrivals defined by* to "Rate schedule" and set the field *Rate schedule* to the schedule that defines the arrivals ("ArrivalSchedule").
2. The *TakeOrder* activity rectangle is implemented as a corresponding processing node (an AnyLogic "Service" element). Its performer role *orderTaker:OrderTaker* (in the diagram above abbreviated by *:OrderTaker*) is turned into an AnyLogic resource pool "orderTakers" with its *Capacity* field set to 2. The "Service" element's *Seize* field is set to "units of the same pool" and its *Resource pool* field is set to the previously defined pool "orderTakers" (by selecting it from the drop-down list). The field *Delay time* is set to an AnyLogic Java expression corresponding to the activity's duration value: *uniform(1,4)*.
3. Likewise, the *MakePizza* activity rectangle is implemented as a corresponding "Service" element with two resource pools "pizzaMakers" and "ovens" modeled after the corresponding performer roles (specifying a resource multipliclity of "2" for "ovens" in the field *Resource sets*). The field *Delay time* is set to *triangular(3,6,4)*.
4. Likewise, the *DeliverPizza* activity rectangle is implemented as a corresponding "Service" element in a similar way as *MakePizza*.
5. Finally, the AnyLogic model is completed by appending a "Sink" element (called *DeliveredOrder*) to the *DeliverPizza* "Service" element. An AnyLogic "Sink" element represents an exit node, which is needed for a Processing Network, while we don't need it in the DPMN Activity Network model of Figure 9.

After performing these steps, the diagram shown in Figure 10 is obtained.

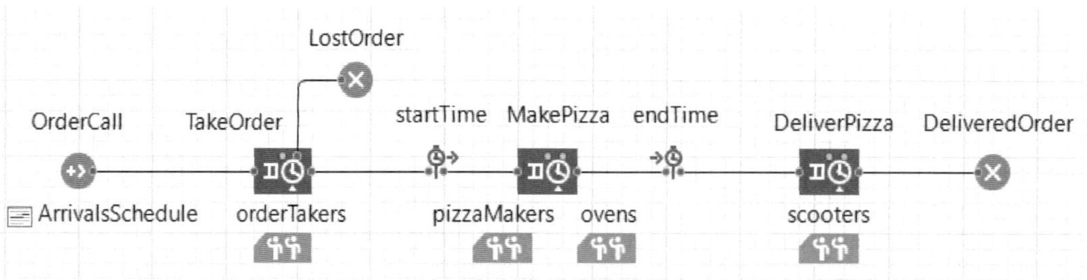

Figure 10 *An AnyLogic process diagram for the Make-and-Deliver-Pizza business process*

Notice that when using a *Service* element for implementing a processing node, all resource pools are individual pools, and the option of using the simpler concept of count pools is not available. A resource object of a pool is by default an instance of AnyLogic's built-in *Agent* Class, if no custom resource type is defined for the pool. Confusingly, resource objects, which are 'agents', are called 'resource units'.

3.2 Implementation with Simio

For implementing the Make-and-Deliver-Pizza model with Simio, we use Simio's *Facility* view and its Standard Library where the PN modelling concept of an *entry node* is called "Source", a *processing node* is called "Server" and an *exit node* is called "Sink". We (1) first drag and drop a "Source" element from Simio's Standard Library and rename it to *OrderCall*, followed by (2) three "Server" elements, renamed to *TakeOrder*, *MakePizza* and *DeliverPizza*, followed by (3) a "Sink" element renamed to *ReceivePizza*. In addition, we define:

1. two individual Resource objects *orderTaker1* and *orderTaker2*, which are placed in a Simio Object List *orderTakers* representing an individual resource pool for the *TakeOrder* activity;
2. six individual Resource objects *pizzaMaker1*, ..., *pizzaMaker6*, which are placed in a Simio Object List *pizzaMakers* representing an individual resource pool for the *MakePizza* activity;
3. two count pools *ovens* and *scooters* in the form of Simio Resource objects with capacities 3 and 10, respectively.

This results in the Simio process diagram shown in Figure 11.

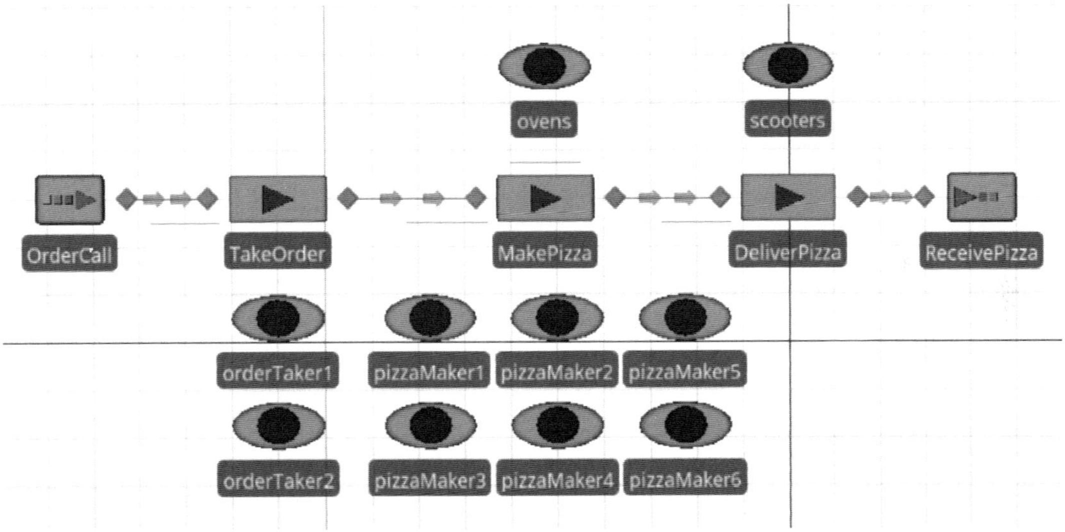

Figure 11 *A Simio process diagram for the Make-and-Deliver-Pizza business process*

It is remarkable that Simio does not have an explicit resource pool concept, which would allow to simplify the model.

REFERENCES

Banks J, Carson J S, Nelson B L, and Nicol D M (2010). *Discrete-Event System Simulation*. 5th ed. Upper Saddle River, New Jersey: Prentice-Hall, Inc.

Banks J (1998). Principles of Simulation. In: Banks J (eds). *Handbook of Simulation*. New York: John Wiley & Sons, Inc.

BPMN (Version 2.0), 2011. http://www.omg.org/spec/BPMN/2.0, accessed 5th July 2021.

Gordon, G. 1961. A general purpose systems simulation program. In *AFIPS '61: Proceedings of the Eastern Joint Computer Conference*, 87–104, New York: Association for Computing Machinery.

Guizzardi, G., and G. Wagner. 2010. Towards an Ontological Foundation of Discrete Event Simulation. In *Proceedings of the 2010 Winter Simulation Conference*, edited by B. Johansson, S. Jain, J. Montoya-Torres, J. Hugan, E. Yücesan. 652–664. Piscataway, New Jersey: IEEE.

Gurevich, Y. 1985. A New Thesis. *Abstracts, American Mathematical Society*, 6(4):317.

Loch, C.H. 1998. Operations Management and Reengineering. European Management Journal, 16, 306–317.

Markowitz, H., B. Hausner, and H. Karr. 1962. *SIMSCRIPT: A Simulation Programming Language*. Englewood Cliffs, N. J.: Prentice Hall.

Pegden, C.D. and D.A. Davis. 1992. Arena: a SIMAN/Cinema-Based Hierarchical Modelling System. In *Proceedings of the 1992 Winter Simulation Conference*, edited by J.J. Swain, D. Goldsman, R.C. Crain, and J.R. Wilson, 390–399. Piscataway, New Jersey: IEEE.

Pegden, C.D. 2010. Advanced Tutorial: Overview of Simulation World Views. In *Proceedings of the 2010 Winter Simulation Conference*, edited by B. Johansson, S. Jain, J. Montoya-Torres, J. Hugan, and E. Yücesan, 643–651. Piscataway, New Jersey: IEEE.

Schruben, L.W. 1983. Simulation Modelling with Event Graphs. *Communications of the ACM* 26:957–963.

Tocher, K. D., and D. G. Owen. 1960. The automatic programming of simulations. In *Proceedings of the Second International Conference on Operational Research*, ed. J. Banbury and J. Maitland, 50–68. London: The English Universities Press Ltd.

Wagner, G. 2022. *Discrete Event Simulation Engineering*. https://sim4edu.com/reading/des-engineering/, accessed 1 December 2022.

Wagner, G. 2021. Business Process Modelling and Simulation with DPMN: Processing Activities. In *Proceedings of the 2021 Winter Simulation Conference*, edited by S. Kim, B. Feng, K. Smith, S. Masoud, Z. Zheng, C. Szabo and M. Loper. Piscataway, New Jersey: IEEE.

Wagner, G. 2020. Business Process Modelling and Simulation with DPMN: Resource-Constrained Activities. In *Proceedings of the 2020 Winter Simulation Conference*, edited by K.-H. Bae, B. Feng, S. Kim, S. Lazarova-Molnar, Z. Zheng, T. Roeder, and R. Thiesing. 45–59. Piscataway, New Jersey: IEEE.

Wagner, G. 2018. Information and Process Modelling for Simulation – Part I: Objects and Events. *Journal of Simulation Engineering* 1:1–25.

Wagner, G. 2017. An Abstract State Machine Semantics for Discrete Event Simulation. In *Proceedings of the 2017 Winter Simulation Conference*, edited by W. K. V. Chan, A.D'Ambrogio, G. Zacharewicz, N. Mustafee, G. Wainer, and E. Page. 762–773. Piscataway, New Jersey: IEEE.

Williams, R.J. 2016. Stochastic Processing Networks. *Annual Review of Statistics and Its Application* 3:1, 323–345.

AUTHOR BIOGRAPHY

GERD WAGNER is Professor of Internet Technology in the Dept. of Informatics, Brandenburg University of Technology, Germany. After studying Mathematics, Philosophy and Informatics in Heidelberg, San Francisco and Berlin, he (1) investigated the semantics of negation in knowledge representation formalisms, (2) developed concepts and techniques for agent-oriented modelling and simulation, (3) participated in the development of a foundational ontology for conceptual modelling, the *Unified Foundational Ontology (UFO)*, and (4) created a new Discrete Event Simulation paradigm, *Object Event Modelling and Simulation (OEM&S)*, and a new process modelling language, the *Discrete Event Process Modelling Notation (DPMN)*. Much of his recent work on OEM&S and DPMN is available from sim4edu.com and dpmn.info.

Proceedings of the Operational Research Society Simulation Workshop 2023 (SW23)
C. Currie and L. Rhodes-Leader, eds.
https://doi.org/10.36819/SW23.005

INTRODUCTION TO HYBRID SIMULATION MODELLING

Sally Brailsford

Southampton Business School
University of Southampton
Southampton SO17 1BJ, UK
scb@soton.ac.uk

ABSTRACT

This tutorial paper, which is an update on a tutorial given at SW18, provides a basic introduction to hybrid simulation modelling as understood by an operational researcher. Hybrid simulation is defined as a modelling approach that combines two or more of the following simulation methods: agent-based simulation, discrete-event simulation and system dynamics. It has gained considerably in popularity in recent years, partly due to the availability of commercial software for developing hybrid models, and partly due to the capability of hybrid models to tackle different aspects of the same problem situation. The session itself will include a hands-on introduction to hybrid simulation modelling in AnyLogic.

Keywords: Hybrid simulation, Agent-based modelling, Discrete-event simulation, System dynamics

1 INTRODUCTION

In this tutorial paper hybrid simulation is defined as a modelling approach that combines two or more of the following simulation methods: agent-based simulation (ABS), discrete-event simulation (DES) and system dynamics (SD). It adopts an operational research (OR) perspective, in other words models developed by operational researchers with the ultimate aim of tackling real-world problems in a pragmatic way, and that use at least two of DES, SD or ABS as they are traditionally described in standard OR textbooks. This definition excludes physics/engineering type simulation models that have both continuous and discrete variables, 'human in the loop' type gaming simulations that combine computer models with human decision-makers, and models that combine simulation with other OR methods, e.g. simulation/optimization. This paper assumes that readers are familiar with the basics of the three individual approaches: for those who are not, see for example Brailsford, Churilov and Dangerfield (2014), which contains introductory tutorial chapters on all three methods. The paper is an update on a tutorial given at SW18 and also includes some of the findings from an EJOR invited review on hybrid simulation (Brailsford et al, 2019).

1.1 Mixing Methods in OR

Mixing OR methods in general, i.e. using more than one method to tackle a particular problem, is far from new. The literature on mixed methods in OR dates back at least to the 1980's and possibly earlier. Jackson and Keys (1984) argue that since all OR methods have different strengths and weaknesses, mixing methods offers the potential to overcome some of the drawbacks of using a single approach. Peter Bennett (1985) discusses three levels at which different OR methods could be applied. The lowest level, Comparison, involved using two methods entirely separately, for the purpose of solving different aspects of a problem which either method used on its own could not tackle. The next level, Enrichment, aims to enhance one method (the main method) by using elements of the other. The highest level, Integration, treats the methods on an equal footing and uses elements of each to generate something totally new. The above definition of hybrid simulation excludes comparison but includes the other two levels.

Combining simulation methods in particular has a strong appeal for OR modellers. Most real-world problems and systems are complex, with different components or subsystems each exhibiting a range of features and characteristics, and rarely is one single simulation approach used on its own ideally suited to capture all of them. The modeller who chooses to use only one method (for the sake of argument, DES) is therefore faced with a dilemma: to model everything in DES, accepting that it won't really work for some parts of the problem, or to model only those parts of the problem for which DES is suitable and simply say that the remaining parts are out of scope? The former approach may lead to a poor model, but equally, it may be neither useful nor sensible to study only one aspect of the system in isolation since the key problem for the decision-maker may be caused by the interaction effects of multiple aspects.

1.2 Structure of this Paper

The rest of this paper is structured as follows. Section 2 provides a condensed history of hybrid simulation. Section 3 presents some frameworks for combining methods while Section 4 summarises the main findings from an EJOR invited review on hybrid simulation (Brailsford et al, 2019). Section 5 briefly presents four case studies in which hybrid simulation is used, and Section 6 is a short introduction to AnyLogic, which will be the main focus of the tutorial session at the conference. Finally, Section 7 discusses the challenges of hybrid modelling and sets out a potential future research agenda.

1.3 Conflict of Interest (or not)

This tutorial paper is definitely NOT an advertisement for AnyLogic, although it may feel like that in places. I have no connection with the company that sell it, and will get no commercial benefit (or indeed any other kind of benefit) from encouraging people to experiment with it.

2 HYBRID SIMULATION MODELLING: A BRIEF HISTORY

DES has been an established tool in the "OR toolbox" since the 1960s. SD has been around for a similar length of time (Forrester, 1961) but was less widely taught on MSc programmes in the 1980s and 1990s and has only gained widespread acceptance in the mainstream OR community within the last twenty years or so. For a long time simulation was synonymous with DES for many OR people, and it is only relatively recently that the scope of the *Journal of Simulation* was extended to include SD and ABS. ABS is definitely the "new kid on the block" as far as mainstream OR is concerned, although it too has been around for many years. In its computerized form it became popular in the 1980s, in the new disciplines of computer science and artificial intelligence. However its origins lie in social science and date back to a time before computers were widely available: Schelling's famous segregation model (Schelling, 1971) was implemented on a real chequerboard.

The burgeoning popularity of SD among the OR community in the early 2000s gave rise to considerable interest in comparing DES and SD. Several authors discussed which approach should be used and when (Brailsford and Hilton, 2001; Brailsford, Churilov and Liew, 2003; Morecroft and Robinson, 2006) while others compared the differences in model-building approaches by users of DES and SD (Tako and Robinson, 2009). I had always thought the idea of actually combining simulation approaches in the field of OR was relatively new, but when writing a paper for the History Track at the 50th Anniversary Winter Simulation Conference on the history of simulation in healthcare (Brailsford, Carter and Jacobson, 2017), I discovered a paper from 1977 (written by modellers who would definitely describe themselves as OR people) which described a hybrid continuous/discrete model of the primary health care system in Indiana (Standridge et al. 1977). Hybrid simulation is clearly not quite such a new idea as Nav Mustafee, Tillal Eldabi and I thought when we started the Hybrid Simulation mini-track in WSC'14, which is now a full track.

Many DES packages provide a limited facility to model continuous processes as well as discrete events, and so (in theory) can be adapted to represent some aspects of SD models. Some also have the capability to model limited aspects of ABS: see example 5.4 below. Most SD software has the capability to employ probabilistic sampling and also includes devices like "conveyors" in Vensim to model activity durations. However despite these valiant efforts, these packages remain essentially either a DES tool with some additional features bolted on, or an SD tool with some discrete or stochastic features. A modeller trying to develop a hybrid model in any of these basically has to force a square peg into a round hole, and make the software do a job it was not really designed to do.

Compared with DES, and even with SD, there are relatively few ABS software packages. Most of the available tools (e.g. Netlogo and Repast) were primarily developed for academic research purposes and hence building models in them involves writing code. While this obviously provides great flexibility (and also means they can be more easily combined with aspects of DES or SD), these packages are rarely taught outside computer science degree programmes. The first, and still the only, commercial software tool that was purposely designed from the start to allow modellers to develop practical hybrid simulation models in all three methods is AnyLogic (www.anylogic.com) which now has a nice graphical interface that allows the user to drag-and-drop icons on the screen and use dialog boxes to enter model parameters, etc.

3 FRAMEWORKS FOR COMBINING SIMULATION METHODS

In the early 2000s, the OR literature on hybrid simulation focused mainly on combining DES and SD. One of the most frequently referenced papers from this period is Chahal and Eldabi (2008), who identify three modes in which DES and SD can be combined. The simplest is the *Hierarchical* mode in which two separate models pass data from one to the other in a sequential manner. In the *Process Environment* mode there are still two distinct models, but the DES model actually sits inside the SD model and models a small section of the system, which then interacts dynamically with the wider SD environment. Finally, in the genuine *Integrated* mode, there is one seamless model with no clear distinction between the DES and SD parts.

More recently, Morgan et al (2015) identified five modes of interaction between simulation methods. While this paper again focused only on DES and SD, these can be extended to include ABS.

a) *Parallel*: this includes Bennett's comparison mode. Two (or more) totally independent models are developed either for direct comparison or to address separate aspects of a problem, and thus even if the results are subsequently combined this does not count as hybrid under our definition.
b) *Sequential*: two or more distinct single-method models that are executed sequentially (but only once), so that the output of one becomes the input to another.
c) *Enriching*: one dominant method with limited use of other method(s). This arguably includes Chahal and Eldabi's Process Environment mode as a special case (i.e. when SD is dominant and the DES component is relatively minor).
d) Interaction: distinct but equal single-method submodels that interact cyclically at runtime. This is in essence Chahal and Eldabi's Hierarchical mode.
e) *Integration*: one seamless model in which it is impossible to tell where one method ends and another begins.

The EJOR review (Brailsford et al, 2019) defines four modes. *Enriching* and *Integration* are the same as Morgan et al's, but in the revised *Sequential* mode the submodels are always executed in a predetermined pattern (e.g. A-B-A-B-A-B-...) whereas in the revised *Interaction* mode the execution order is determined dynamically at runtime depending on the system state, and hence cannot be specified in advance.

4 KEY FINDINGS FROM THE EJOR REVIEW

139 papers met the inclusion criteria and of these, 109 described an actual model (the other 30 were purely conceptual or theoretical). Table 1 presents the breakdown of the 109 models by method and hybridization mode. 'Other' means that it was not possible, based on the information provided, to determine how the submodels were connected; and for two papers, it was not entirely clear what simulation methods were used!

Table 1 *Type of hybridization, by methods used. Source: Brailsford et al (2019)*

Form of Hybridization	Simulation methods used					
	ABS+DES	ABS+DES+SD	SD+ABS	SD+DES	Unclear	Total
Interaction	15	11	16	21	1	64
Sequential	2	1	4	19	-	26
Enriching	-	1	2	4	-	7
Integration	2	-	1	1	-	4
Other	1	1	2	3	1	8
Total	20	14	25	48	2	109

Of these 109 models, only 11 were coded from scratch in a standard programming language. By far the most popular tool was AnyLogic (47 models) with the next most popular, Arena, some way behind (13 models). Three distinct methods of combining methods were identified:

- *Manual*: the modeller literally types in (or copies & pastes) the output from one software package into another [5 models];
- *Intermediary tools*: two (or more) different software packages are linked via a common interface which enables them to pass data from one to another, e.g. MS Excel, distributed simulation software, programming languages and databases for information exchange [21 models];
- *Automated*: the entire model is contained within one single software package that handles all aspects of integration [67 models]. All AnyLogic models fall into this category, although as noted above other packages offer some limited kind of of automation.

In terms of application areas, healthcare was the most popular (22%) closely followed by supply chain logistics (19%) and manufacturing (17%). This is perhaps unsurprising given the interests of the leading early researchers in hybrid simulation, but also reflects the complexity and interconnected nature of problems in these areas. It is notoriously difficult to define sensible model boundaries for healthcare problems and by definition, supply chains involve complex cross-organisational networks. As for my personal 'hobby horse', real-world implementation of model findings, as expected for a relatively new approach, this was disappointingly low with only 3% of papers reporting any kind of practical use. However the review only considered papers published in the academic literature up to December 2016. By the time of this conference, given the rapid growth in publications since then, it is likely that many more hybrid models have been used in practice. The consultancy pages on the AnyLogic website certainly suggest this is the case!

5 ILLUSTRATIVE EXAMPLES

As a rule, SD is used for the "big picture" system-wide aspects, or for aspects of a problem where variability is less important; DES is used for detailed sub-models (typically, stochastic resource-constrained queuing systems) where individual variability matters; and ABS is used to capture behavioural or spatial aspects.

However in the first example, which uses all three methods, SD was used at the lowest level (i.e. inside each patient agent) to model disease progression.

5.1 The AMD Model (SD + ABS + DES)

Age-related macular degeneration (AMD) is a serious but fairly common eye disease which affects older people. Until about 15 years ago it was untreatable, and patients with AMD were left to face blindness and the consequent loss of independence, which increased their need for social care support and often led to placement in residential care. AMD still cannot be cured, but its progression can now be delayed by monthly injections into the eyes which must be given in a hospital outpatient clinic. This, combined with the ageing population, has led to huge demand for AMD clinics. Overcrowded clinics can lead to patients having to leave without being treated, not because they are unwilling to wait but because they are often dependent on hospital transport to take them home. This highlights the lack of coordination between the health and social care systems in the UK, where health care and social care are provided by different organisations and are funded differently. In the case of AMD, the medical treatment is provided and paid for by the NHS, but the main financial benefit is felt by the social care system as patients can now live independently in their own homes for longer. So who should pay for the extra AMD clinics?

Viana et al (2012) developed a hybrid simulation model in AnyLogic which combines an agent-based model representing individuals with AMD with a DES model of the outpatient clinic and an SD model of disease progression. Patient agents contain two simple embedded state transition models which represents the progression of AMD in each eye. Treatment slows down the disease process. Each patient has a social care need status, and this, in conjunction with their level of social care provision, affects their probability of clinic attendance. Social care provision is represented by a statechart which consists of the three states: not required, partly met and fully met. The agents enter the DES model when the scheduled time of their clinic appointment arrives. The patient may (or may not) attend the clinic, and may (or may not) receive treatment, depending on waiting times in the clinic and the overall performance of the clinic.

The original plan was to include the geographical location of the real-world patients in the area served by the hospital, which would also affect their probability of attendance, but it was not possible to obtain the necessary data to do this so the agents were just placed at random. Nevertheless, the model provided a neutral 'sandpit' that enabled stakeholders to explore the consequences of different levels of social care provision.

5.2 The Chlamydia Model (SD + DES)

The sexually transmitted infection (STI) chlamydia is common in young people aged 16-24 and is often asymptomatic. Hence it can be unknowingly transmitted, although once detected it can be easily (although only temporarily) treated by a short course of antibiotics. It is an important public health issue because repeated infections can lead to serious and even life-threatening complications, such as ectopic pregnancy.

Viana et al (2014) present a hybrid simulation in which an SIR (susceptible-infected-recovered) model of the spread of chlamydia is combined with a model of the hospital STI clinic in which patients get treated. Clearly, in reality these two aspects are closely interconnected and affect each other. If the clinic is overcrowded the queues get long: people may be unwilling to wait, and leave without being tested or treated. This then increases the proportion of infected people in the community, leading in turn to an increased number of new infections. DES captures the stochastic nature of the clinic operations and the impact on overall clinic performance of adding extra resources, while SD is a natural choice for population level SIR models. The SD and DES models are totally separate and can in fact be run independently. They were developed in Vensim and Simul8 respectively, linked by an Excel interface. The SD model first runs for one month and generates the demand for that month. This monthly demand is then automatically exported from Vensim into Excel, which disaggregates the demand into time-dependent inter-arrival rates based on historical data, to be used as arrival distributions in Simul8. The DES clinic model then performs 20 iterations of one day, and the average number of people treated per day is automatically exported from

Simul8 through Excel to the SD model. The SD model advances a time step and the whole process starts again. This is therefore a *sequential* type hybrid model; the two submodels run alternately in a predetermined pattern.

In both this model and the AMD model, DES is appropriately used to model resource-constrained queuing systems (hospital outpatient clinics) embedded within, and interacting with, wider complex systems outside the hospital for which DES is less well suited. In both cases, clinic performance has wide-reaching effects. Traditional DES models of outpatient clinics or Emergency Departments, of which there are many hundreds in the academic literature, often treat these as closed systems; the objective is to maximise throughput while minimising resource use. However, in these two examples long waits and low throughput have knock-on effects outside the hospital and can ultimately increase demand. Ignoring these wider system effects only tells half the story, and hence a hybrid approach is useful.

5.3 The Wessex Dementia Model (ABS + SD)

Evenden et al (2021) developed a hybrid model in AnyLogic to predict future numbers of patients with dementia, and the associated care costs, in the Wessex region of southern England. SD is used to represent the total population aged over 65, split into 5-year age bands. In each age band there are two stocks; cognitively normal (CN) and people with dementia (PWD). The flow from CN to PWD is based on population-level rates of dementia onset in that age band. As soon as the cumulative value of this flow crosses an integer boundary (e.g. from 25.8 to 26.1) a new patient agent is created in the ABS part of the model, which represents disease progression at individual patient level. New agents are assigned a specific age, a rate of severity progression (fast, medium or slow) sampled from published data, and a sampled time to death from other causes. Their dwelling times in each state (mild, moderate and severe) are subsequently sampled. As each agent progresses through the various disease stages, their quality of life and resource use are calculated. If an agent ages into the next age band, the relevant PWD stocks in the SD part are adjusted accordingly and if an agent dies (either from dementia or from other causes) the corresponding PWD stock is then reduced by 1. The model was used to evaluate three scenarios: do nothing, a hypothetical medical treatment to slow progression, and the potential impact of an intervention that encourages healthier lifestyles in middle age, which both delays onset and slows progression.

SD is the method of choice for modelling "ageing chains" in large populations. The CN stocks in this model contain many thousands of people as the study region is a popular retirement destination. However dementia progression rates vary considerably between patients, and hence a stochastic individual-level approach is required after the onset of dementia. A hybrid approach is therefore beneficial.

5.4 The Telecare Dementia Model (DES + ABS)

In this context, telecare means the use of personal alarms and devices such as fall detectors or environmental sensors to help people live independently in the early stages of dementia. Such devices monitor for changes (such as a patient leaving their house or letting a saucepan boil dry) and either warn the patient themselves, their carer or another family member, or raise an alert at a control centre. Penny et al (2022) developed a hybrid model in an experimental extended version of the DES software Simul8 that incorporates elements of ABS. The aim of the model was to explore the extent to which telecare enables patients to continue living in their own homes for longer, reducing demand for residential care beds, and improving quality of life for both the individual and their carer.

The DES part represents the flow of patients through the care system, starting from living at home through to admission to residential care. As in real life, periodic assessments are made to check whether the patient needs to be referred to a new service due to a change in their dementia state. Disease progression, together with the 'coping' status of the patient's carer (if they have one) is modelled using statecharts. A change in dementia status can result in their carer no longer being able to cope, and triggers an urgent referral for additional support, since this cannot wait until the next routine assessment.

Here, a hybrid approach overcomes the practical challenge of modelling the impact of disease state change on real-world process type activities. It is possible to use DES to model disease progression by representing the dwelling times in each health state as 'activities'. There are therefore two distinct types of activity in such models; process activities (treatment and investigations) and health status activities. Unfortunately, interrupting an activity once it has started can be technically difficult in some DES software tools. If process activities are relatively short in duration, e.g. in a model of an Emergency Department, it is a reasonable approximation to reality to wait until the end of an activity to check the patient's health status, and then route them to the appropriate place in the care pathway. However if (as here) process activities are measured in months or even years, waiting until such activities end before reassessing the patient's care needs is not a realistic way to represent what actually happens in practice. Hence statecharts lend themselves particularly well to long-term conditions like dementia.

6 ANYLOGIC

Like Netlogo and Repast, AnyLogic has its roots in computer science. It was originally developed by Russian computer scientist Andrei Borshchev in the early 1990s, but has only really gained widespread popularity for OR modelling in the last 10-15 years. A free PLE (Personal Learning Edition) version of AnyLogic is now available for anyone to download and as a result AnyLogic is increasingly being used by students and researchers alike. It is obviously way beyond the scope of this tutorial to provide anything more than a cursory introduction to AnyLogic, but anyone interested in having a play with it is recommended to download the PLE, and work through a few of the examples provided in the free textbook "*AnyLogic in Three Days*" by Ilya Grigoryev (2014), which is free to download.

In the tutorial session, I will lead a hands-on session to demonstrate AnyLogic. The software runs on MS Windows, Mac OS and Linux. Before the session (or even during!) please download and install the free PLE version from https://www.anylogic.com/downloads/personal-learning-edition-download/. It is quite a large file, so downloading *before* the session is advisable…

7 CHALLENGES IN USING HYBRID SIMULATION

Combining simulation methods has many advantages in modelling real-world problems. However, these benefits may come at a price. Experienced users of any one method (or software tool) will have a natural preference to choose the method they know best. It is not quite as simplistic as "when all you have is a hammer, every problem is a nail" but DES experts, say, will tend to conceptualise a problem as a stochastic queuing network, whereas SD experts will conceptualise the same problem as a feedback system. AnyLogic is the first tool which has started from an "agnostic" basis and allows the user to use DES, ABS and SD all within the same environment. AnyLogic embodies a philosophy in which the modeller does not have to begin by thinking "Is this a DES problem, an ABS problem or an SD problem?" Of course there are not "DES problems" in the real world, just problems …

The EJOR review identifies several broad areas where further research is needed. One of these is conceptual modelling. Standard diagramming methods exist for conceptual modelling in ABS, DES and SD, but each is bespoke to that method and there is no overarching general approach that is agnostic to simulation method for conceptual modelling in hybrid simulation. Likewise, there is no recognised standard approach for representing links and modes of information exchange between submodels when combined in a hybrid model. Hence there is a need for a new methodology to capture the problem situation in terms of the subsystems that comprise it, and to provide a rigorous and systematic way to identify the characteristics of each subsystem that indicate the use of a given simulation method.

Another area needing further research is validation and verification (V&V). There are established quantitative methods for DES, but while the use of statistical methods in SD is increasing, V&V of SD models is often qualitative (face validity with domain experts, involving stakeholders in group model building) or fairly basic (dimension checking). For ABS, V&V is even more difficult and is mainly 'black-box' (checking consistency of aggregated results with observed data). There are no established methods for validating the links between submodels, data exchange mechanisms etc. Such methods are essential if problem owners (clients) are to have trust in hybrid models.

8 CONCLUSION

The 2018 version of this paper concluded with some cautious advice to prospective modellers. *'Even today, the decision to develop a hybrid model should not be taken lightly, especially in a consultancy setting where there is a client who needs a quick answer. Suppose you are faced with modelling a problem which has features which suggest DES alone may not be capable of capturing everything. If you are an experienced DES modeller and are very familiar with one particular software tool, try that method alone first and only use a hybrid if you find it is impossible to capture some aspects in a sensible way. Given the relative maturity of DES software, there is a reasonable chance that you will be able to model the required features within the DES environment, using the facilities of the software and with only a small amount of additional coding. Moreover, the full battery of analysis tools standard in such packages - optimization, automatic calculation of the number of iterations, variance reduction, output display tools and so on - will be at your disposal. If you are an experienced SD or ABS modeller, the same applies (although to a lesser extent)*'. While much of this is still true, six years on I would be far less cautious! AnyLogic has become much more user-friendly and has many additional features that were lacking in earlier versions; there is also now a much larger user community.

The 2018 version also stated: '*Software maturity is highly relevant. As yet, hybrid simulation is not routinely used by practitioners: most models are developed by academic researchers, sometimes for genuine practical reasons but often just out of curiosity. AnyLogic is gradually changing this, as anyone who visits their website and looks at their list of clients can see, but it will not happen overnight'*. While it is still true that most published hybrid models are developed by academics, the same is also true for single-method models. There is now clear evidence that hybrid simulation is starting to be used more widely.

I strongly believe that in the long run, hybrid simulation will lead to more useful models that better represent real-world problems and provide better solutions. In conclusion, and bearing in mind that I am not getting paid for promoting AnyLogic, it is definitely worth giving the free version a go!

REFERENCES

Bennett PG (1985). On linking approaches to decision aiding: issues and prospects. *Journal of the Operational Research Society*, **36(7)**: 659-669.

Brailsford SC and Hilton N (2001). A comparison of discrete event simulation and system dynamics for modelling health care systems In, Riley, J. (eds.) *Planning for the Future: Health Service Quality and Emergency Accessibility*. Operational Research Applied to Health Services, Glasgow Caledonian University.

Brailsford SC, Churilov L and Liew S (2003). Treating ailing emergency departments with simulation: An integrated perspective. In: Anderson J (ed). *Proceedings of the 2003 Western Multiconference on Health Sciences Simulation*. Society for Modelling & Simulation International, San Diego, CA.

Brailsford SC, Churilov L and Dangerfield BC (2014). *Discrete-Event Simulation and System Dynamics for Management Decision Making*. John Wiley & Sons, Chichester.

Brailsford SC, Carter MW and Jacobson SH (2017). Five decades of healthcare simulation. In: *Proceedings of the 2017 Winter Simulation Conference*, Las Vegas, NV. W. K. V. Chan, A. D'Ambrogio, G. Zacharewicz, N. Mustafee, G. Wainer, and E. Page, eds.

Brailsford SC, Eldabi T, Kunc M, Mustafee N, Osorio AF (2019). Hybrid simulation modelling in operational research: A state-of-the-art review. *European Journal of Operational Research*, 278(3), 721-737.

Chahal K and Eldabi T (2008). Applicability of hybrid simulation to different modes of governance in UK healthcare. *Proceedings of the 2008 Winter Simulation Conference, Miami, Florida.* S. J. Mason, R. R. Hill, L. Mönch, O. Rose, T. Jefferson, and J. W. Fowler, eds.

Evenden DC, Brailsford SC, Kipps CM, Roderick PJ & Walsh B (2021) Hybrid simulation modelling for dementia care services planning, *Journal of the Operational Research Society*, 72:9, 2147-2159.

Forrester JW (1961). *Industrial dynamics*. MIT Press, Cambridge, MA.

Grigoryev I (2014). *AnyLogic in Three Days*. www.anylogic.com/resources/books/free-simulation-book-and-modeling-tutorials/. Accessed 29.11.17.

Morecroft, J. and Robinson, S. 2006. Comparing discrete event simulation and system dynamics: modelling a fishery. *Proceedings of the 2006 OR Society Simulation Workshop SW06*, 137-148.

Morgan JS, Howick S and Belton V (2014). A toolkit of designs for mixing discrete event simulation and system dynamics. *European Journal of Operational Research*, **257(3)**:907–918.

Penny KEE, Bayer SC & Brailsford SC (2022). A hybrid simulation approach for planning health and social care services, *Journal of Simulation*, DOI: 10.1080/17477778.2022.2035275

Schelling TC (1971). Dynamic models of segregation. *Journal of Mathematical Sociology* **1(2)**: 143–186.

Standridge C, Macal C, Pritsker AB, Delcher H, and Murray R (1977). "A Simulation Model of the Primary Health Care System in Indiana". In *Proceedings of the 1977 Winter Simulation Conference*, 349-358. Piscataway, New Jersey: Institute of Electrical and Electronics Engineers, Inc.

Tako AA and Robinson S (2009). Comparing discrete-event simulation and system dynamics: users' perceptions. *Journal of the Operational Research Society*, **60**:296-312.

Viana J, Rossiter S, Channon AR, Brailsford SC and Lotery AJ (2012). A multi-method, whole system view of health and social care for age-related macular degeneration. *Proceedings of the 2012 Winter Simulation Conference, Berlin, Germany.* C. Laroque, J. Himmelspach, R. Pasupathy, O. Rose, and A.M. Uhrmacher, eds.

Viana J, Brailsford SC, Harindra V and Harper PR (2014). Combining discrete-event simulation and system dynamics in a healthcare setting: a composite model for Chlamydia infection. *European Journal of Operational Research*, **237**:196–206.

AUTHOR BIOGRAPHY

SALLY C. BRAILSFORD is Professor of Management Science at the University of Southampton, UK. She received a BSc in Mathematics from the University of London, and MSc and PhD in Operational Research from the University of Southampton. Her research interests include hybrid simulation modelling methodologies, system dynamics, health service research and disease modelling, and the modelling of human behaviour in healthcare systems. From 2010-19 she was chair of the European Working Group on OR Applied to Health Services (ORAHS) and Editor-in-Chief of the journal Health Systems. She is on the editorial boards of the EURO Journal of Decision Processes, Health Care Management Science and Operations Research for Health Care. Her email address is s.c.brailsford@soton.ac.uk.

Proceedings of the Operational Research Society Simulation Workshop 2023 (SW23)
C. Currie and L. Rhodes-Leader, eds.
https://doi.org/10.36819/SW23.006

WORKSHOP: THE SUPPLY CHAIN BUSINESS GAME

Dr David Exelby

Decision Analysis Services Ltd
Grove House, Lutyens Close, Chineham Court,
Basingstoke, Hampshire, RG24 8AG
DaveExelby@DAS-Ltd.co.uk

Dr Siôn Cave

Decision Analysis Services Ltd
Grove House, Lutyens Close, Chineham
Court, Basingstoke, Hampshire, RG24 8AG
SionCave@DAS-Ltd.co.uk

1 BACKGROUND

The Supply Chain Business Game, also called the Beer Game, is a hands-on, interactive and fun simulation representing the dynamics of a typical supply chain. The purpose of the workshop is to demonstrate how systems cause the behaviour they exhibit, and how unintended consequences can result from system design. The lessons from the beer game are relevant to anyone who relies on supply chains for delivery, from manufacturing industries, technology companies to workforce planners.

The Beer game was created by a group of professors at MIT Sloan School of Management in early 1960s to demonstrate a number of key principles of supply chain management and introduces the players to the concept of the "bullwhip effect", how structure produces behaviour and the importance of information sharing and collaboration.

Papers describing the Beer game (Sterman, 1989; Thompson and Badizadegan, 2015) are listed in the References section. More information is also available here:

- https://systemdynamics.org/products/beer-game/
- https://systemdynamics.org/product/supply-chain-game-the-beer-game-complete-game-set/

2 APPROACH

The Beer Game is a supply chain simulation that lets the participants experience the pressures of playing a role in a production/distribution supply chain. The simulation is played via a large board game with teams of at least four players. Players move pieces around the board and determine what they want to make or purchase and how much stock to hold, all within a fun and competitive environment. The game is facilitated by experts in systems modeling who provide information about the key insights generated through the session.

3 IMPACT

DAS is expert in the delivery of interactive business games, and has run sessions for clients as diverse as diamond merchants, technical consultancies and manufacturing industries and at international academic conferences.

Key learning points from the game include:

- The importance of using real world data to support decision making.
- The need for transparency across the supply chain.
- How a system creates systemic behaviour
- What the "bull whip effect" is and how to recognise it
- The importance of supplier-customer relationships
- How an individual's decisions can affect the system as a whole.
- The complexities of supply chain planning
- Effective decision making and learning require tools to better understand how the structure of
- complex systems influences the behavior of individuals within them and the overall outcomes.

- Use learning to improve culture and have fun as a teambuilding exercise

REFERENCES

Sterman,J. (1989) Modeling managerial behavior: Misperceptions of feedback in a dynamic decision making experiment, *Management Science* 35(3), 321-339

K. M. Thompson and N. D. Badizadegan, (2015) Valuing Information in Complex Systems: An Integrated Analytical Approach to Achieve Optimal Performance in the Beer Distribution Game, *IEEE Access* (3), 2677-2686

Proceedings of the Operational Research Society Simulation Workshop 2023 (SW23)
C. Currie and L. Rhodes-Leader, eds.
https://doi.org/10.36819/SW23.007

METAMODELLING: POWER, PITFALLS, AND MODEL-FREE INTERPRETATION

Russell R. Barton

The Pennsylvania State University
413 Business Building, University Park, PA 16802 USA
rbarton@psu.edu

ABSTRACT

Simulation metamodels have been a valuable tool since before the term was coined. They can provide insight on system behaviour that is not obvious from the simulation model and are a computationally inexpensive proxy for the simulation model that they approximate. But fitting and using metamodels requires care. This tutorial explains the power of metamodeling, while highlighting potential pitfalls in fitting and using various metamodel technologies. Finally, a particular advantage of metamodeling is illustrated: the counterintuitive ability to provide model-free interpretation of the simulation input-output behaviour.

Keywords: Metamodel, Response Surface, Surrogate Model, Response-Scaled Design Plots

1 INTRODUCTION

Metamodels are models of models. That is, they are a mathematical approximation to the implicit input-output function of a simulation model. The simulation model might be discrete-event, agent-based, system dynamics, or one of a variety of engineering simulation models including finite-element models and microwave circuit simulation models. The term originated in the discrete-event simulation context, and was coined by Robert Blanning (Blanning, 1974; 1975) and popularized and developed by Kleijnen (Kleijnen, 1975; 2015). In fact, the use of metamodels is much older, inspired by response surface models in Box and Wilson (1951), and applied in simulation eight years before Blanning by Burdick and Naylor (1966).

Simulation metamodels take as arguments one or more parameters that define the simulation model. For a discrete-event simulation, such parameters might include the mean interarrival time of customers to a service counter, the number of servers, the maximum capacity of the waiting area the probability distribution parameters for service time, and so forth. Given specific values for the argument, a metamodel gives a value that approximates some characterization of the simulation output when operated using the specific parameter values in the argument. For example, a metamodel output might be the mean time in system, or the mean number of busy servers, or the 90th percentile of waiting time.

Metamodels provide valuable information to the simulation practitioner. Simple linear regression metamodels give insight on the first-order impact of simulation model parameters on system performance. For more complex input-output relationships, metamodel structure may not provide insight on system behaviour, as we will see below. But fitted metamodels allow prediction of system performance on a multidimensional grid of system parameter values, and we will see that this permits a graphical representation that allows model-free interpretation of the model behaviour. Metamodels, either local or global, can assist in optimization of some system output characteristic as a function of system design parameters.

The power of metamodels has two dimensions: speed and insight. When the simulation model is time consuming to run, a metamodel allows virtually instantaneous "what-if" analyses by the decision maker. This can result in better decisions than enduring the delays in computed response of the base simulation model, even when the delay in simulation runs is only one second (Simpson et al, 2007). The dimension of insight itself has two parts: efficiency in determining global behaviour (i.e., over all possible parameter values) from a small number of runs at different parameter settings, and characterization of behaviour based on metamodel form or on exercise of the metamodel (e.g., relative importance of different parameters, convexity in response, maximum and minimum values). Metamodels for job completion time as a function of system state can predict completion time quantiles to within a few percent based on a fitting experiment using only a fraction of the large number of possible system states (Pedrielli and Barton, 2019). Further, these models use simple second-order polynomial regression, and the coefficients directly characterize sensitivity, interactions, and nonlinearity.

Combining speed and insight, metamodels permit optimization of the response as a function of the parameter values. For metamodels with simple form, analytic optimization methods can be used. For more complex metamodels, black-box optimization methods can perform efficiently since the function evaluations (metamodel evaluations) are inexpensive computationally (Kleijnen, 2015).

In this tutorial we focus on global metamodels: approximations that have as domain the full feasible range of possible parameter values. Local metamodels are also used in simulation optimization, but this optimization approach is often referred to as response surface methodology. See, for example, Kleijnen (2015), Law (2014) or Myers et al (2009).

While metamodels provide a powerful tool in the simulationist's toolkit, care is required to gain the potential benefits. This tutorial provides some guidance on effective use of simulation metamodels. It begins with a summary of notation and an overall metamodeling process. Next, several metamodel types are constructed to approximate the average number in system for a queue, simple metamodels with a single parameter. Metamodel predictions and the R code for generating the metamodels and plots are included in an appendix. Potential pitfalls are apparent even for this simple case. The next section raises additional pitfalls, discussed in more detail in Barton (2021). Finally, insight is often difficult, even when using second-order linear regression models. The last section illustrates the issue and provides a remedy: response-scaled design plots based on metamodel predictions on a factorial grid (Barton, 1998). All of the examples in this tutorial were programmed in R using RStudio on a PC. A post on my personal page https://sites.psu.edu/russellbarton/ has a link to the code.

2 METAMODELING: FORM AND PROCESS

Mathematical notation is important to clarify the important concepts in metamodeling and in the metamodeling process. The metamodel is represented as a mathematical function of one or more simulation model design parameters. For this tutorial, the output is univariate; multi-output metamodels are either multiple univariate output models, fitted separately, or if there is dependence of coefficients across outputs or dependence of outputs, the resulting complexity is beyond the scope of this tutorial.

2.1 Functional Form

A metamodel is a function denoted f, that takes as its argument simulation model design parameters, represented here by a vector x, to produce an output, $f(x)$. Possible design parameters include input probability distribution parameters such as arrival and service rates, time to failure and time to repair distributions, and routing probabilities. Design parameters can also include system configuration parameters, such as the number of servers, service priority, operational protocols, and buffer capacity. For now, assume that any design parameter can be coded numerically, even if only as a 0-1 variable.

The metamodel $f(x)$ produces a (univariate) approximation to some characterization of a simulation output, denoted Y. Examples of simulation outputs are time in the system for a set of jobs or customers, utilization of a particular resource (e.g., operator, machine), or perhaps net revenue over a specific time period. Generally, these outputs are averaged over the length of a simulation run, but the averages vary randomly from run to run. If the value of the output is $Y(x)$ for an actual simulation run with the design parameters set to the values in x, then we represent the fitted metamodel approximation $f(x)$ as:

$$h(Y(x)) \approx f(x), \qquad (1)$$

where h represents a (univariate) characterization of the random variable Y, such as its mean, standard deviation, or a quantile. To simplify notation, we let y represent observed values of the random variable $h(Y(x))$. Multivariate f and h are possible, but beyond the scope of this tutorial.

2.2 Metamodeling Process

The metamodeling process is part of the overall process of building, validating, and using simulation models for decision support. Metamodeling typically occurs after a simulation model has been validated, although metamodel insights can be helpful in the validation process as well. This tutorial gives a seven-step process:

1. Determine Purpose(s) for Metamodeling,
2. Identify Design Parameter(s), Output(s), and Characterisation, h,
3. Choose Metamodel Type,
4. Based on Metamodel Type and on Purpose, Choose Experiment Design to Fit Metamodel,
5. Conduct Simulation Runs Specified by the Experiment Design; Fit Metamodel,
6. Validate Metamodel Adequacy: If Unsatisfactory Return (usually) to Step 3, and
7. Use Metamodel for Intended Purposes.

As for the simulation models they emulate, metamodels should be designed with specific purpose in mind. What kind of decision support with they provide? This information will guide the choice of important design parameters and their ranges of interest, and the particular simulation output or outputs of interest, along with the characterization function, h. In choosing the metamodel type, simpler, interpretable metamodels such as first- and second-order linear regression are preferred, if the output characterization (perhaps after transformation) is well-approximated by a linear or quadratic function. In cases where this does not obtain, more complex nonlinear models such as Gaussian process (GP) or artificial neural network (ANN) regression methods might be appropriate.

The set of parameters and the metamodel type will affect the choice of experiment design. For example, a factorial experiment design is not appropriate for fitting a GP metamodel (Barton, 1994). The experiment design consists of a set of two or more parameter vector (x) settings, each unique setting called a *design point*. The simulation model is run, possibly with multiple replications, at each of the design points. If the random number streams are independent across design points, the usual statistical analysis methods are available for linear regression models. Sometimes improvements in precision are possible by the use of common random number streams across multiple design points. The random number assignment strategy and analysis method for this case are complex, beyond the scope of this tutorial. The interested reader can find more information in Nozari et al (1987), Schruben and Margolin (1978), and Tew and Wilson (1992). Next, the metamodel should be validated before use. For simple linear regression models, lack of fit, R^2 and statistical significance of model coefficients all provide measures of validity and adequacy. Alternatively, for all metamodel types, prediction accuracy can be tested at a number of randomly chosen points in design space that were not used in fitting the metamodel. These out-of-sample tests compare the model prediction with the simulation result and give an empirical characterisation of metamodel predictive performance.

While the process is simple, there are pitfalls to avoid. In the next section the model construction process is illustrated for a number of different model forms. The emphasis is on steps 3 and 6.

3 METAMODELS FOR A SIMPLE CASE

The metamodels constructed in this section approximate the long-term average number in an $M/M/1$ queueing system as a function of arrival rate. The structure is shown in Figure 1. Customers arrive at rate λ with exponentially distributed interarrival times. If the server is busy, they wait in queue, under a first-in-first-out priority until they are served. The service rate (and mean) is $\mu = 1$. After service they depart the system.

Although it is typical for metamodels to have multiple elements in the argument vector, the ability to see the nature of the fitted metamodel is instructive, and so the argument has only one parameter: the arrival rate. The service rate μ is set to one in the example. As a consequence, the arrival rate λ and the utilization ρ are equivalent. Suppose that the region of interest for prediction is $.75 \leq \lambda \leq .95$. Of course, for this case an analytic form for average number in system is known:

$$E(\textit{Number in System}) = L = \rho/(1\text{-}\rho). \qquad (2)$$

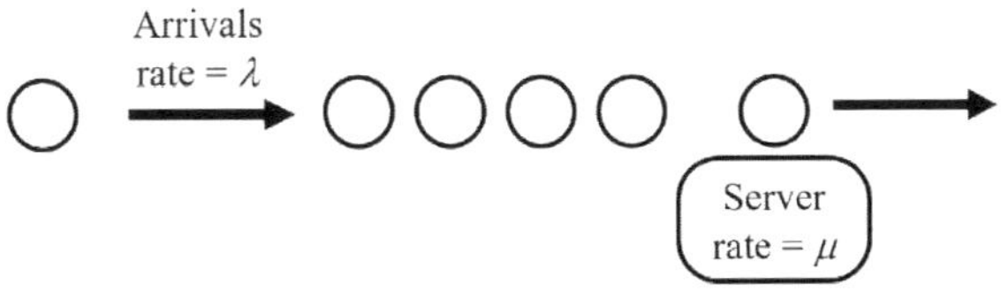

Figure 1 *An M/M/1 Queue*

For this example $\rho/(1\text{-}\rho)$ is equivalent to $\lambda/(1\text{-}\lambda)$. This relationship will be important in choosing a scaling that allows a simple yet high-fidelity metamodel, demonstrated later in this section.

The experiment design used in fitting the metamodel is the same for all metamodel types:

$$\{x \textit{ design points}\} = \{.75, .8, .85, .9, .95\}. \qquad (3)$$

Then the performance of different metamodels can be compared based on identical fitting data. Three forms of metamodels are compared: linear regression, Gaussian process regression, and neural network regression. Although the analytic solution is known, here we will pretend that the expected number in system is estimated by computing the average number in system using a simulation model of an *M*/*M*/1 queue. Four replications of 10,000 time units each, starting empty and idle, produced the simulation model results in Table 1. Note that since the arrival rate varies, the number of customers served will vary both stochastically and across design points. Further, we will soon see that, as often happened in simulation metamodeling, variance is heterogeneous over design points. The data will be used to fit the metamodels, and represented with lowercase y since they are observations of the random variable Y.

Table 1 *Simulation Results for Each Design Point (Arrival Rate)*

Design Point (x)	Results (y = Avg # in System)
.75	2.73 2.77 3.13 3.53
.80	4.68 3.54 3.57 3.93
.85	5.50 8.35 5.69 5.05
.90	9.32 8.48 7.42 9.32
.95	23.68 20.29 28.35 33.67

3.1 Linear Regression Metamodels: First-, Second- and Third-Order

Linear regression metamodels have long history and have many advantages. They are easy to fit; and experiment designs that produce desired performance are known, and generated automatically by commercial and public domain software, for example, the `skipr` package in R (Morgan-Wall and Khoury, 2021).

The first-order linear regression metamodel fit is shown in Figure 2. Code for fitting the model and generating figures is in the appendix. For Figure 2, the metamodel corresponding to (1) has $Y \equiv$ average number in system over 10,000 time units (starting empty and idle), $h(Y) \equiv$ the expected value of this run average, i.e., $E(Y)$; $x \equiv$ arrival rate (λ), and $f(x) = -78.10 + 103.24x$. Although the fit has problems, statistical significance for the intercept and the slope are high, and the adjusted $R^2 = .64$ is good. The positive relationship between arrival rate and average number in system is an insight captured by this model, but the implied increase in average number in system from $\lambda = .75$ to $\lambda = .95$ of ≈ 21 is not a useful insight.

The *linear* in linear regression refers to the model being a linear combination of functions of x, but the functions of x need not be linear. The second-order linear regression metamodel adds a quadratic term in x.

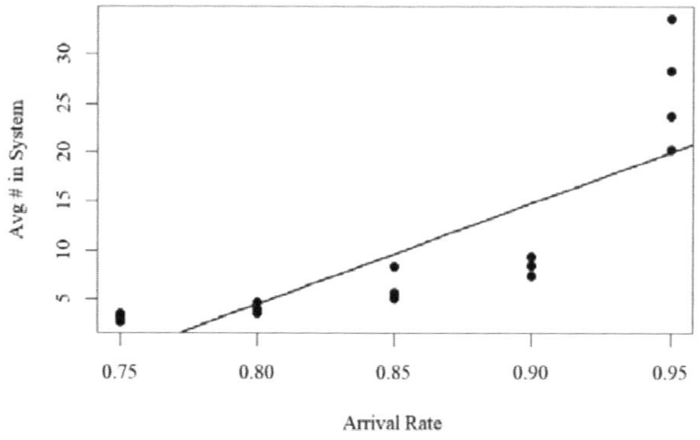

Figure 2 *First-order Linear Regression Metamodel*

Pitfall #1 – failing to scale predictors: to estimate linear and quadratic terms independently of each other, it is necessary to rescale the range of x to +/-1. To further remove the confounding of the quadratic term with the intercept requires orthogonal polynomial representation for the terms; more on that when the cubic model is fitted. The results are shown next.

The fitted quadratic metamodel is $f(x') = 4.76 + 10.32x' + 9.78(x')^2$ where x' represents the rescaled value $(x - x_{min}) / .5(x_{max} - x_{min})$. The insight is provided by the second-order regression metamodel is improved: there is increasing sensitivity to arrival rate at higher levels. Summing the three coefficients for the scaled variable gives the prediction for x' at its highest value $x' = 1$ ($x = .95$). The intuition provided by scaling is useful in many settings. Since the midrange value is now zero and the upper value is one, the coefficients say how much the response will change moving from midrange to the highest value. The resulting fit is shown in Figure 3.

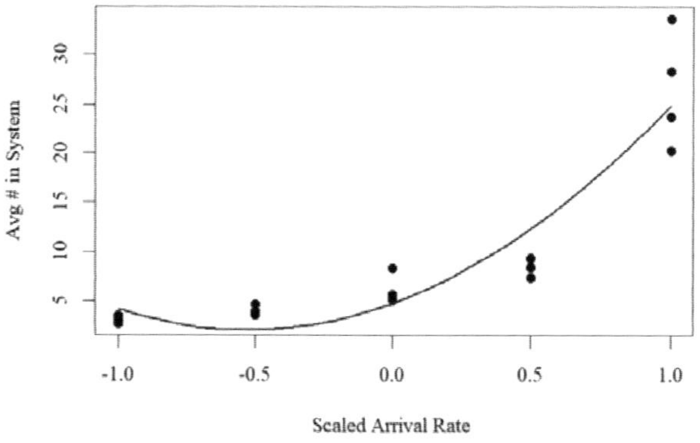

Figure 3 *Second-order Linear Regression Metamodel, x Scaled to ± 1*

The model can be expressed in original units. Then it is $f(x) = 623.4 - 1558.8x + 977.7(x)^2$. The fit is identical, but the intuition is less clear, with great cancellation of linear and quadratic effects to provide the moderate levels seen in the figure. For either scaling, the statistical significance of the intercept, slope and curvature are all high, and the adjusted R^2 at .86 is higher than for the first-order model. But there are still problems: the average number in system decreases initially as x increases, which is not correct. Evaluating the derivative of either model at $x' = -1$ or $x = .75$ shows negative value. The slope is negative, but we expect everywhere an increasing mean number in system as the arrival rate increases. Also, there are problems with bias at all but the highest value of arrival rate.

Pitfall #2 – using linear regression with too-high order: it is natural to try even higher order polynomial models to better approximate the response, but higher order models are more complex and it is hard to validate expected behaviour. Further, for evenly spaced design points, high-order polynomial regression models can exhibit increasingly bad behaviour (Barton, 1992). For this tutorial one further order is explored, a cubic regression again using scaled arrival rate. The fitted regression model with scaled x is $f(x') = 4.76 + 2.36x' + 9.78(x')^2 + 9.34(x')^3$ where x' is the scaled version of x. Figure 4 shows the resulting fitted metamodel, replacing x' by the corresponding x on the horizontal axis. The model adjusted R^2 is yet higher at .91, but strangely, the linear term is not significant now. Even with scaled x, the linear and cubic functions used in the model are not orthogonal, causing the lack of significance in the hypothesis test above. As for the second-order model with scaled x, the quadratic function never drops below zero, it is not orthogonal to the constant (intercept) term.

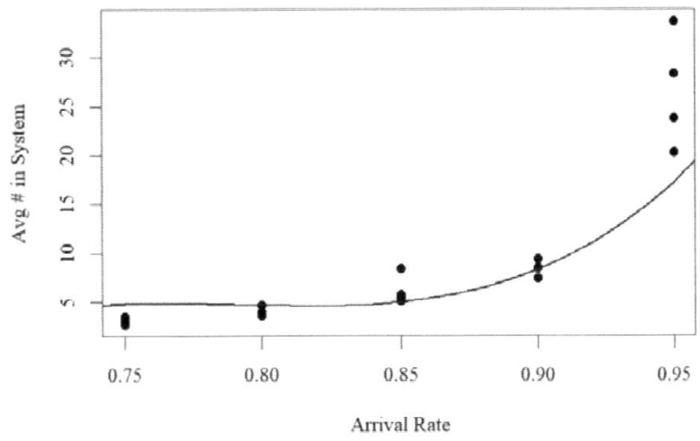

Figure 4 *Third-order Linear Regression Metamodel, Original x*

To fix these problems, it is necessary to replace the predictor terms x, x^2, x^3 by their orthogonal polynomials. Figure 5 shows the original power terms over [-1, 1] in the left plot. The figure on the right shows the corresponding orthogonal polynomial functions: quadratic with an adjusted mean, and cubic removing the linear part. Now these effects can be estimated independently, and the resulting coefficient table shows all are statistically significant.

Denoting these functions as $op_1(x)$, $op_2(x)$ and $op_3(x)$, the fitted metamodel can be constructed. It is $f(x') = 9.65 + 32.65op_1(x') + 18.29op_2(x') + 8.89op_3(x')$. Now, as one expects, the linear component is largest, followed by the quadratic, then the cubic. Note that there is no change to the metamodel predictions, R^2, etc. so the plot remains as in Figure 4. While the insight is aligned with expectations and the R^2 is improved over the second-order model, serious inadequacies remain. The bias at the highest arrival rate remains very large. Higher order models will not correct the shortcomings. There are two alternatives to explore: applying nonlinear transformation to the variables, or fitting a nonlinear regression model.

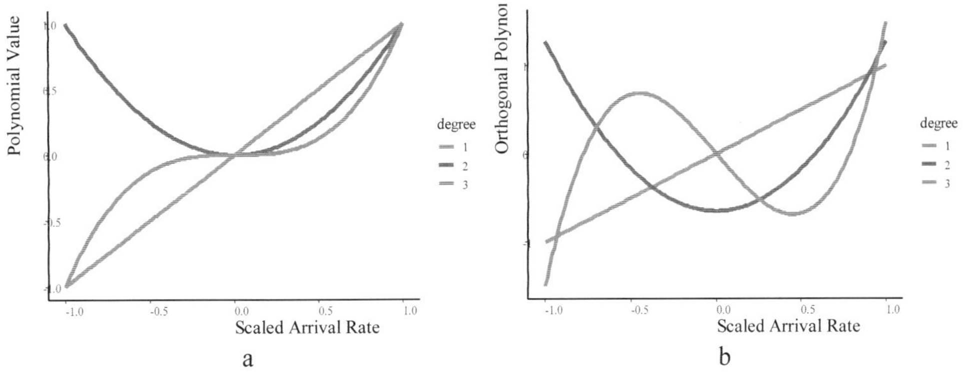

Figure 5 *Linear, Quadratic and Cubic Functions (a) and their Orthogonal Polynomial Form (b)*

3.2 Transforming the Response

There are three alternatives to explore when transforming variables prior to fitting regression models: transforming the observed y to stabilize the variance across different x values, transforming the observed y based on a known or approximately known nonlinear relationship between x and $h(Y(x))$, and/or transforming x. All three of these are discussed in detail by Box and Cox (1964), to be used separately or in combination. This section will focus on the first two.

3.2.1 Variance Stabilizing Transformations

The usual regression form assumes the variance of the response to be the same for all values of x. The data in Figures 2 – 4 show increasing variance at higher arrival rates. Regression models for such data often employ a variance stabilizing transformation (VST). A VST is a (nonlinear) function applied to the y values, e.g., $y' = \log(y)$, so that the transformed values exhibit homogeneous variance across x. The resulting x, y' relationship will often be a simpler, linear or quadratic function. Bartlett (1947) gave an extensive coverage of common transformations. Box and Cox (1964) provided a parametrized family of transforms $y' = (y^{\lambda}-1)/\lambda$ if $\lambda \neq 0$ and $= \log(y)$ if $\lambda = 0$. Member of this family are called Box-Cox transformations. In R, the `boxcox` function in the MASS Library (Venables and Ripley, 2022) identifies the optimal λ and uses that transform in fitting the regression. The result for the queueing simulation data is $\lambda = -.7$ so $y' = (y^{-.7} - 1)/-.7$. The fitted model (to y' not y) is $f(x) = -1.13 + 2.52x$. Because the model is first-order, it is not necessary to use scaled x values. The R^2 is .94 for the transformed y.

Figures 6 and 7 show the resulting fits after applying the optimal Box-Cox transform to y, in transformed and untransformed y. The first-order model is wholly adequate for modelling y', and the plot in the untransformed domain shows good fit. Yet from Figure 7 it is clear that the metamodel is not fully capturing the explosion in expected number in system as the arrival rate approaches 1.

Pitfall #3 – using a variance stabilizing transformation that destroys model simplicity: in our example, the transformed y values exhibit homogeneous spread over arrival rate, and the response function is simplified. This is not always the case. The issue was recognized 75 years ago (Bartlett, 1947), and recently in metamodels for job completion time quantiles (Pedrielli and Barton, 2019).

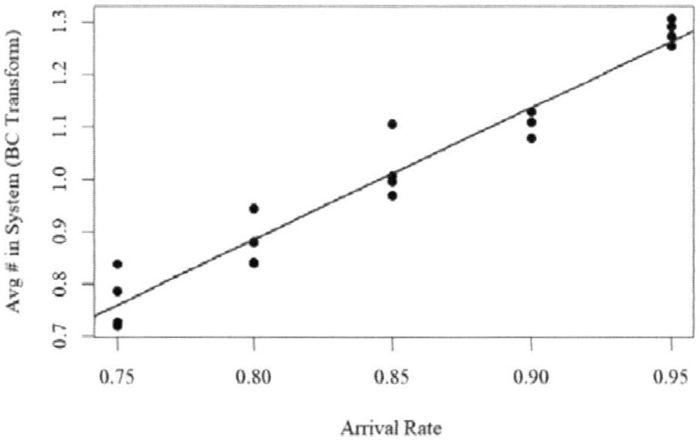

Figure 6 *Linear Regression for Box Cox Transformed y' with* $\lambda = -.7$

3.2.2 When Congestion Drives System Performance – Use Transforms based on x

None of the models considered so far capture the explosive growth of average number in system as the utilization approaches one. For systems where congestion drives behaviour, researchers have found transformations based on general queueing model behaviour that simplify the response relationship. Cheng and Kleijnen (1999) developed models for waiting time in queueing systems based on the simple $M/M/1$ relationship between utilization and waiting time: $1/(1-x)$, an extension of the strategy described in Cheng (1990). This strategy was further generalized and exploited by Yang and co-authors (Yang et al, 2007; 2008) who proposed an expected waiting time metamodel as a polynomial regression divided by a power (p) of $(1 - x)$:

$$\mu_t(x, b, p) = \frac{\sum_{l=0}^{t} b_l x^l}{(1-x)^p}. \quad (3)$$

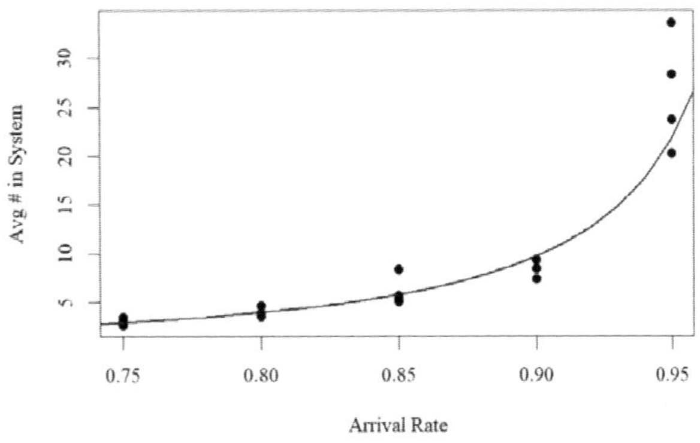

Figure 7 *Linear Box Cox Regression with Untransformed y Scale*

As a consequence, transforming y to permit a linear regression metamodel fit requires its multiplication by the denominator in (3). For our M/M/1 example, the form for expected number in system is given by (2), suggesting the transform $y' = y(1 - x)$. The fitted model is $f(x) = -1.12 + 2.41x$. As for the Box-Cox model, untransformed x can be used because the model again is first-order, and the model predicts y' not y. The R^2 is further improved to .94 for the transformed y. Based on the transformation, one would expect an intercept of zero and a slope of 1 for the coefficients. There is some deviation (not statistically significant) for the intercept but the confidence interval for the true value includes zero. On the other hand, the lower limit of a 95% confidence interval for the fitted slope does not include 1 (lower limit:1.125). But remember that the simulation experiments did not generate true steady-state values for number in system, so some discrepancy is expected. Note the relatively low R^2. This is understandable since a large part of the functional variation is removed by the transformation.

Figures 8 and 9 show the resulting models in transformed and untransformed scales. There is improvement in stabilizing the variance, but model nonlinearity is apparent in Figure 8 and some bias remains at high utilization as seen in Figure 9.

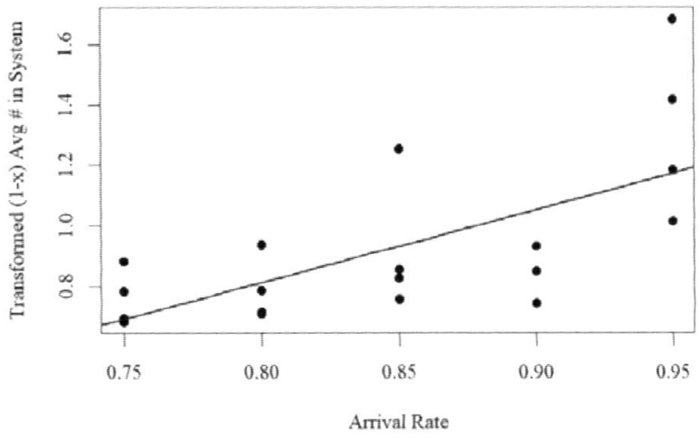

Figure 8 *Linear Regression for Transformed $y' = y(1 - x)$*

The model (3) as discussed by Yang et al. (2007) permits both the form of the linear model in the numerator and the choice of exponent in the denominator to be chosen based on empirical fit. Keeping the first-order regression but changing the exponent to $p = 2$ provides a better result in terms of fit, but the regression slope is not interpretable in the usual way: it is negative. The transformation trades off insight (the regression portion shows negative slope!) for better fit. Figures 10 and 11 show the resulting model in transformed and untransformed scale.

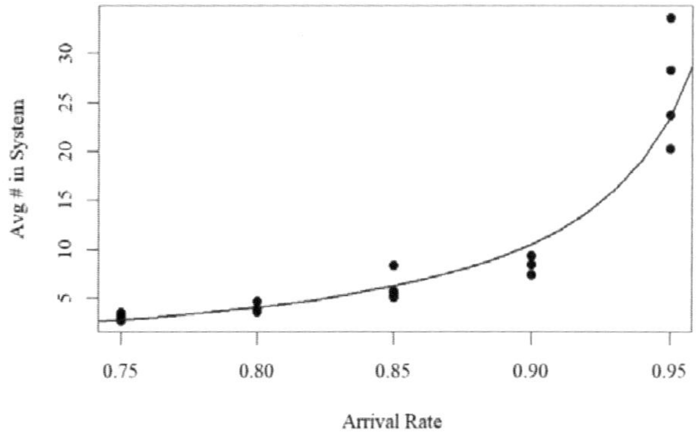

Figure 9 *Linear Regression for Transformed y' = y(1 – x) with Untransformed y Scale*

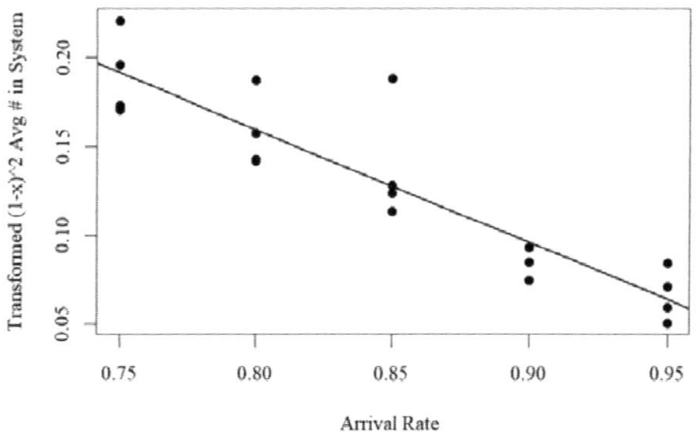

Figure 10 *Linear Regression for Transformed y' = y(1 – x) ^2*

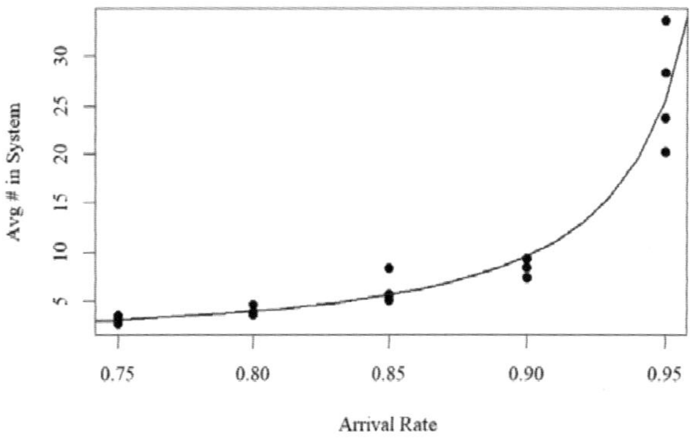

Figure 11 *Linear Regression for Transformed y' = y(1 – x)^2 with Untransformed y Scale*

3.3 Nonlinear Regression Metamodels for the Queueing Example

The difficulty in finding an appropriate linear regression metamodel for this simple example is discouraging. Remember that in realistic metamodel settings, the ability to discern quality of fit is greatly reduced: there will not be a simple plot of y vs. x with data superimposed. But beware the temptation to assume that flexible nonlinear regression models will handle all such difficulties. Problems arise in both fit and interpretation.

3.3.1 Gaussian Process Metamodels

Gaussian process models arose from Kriging models. The simplest Kriging model is:

$$Y(x) = \beta_0 + M(x), \qquad (4)$$

where M is the realization of a mean zero random field, i.e., a function drawn at random from all functions whose nearby values are correlated according to some spatial correlation function. For that reason, these models are also called spatial correlation models. When the randomness is assumed to be Gaussian, the models are called Gaussian Process (GP) regression models. GP models can approximate deterministic response functions, since once the realization occurs, the model (4) has no intrinsic randomness. Other regression-type terms might be added, but typically the intercept β_0 and $M(x)$ are all that is needed for a good fit. For stochastic GP models, a term $\varepsilon(x)$ is added to (4). Ankenman et al. (2010) modelled the variance of $\varepsilon(x)$ by a second spatial correlation model. Popular GP modelling software implements a version of stochastic kriging, with diagonal $\hat{\Sigma}_{\varepsilon}$, characterized as the *nugget*. The nugget can have equal values on the diagonal, or they may vary across the design points.

Because GP metamodels have great flexibility to fit arbitrary functions, they are an attractive alternative to high-order linear regression models. This comes at a cost. First, the model is more complex to fit, and GP modeling capability is not available in some statistical packages. Second, fitted model coefficients give some indication of how rapidly the response changes as components of x change, but the detailed insight available from a fitted linear regression model cannot be obtained. Further, if experimental run conditions are scarce, predictions in design parameter space between experimental runs can be significantly in error due to mean reversion. Mean reversion will be a minor problem below, but see Figure 1 in Erickson et al (2018).

Figures 12 and 13 show stochastic GP models fitted to the M/M/1 data. For Gaussian process regression, the R package `mlegp` (MLE-fitted Gaussian process) is used (Dancik and Dorman, 2008). The stochastic variance terms can be estimated directly at each design point using the replication sample variance, provided to `mlegp;` otherwise `mlegp` uses a pooled estimate, identical at every design point.

Figure 12 uses mlegp with the option of different variances for each design point, estimated by the four replications at each point. It shows a non-monotonic structure, and mean reversion below .75 and above .95. Figure 13 uses the common variance estimate across all design points. The result is a smoother fit but still nonmonotonic, and with failure to capture the explosive growth of average number in system as the utilization approaches one. Neither of the GP metamodels provides adequate fit.

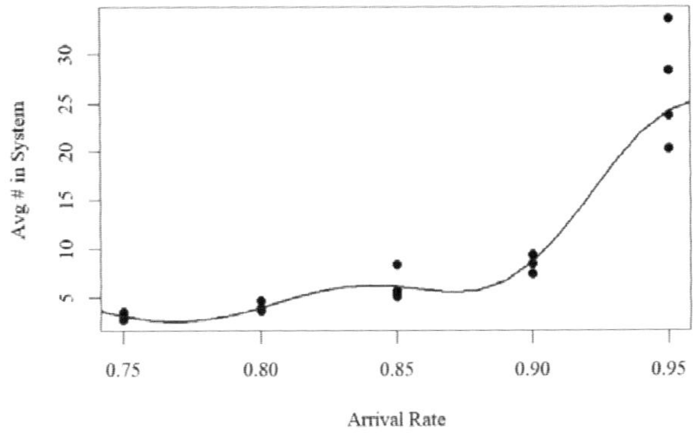

Figure 12 *Stochastic GP Metamodel (mlegp), Local Nugget*

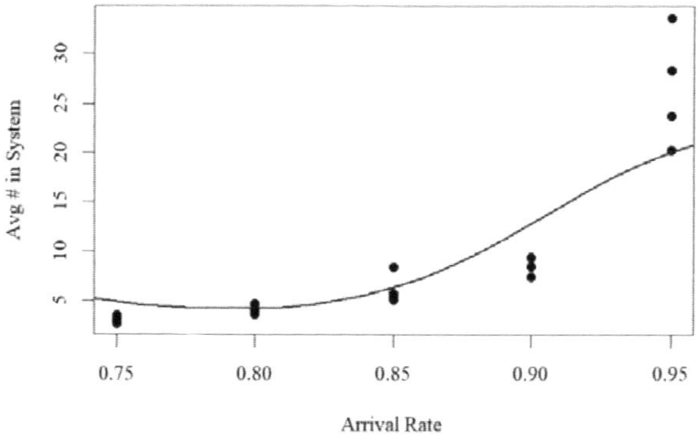

Figure 13 *Stochastic GP Metamodel (mlegp), Global Nugget*

Pitfall #4 – using a stochastic GP model when specific response properties are known: in our example, the expected number in system is a monotonic function of utilization, but the GP models failed to preserve that property. There have been attempts to address this shortcoming (Kleijnen and van Beers, 2013). In high dimensions the systematic errors of a GP metamodel will be hard to detect.

3.3.2 Artificial Neural Network Metamodels

Today, many researchers and practitioners are enthusiastic about deep learning, reinforcement learning, and other complex neural network formulations. These are largely applied in classification settings, but neural networks are also used for regression. Artificial neural network (ANN) metamodels develop the approximation function $f(x)$ as a composition of networked simple nonlinear functions with adjustable coefficients. The ANN model can be thought of as a nested form of ordinary regression models, where the output for each node, a weighted sum y of the immediate predecessor nodes, is post-processed by an activation function, e.g., $g(y) = 1/(1+e^{-y})$. There is one input node for each component of x. Layers between the input and output nodes are called hidden layers. For example, a neural network with five input components and two hidden 3-node layers (with nodes indexed by i and h) would have mathematical form:

$$f(x) = g(w_{30} + \textstyle\sum_{h=1}^{3} w_{3h} g(w_{2h0} + \sum_{i=1}^{3} w_{2hi} g(w_{1i0} + \sum_{j=1}^{5} w_{1ij} x_j))), \qquad (5)$$

where the w values in (5) are analogous to b values in a fitted regression model, and would be found via least squares. ANN fidelity and flexibility make the model type attractive for complex metamodeling. But ANN metamodels in high dimensions are difficult to validate, and performance can be sensitive to the choice of network structure and other issues. In their simulation modelling study, Fonseca et al (2003) stated "the ANNs capabilities of effectively learning were highly restricted by the selected codification scheme." This assessment is further supported by Al-Hindi (2004) and Can and Heavey (2012) who explored the predictive ability of neural networks for simple (s, S) DES models.

An ANN for the queueing example can be quite simple: only two hidden nodes are needed. Then the form in (5) reduces to

$$f(x) = g(b_0 + \textstyle\sum_{i=1}^{2} w_i' g(b_i + w_i \lambda)), \qquad (6)$$

with hidden nodes indexed by i. Figure 14 shows the resulting fit, using the R package `nnet`. The fitted metamodel appears monotonic but, perhaps surprisingly, fails to capture the explosive growth of average number in system as the utilization approaches one.

When used for regression as opposed to classification, neural network software typically offers the option to report the final node output directly, rather than scaled by the activation function whose value is limited to (0, 1). One might expect a better fit at high utilization using this option, but in fact the opposite holds. Figure 15 shows the metamodel fit with a linear output option. The fit is virtually identical to Figure 14 for utilization values up to .9, but inferior to the previous version for higher utilization. Note that for both ANN models the average number in system simulation output data were scaled by 1/30 so that predicted values no larger than one would be needed; adequate for capture by the ANN with activation function output.

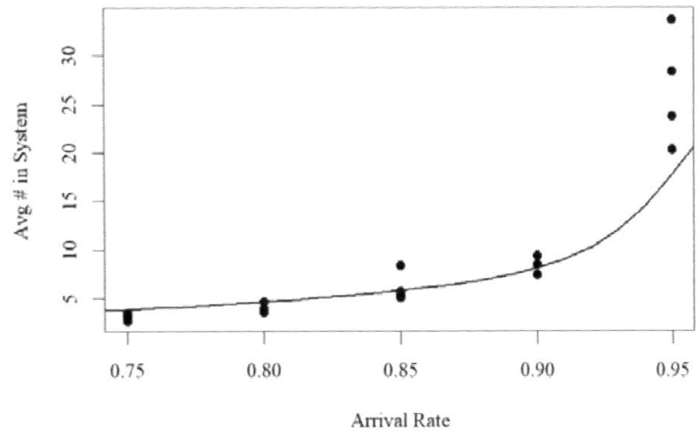

Figure 14 *Artificial Neural Network Metamodel, Logistic Activation Function Output*

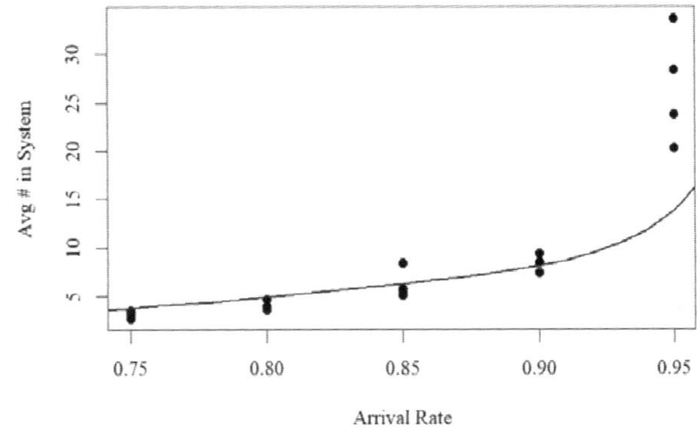

Figure 15 *Artificial Neural Network Metamodel, Linear Function Output*

Pitfall #5 – trusting the universal fit property of neural network models fitting stochastic responses with replications: neural networks are universal approximators to continuous nonconstant functions (Hornik et al, 1989). But in our situation, two aspects fail: first, the number of experimental points is finite; in fact small. Second, the response is stochastic. As the number of design points becomes dense, the function never becomes continuous due to stochastic variation in response (unless we fit to the replication means and allow the number of replications to become infinite). Also be warned: in high dimensions the systematic errors of an ANN metamodel will be hard to detect.

Artificial neural networks have an additional drawback: intuition is hard to gain from the model coefficients. Figure 16 shows the fitted coefficients for the ANN metamodel in Figure 14.

3.4 Learnings from the *M*/*M*/1 Queue Example

This very simple metamodeling scenario has exposed many important issues in choosing metamodel type and validating metamodel fit. The one-dimensional scenario allowed easy view of the shortcomings of each metamodel. Overall, what can one conclude? First, that complex model form such as GP or ANN can promise more than they deliver when the response function is stochastic. Second, if the simulation output is driven by a congestion or utilization phenomenon, then identifying the approximate congestion vs. output function is necessary. The best metamodels in this section used the structure in (3) to reduce the nonlinear behaviour. Then simple linear regression was sufficient.

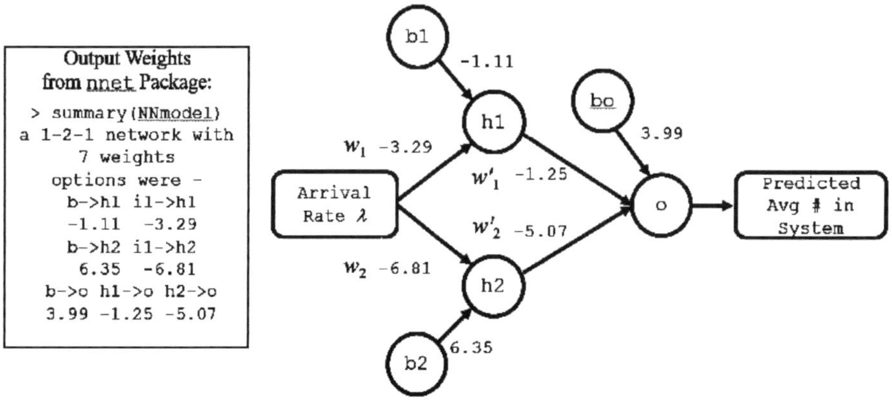

Figure 16 *No Intuition from Artificial Neural Network Weights for Metamodel in Figure 15*

Not all metamodeling issues were uncovered in this section. The choice of design points and design size for the experiment is also important. ANOVA-type compositions for high-dimension models are important for validating behavior and gaining insight. For more on these topics, see Santner et al (2018) and Law (2014).

4 WHEN MODEL INSIGHT FAILS: MODEL-FREE INTERPRETATION

Even simple regression models can fail to provide insight when interaction terms dominate. But regardless of the model type or experiment design, once the metamodel is fitted, its predicted response can be evaluated on a factorial grid, and a response-scale design plot can give model-free direct insight on the system behaviour (Barton, 1998). Figure 17 shows such a plot for the metamodel of a system dynamics simulation of wolf vs. sheep in Law (2014), which contains the full list of the fitted regression metamodel coefficients. They include four significant main effects, five significant two-factor interactions, and one three-factor interaction. The response is the average number of wolves over 2000 time epochs, as a function of sheep energy gain, sheep reproduction rate, wolf energy gain, wolf reproduction rate, and the rate at which grass re-grows. The behaviour is difficult to determine from the coefficients alone: the regression model fails to provide full insight. On the other hand, the response scaled design plot in Figure 17 makes behaviour clear.

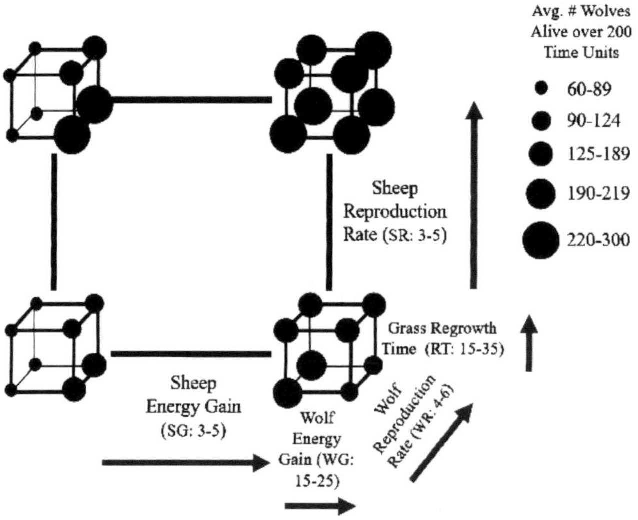

Figure 17 *Response Scaled Design Plot for Wolves vs. Sheep Metamodel*

Figure 17 shows behaviour consistent with expectations except in one case, which would not be apparent by examining the coefficients. When sheep energy is low and wolf energy high, or sheep energy high and wolf energy low, why does slow grass regrowth result in more wolves?

5 CONCLUSION

Metamodels can be useful in understanding simulation model input-output behaviour and can serve as a computationally efficient proxy for exploration of the behaviour of a simulated system. Care must be taken in the metamodeling activity to avoid hard-to-detect model misfit. For models involving more than two or three predictor variables, response-scaled design plots are key for validation and insight.

Of course, one must also avoid pitfalls in building the simulation model itself. Advice may be found in Law and McComas (1989).

ACKNOWLEDGMENTS

I have had the good fortune to benefit from the wisdom of the authors cited in this paper. I thank the anonymous reviewers for suggestions to improve this tutorial.

REFERENCES

Al-Hindi H (2004). Approximation of a discrete event stochastic simulation using an evolutionary artificial neural network. *Journal of King Abdulaziz University-Engineering Sciences* **15**: 125-138.

Ankenman, B, Nelson B L and Staum J (2010). Stochastic Kriging for simulation metamodeling. *Operations Research* **58**: 371–382.

Bartlett M S (1947). The use of transformations. Biometrics **3**: 39-52.

Barton R R (1992). Metamodels for simulation input-output relations. *Proceedings of the 1992 Winter Simulation Conference*. IEEE: Piscataway, NJ USA, pp 289-299.

Barton R R (1994). Metamodeling: a state of the art review. *Proceedings of the 1994 Winter Simulation Conference*. IEEE: Piscataway, NJ USA, pp 237-244.

Barton R R (1998). Design-plots for factorial and fractional-factorial designs. *Journal of Quality Technology* **30**: 40-54.

Barton R R (2021). Some formulation issues in constructing metamodels. *Proceedings of the 10th Operational Research Society Simulation Workshop*. The Operational Research Society: Birmingham, UK, pp 287-294.

Blanning R W (1974). The sources and uses of sensitivity information. *INFORMS Journal on Applied Analytics* **4**: 32-38.

Blanning R W (1975). The construction and implementation of metamodels. *Simulation* **24**: 177-184.

Box G E P and Cox D R (1964). An analysis of transformations. *Journal of the Royal Statistical Society Series B* **26**: 211-252.

Box G E P and Wilson K B (1951). On the experimental attainment of optimum conditions. *Journal of the Royal Statistical Society Series B* **13**: 1-45.

Burdick D S and Naylor T H (1966). Design of computer simulation experiments for industrial systems. *Communications of the ACM* **9**: 329-339.

Can B and Heavey C (2012). A comparison of genetic programming and artificial neural networks in metamodeling of discrete-event simulation models. *Computers & Operations Research* **39**: 424-436.

Cheng R C H (1990). Fitting parametric models by conditional simulation. *Proceedings of the 1994 Winter Simulation Conference*. IEEE: Piscataway, NJ USA, pp 333-336.

Cheng R C H and Kleijnen J P C (1999). Improved design of queueing simulation experiments with highly heteroscedastic responses. *Operations Research* **47**: 762-777.

Dancik G M and Dorman K S (2008). mlegp: statistical analysis for computer models of biological systems using R. *Bioinformatics* **24(17)**: 1966-1967.

Erickson C B, Ankenman B E and Sanchez S M (2018). Comparison of Gaussian process modeling software. *European Journal of Operational Research* **266**: 179-192.

Fonseca D J, Navaresse D O and Moynihan G P (2003). Simulation metamodeling through artificial neural networks. *Engineering Applications of Artificial Intelligence* **16**: 177-183.

Hornik K, Stinchcombe M and White H (1989). Multilayer feedforward networks are universal approximators. *Neural Networks* **2**: 359-366.

Kleijnen J P C (1975). A comment on Blanning's 'Metamodel for sensitivity analysis: the regression metamodel in simulation.' *INFORMS Journal on Applied Analytics* **5**: 21-23.

Kleijnen J P C (2015). *Design and Analysis of Simulation Experiments*, Springer: Berlin.

Kleijnen J P C and van Beers W (2013). Monotonicity-preserving bootstrapped Kriging metamodels for expensive simulations. *Journal of the Operational Research Society* **64**: 708-717.

Law A M (2014). *Simulation Modeling and Analysis*. McGraw-Hill: New York.

Law A M and McComas G (1989). Pitfalls to avoid in the simulation of manufacturing systems. *Industrial Engineering* **31**: 28-31.

Morgan-Wall T and Khoury G (2021). Optimal design generation and power evaluation in R: the skpr package. *Journal of Statistical Software* **99**: 1-36.

Myers R H, Montgomery D C and Anderson-Cook C M (2009). *Response Surface Methodology*. John Wiley and Sons: Hoboken, NJ.

Nozari A, Arnold S F and Pegden C D (1987). Statistical analysis for use with the Schruben and Margolin correlation induction strategy. *Operations Research* **35**: 127-139.

Pedrielli G and Barton R R (2019). Metamodel-based quantile estimation for hedging control of manufacturing systems. *Proceedings of the 2019 Winter Simulation Conference*. IEEE: Piscataway, NJ USA, pp 452-463.

Santner T J, Williams B J and Notz W I (2018). *The Design and Analysis of Computer Experiments*. Springer-Verlag: New York.

Schruben L W and Margolin B H (1978). Pseudorandom number assignment in statistically designed simulation and distribution sampling experiments. *Journal of the American Statistical Association* **73**: 504-520.

Simpson T W, Barron K, Rothrock L, Frecker M, Barton R R and Ligetti C (2007). Impact of response delay and training on user performance with text-based and graphical user interfaces for engineering design. *Research in Engineering Design* **18**: 49-65.

Tew J D and Wilson J R (1992). Validation of simulation analysis methods for the Schruben-Margolin correlation-induction strategy. *Operations Research* **40**: 87-103.

Venables W N and Ripley B D (2022). boxcox: Box-Cox transformations for linear models in MASS. https://rdrr.io/cran/MASS/man/boxcox.html. Accessed 25th November, 2022.

Yang F, Ankenman B E and Nelson B L (2008). Estimating cycle time percentile curves for manufacturing systems via simulation. *INFORMS Journal on Computing* **20**: 628-643.

Yang F, Ankenman B E and Nelson B L (2007). Efficient generation of cycle time-throughput curves through simulation and metamodeling. *Naval Research Logistics* **54**: 78-93.

AUTHOR BIOGRAPHY

RUSSELL BARTON is Distinguished Professor of Supply Chain and Information Systems in the Smeal College of Business and Professor of Industrial Engineering at the Pennsylvania State University. He is Chair of the Computer Simulation Archive Advisory Committee, housed at the NC State University Libraries; see https://d.lib.ncsu.edu/computer-simulation/. His research interests include applications of statistical and simulation methods to system design and to product design, manufacturing and delivery. He is a Fellow of IISE and a member of IEEE, INFORMS, and the Operational Research Society. https://sites.psu.edu/russellbarton/

Proceedings of the Operational Research Society Simulation Workshop 2023 (SW23)
C. Currie and L. Rhodes-Leader, eds.
https://doi.org/10.36819/SW23.008

AN AGENT-BASED MODELLING APPROACH FOR ASSESSING ENTREPRENEURIAL INTENTIONS AND DEVELOPMENT STRATEGIES IN THE RURAL HOSPITALITY INDUSTRY

Yunfei Gu

University of Southampton
Y.Gu@soton.ac.uk

Prof. Bhakti Stephan Onggo

University of Southampton
B.S.S.Onggo@soton.ac.uk

Prof. Martin Kunc

University of Southampton
M.H.Kunc@soton.ac.uk

Dr. Steffen Bayer

University of Southampton
S.C.Bayer@soton.ac.uk

ABSTRACT

Rural hospitality and tourism (RHT) play a key role in rural revitalization, especially due to the impact of COVID-19, with more citizens choosing to travel to the countryside for a staycation. Local SMEs, especially family-owned enterprises, make up the majority of the RHT sector, not only providing services and products to satisfy tourists, but also helping with local employment. However, entrepreneurs operating in rural areas face many challenges in terms of capital, skills and education. Hence, it is important to explore the entrepreneurial intention (EI) of local people and how policies can support or change their behaviours. Current research on the RHT industry, rarely study the EI of local people, and the literature on rural entrepreneurship concentrates on developed countries. This study therefore uses agent-based modelling to explore how locals' EI in Chongming island (China) respond to the current impact of COVID-19, and whether policies will bring about changes on the supply side of RHT sector.

Keywords: Rural tourism, Rural entrepreneurship, Entrepreneurship intention, Agent-based modelling

1 INTRODUCTION

Rural tourism has been recognized as a tool to promote socio-economic growth, especially in a period of transition when traditional rural industries and agriculture were in recession (Chen, 2019; Iorio & Corsale, 2010; Oppermann, 1996). Supporting rural tourism is gradually becoming a popular choice for both developed and developing countries, as tourism can help to tackle local unemployment, increase income and economic diversification, and help to preserve local culture as well as increase local well-being (Rosalina et al., 2021; Sharpley, 2002; Su et al., 2019). As a host and provider of services to urban visitors, the rural hospitality and tourism sector (RHT) plays a key role in rural revitalisation (Reichel et al., 2000). Especially due to the impact of COVID-19, there has been a shift in the choice of tourist destinations, and rural tourism has become an important alternative during this period (Vaishar & Šťastná, 2022; Wen et al., 2020; Zhu & Deng, 2020). It can satisfy the public's need to travel during the epidemic and can also alleviate their anxiety about travelling (Zhu & Deng, 2020). For example, the scale of China's rural tourism reached 867 million visitors from January to May 2021, an increase of 55.5% from the previous year, making RHT the most significant contributor to the domestic tourism economy (China Tourism Academy, 2022).

While rural entrepreneurship is actively encouraged in countries around the world, businesses operating in rural areas, particularly in the tourism industry, face challenges in terms of capital, skills and education compared to similar urban businesses (Badulescu et al., 2016; Chen, 2019; Su, 2011). At the same time, local small and medium-sized enterprises (SMEs), especially family-owned enterprises, make up the majority of the RHT sector (Ilbery et al., 1998). Hence, to revitalize the rural economy, it is important to understand the *entrepreneurial intention* (EI) of local people and how policies can support or change their behaviours. However, there is little literature on rural tourism entrepreneurship. As a result, tourism policy makers and operators are poorly informed about entrepreneurship in rural areas, especially on the supply side, with less emphasis on local EI and barriers to its operation (Ali & Yousuf, 2019; Pato & Teixeira, 2016; Solvoll, 2015). Furthermore, the existing literature mainly concentrates on developed countries such as UK, Spain, and Finland (Pato & Teixeira, 2016; Thirumalesh Madanaguli et al., 2021).

This research focuses on Chongming island, the third largest island in China, which has historically been a rural and remote area. In the early 21st century, Chongming Island began building an eco-island to promote rural eco-tourism and upgrade its traditional agro-industrial industries (Xie et al., 2019). The Covid-19 pandemic presented an opportunity for Chongming Island, as China's travel restrictions led to a surge in suburban and rural tourism (Zhu & Deng, 2020). Tourist numbers to Chongming Island increased by 199% in 2021 compared to 2018, while local businesses undertaking homestays and catering increased by 59% (Chongming District Bureau of Statistics, 2021). Hence, the aims of this research are to explore the impact of the pandemic on local EI, including whether it served as a catalyst or a hindrance to EI and the extent to which rural EI is influenced by community and peers, and to investigate the role of policy adjustments in encouraging entrepreneurship in RHT sector.

The objective of this paper is to present the conceptual model of an agent-based model (ABM) that can be used to explore the impact of pandemic, community and peer on rural EI. The design of the ABM is based on Esfandiar et al.'s (2019) model of entrepreneurial intention. We will also discuss the questionnaires that we will use to collect local attitudes towards entrepreneurship and other parameters needed for the ABM. As ABM is becoming increasingly popular in business and management research, particularly in innovation and family business studies (Onggo and Foramitti, 2021), it offers a unique advantage in studying heterogeneous populations. ABM allows agents to follow internal logic, interact in a dynamic system, and learn from the external environment (Onggo and Foramitti 2021; Bonabeau, 2002; Swinerd & McNaught, 2012; Vinogradov et al., 2020). This study contributes to the literature on rural entrepreneurship by examining the local EI under the influence of Covid-19 and assisting rural tourism planners in designing strategies to support local SMEs. It also offers new perspectives on the supply side of rural tourism, particularly in developing countries, by contributing to the development of the RHT sector and revitalizing rural economies.

2 LITERATURE REVIEW

2.1 Rural Entrepreneurship

Rural entrepreneurship has been a subject of interest for over a century, but researchers have not yet agreed on a universal definition (Kulawiak et al., 2022). Early research has typically defined rural entrepreneurship in two perspectives: as the image of the traditional entrepreneur with characteristics such as independence and innovation, and as the creation of new firms and jobs (Vaillant & Lafuente, 2007; Wortman, 1990). However, these two perspectives fail to account for the complexity of the rural environment and its differences from urban entrepreneurship (Fortunato, 2014). Instead, a more representative definition views rural entrepreneurship as the creation of a new organization that introduces a new product, serves or creates a new market, or utilizes a new technology in a rural environment (Wortman, 1990). Researchers increasingly recognise that entrepreneurship should be situated within the nature of localised business activities (Schoonhoven & Romanelli, 2001). Rural entrepreneurs are seen as living in rural settings and communities, and they are influenced by rural social characteristics and networks (Stathopoulou et al., 2004).

Rural entrepreneurship has been researched for over twenty years. It has been dominated by economic-related studies especially on the macro policy level, analysing obstacles and challenges (Kulawiak et al., 2022). However, research from a micro perspective is lacking, with existing studies focusing on demographic and psychographic profiles of entrepreneurs, their skill levels, and rural firm characteristics (Pato & Teixeira, 2016). Furthermore, research has primarily focused on the United States and European countries, with less attention paid to low-income countries (Pato & Teixeira, 2016).

Besides, when it comes to RHT sector, entrepreneurship has received relatively limited attention (Thirumalesh Madanaguli et al., 2021). Therefore, more research is needed on the supply side of RHT, especially from the perspective of individuals to analyse barriers and support for development, the characteristics of entrepreneurs, and business performance (Thirumalesh Madanaguli et al., 2021).

2.2 Entrepreneurship Intention

As a crucial stage in the formation of business activity, EI has been studied for over 35 years (López-Fernández et al., 2016). Entrepreneurship arises from individual intention and subsequent actions, so entrepreneurial activities are also considered intentionally planned behaviours (Krueger & Carsrud, 1993; McMullen & Shepherd, 2006). Hence, intention-based models have gained more attention because they are theoretically grounded and explain individual thinking and motivation (Esfandiar et al., 2019). Among the various behavioural models, two widely recognised ones are Shapero's model (Shapero & Sokol, 1982) and Theory of Planned Behaviour (TPB) (Ajzen, 1991). Shapero's entrepreneurial event suggests that launching a new venture requires three key prerequisites, namely propensity to act and perceptions of desirability and feasibility (Shapero & Sokol, 1982). In the TPB model, intentions are triggered by an attitude toward the behaviour, subjective norms and perceived behavioural control (Ajzen, 1991). Both of these behavioural theories have been widely used in different industries over the years.

To describe EI more comprehensively, Krueger (2009) developed a comprehensive model of EI by combining the TPB and Shapero's model (Figure 1). This model includes a new metric, collective effectiveness, which accounts for the influence of the surrounding collective environment and colleagues on intentions, as individual capabilities may sometimes be insufficient to achieve success (Esfandiar et al., 2019; Krueger, 2009). To further understand the relationship between intention and action, Esfandiar et al. (2019) developed a new Entrepreneurial implementation intention model (see Figure 2). In their model, personal desirability is described in terms of attitudes toward entrepreneurship. Intention is described as entrepreneurial goal intentions (EGIs), which capture whether they have goals in the business, while the implementation intention is their determination to start a business. Their studies shows that there is a strong impact of EGIs on implantation intention. Also, social norms are found not to influence the EGI, whereas desirability is the main determinant of EGI, followed by self-efficacy, feasibility, opportunity, attitude, and collective efficacy. While research on EI has been studied for years, the analysis is centred around students' EI (Dolhey, 2019; Gurel et al., 2010). However, little is known about local EI in rural areas, particularly in the RHT industry. Given the significance of revitalising the RHT sector for the local economy, it is crucial to comprehend how local EI is affected by the COVID-19 pandemic.

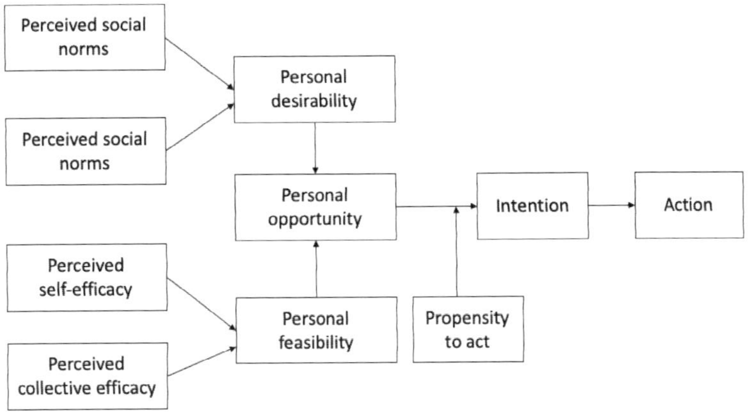

Figure1 *Entrepreneurial intention model (adopted from Krueger, 2009)*

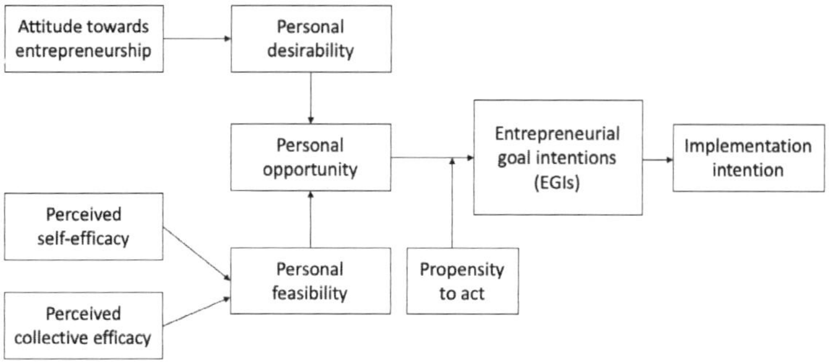

Figure 2 *Entrepreneurial implementation intention model (adopted from Esfandiar et al., 2019)*

3 RESEARCH METHOD

3.1 Agent-based Modelling and Simulation

ABM is a computational method for simulating the interaction of a group of agents to explore dynamic behavioural changes in complex systems (Siebers et al., 2010). Agents in ABM can make decisions based on a series of rules and logic, agents interact with each other, and their behaviour is also influenced by the environment (Bonabeau, 2002; Swinerd & McNaught, 2012). Hence, ABM has the ability to capture the emergent phenomena generated by individual entity interactions, and can simulate the behaviour of complex individual agents to make predictions from a microscopic perspective (Bonabeau, 2002; North & Macal, 2007). These characteristics make ABM appropriate for studying entrepreneurship in tourism, where agents (i.e., entrepreneurs and potential entrepreneurs) can exhibit heterogeneity in terms of their level of education, attitude to risk perception, and financial situation. ABM is widely used in business and management (Onggo & Foramitti, 2021), with applications in supply chain (Utomo et al., 2018), marketing (Negahban & Yilmaz, 2014) and human resource management (An, 2012). In the context of the impact of an epidemic, ABM is a powerful simulation tool that can provide decision analysis (Currie et al., 2020).

Moreover, ABM is increasingly being applied in the tourism industry to analyse the behaviour of tourists and the interactions between individuals (Baktash et al., 2022). ABM has been used to analyse visitor decision-making (Alvarez & Brida, 2019), risk management (Fan et al., 2019) and destination management (Pizzitutti et al., 2014; Student et al., 2020). One advantage of ABM over other modelling and data collection methods is that it can be integrated with theory in modelling and inform tourism decision makers and stakeholders through experimentation (Baktash et al., 2022; Lindkvist et al., 2020). However, most of the applications of ABM in the tourism industry have been cut from the perspective of tourists, with fewer studies analysing the supply side. Thus, this study will use ABM from a supply perspective to provide more information in destination management. The study will also follow the guidelines provided by Monks et al. (2019) on how to document a simulation study. Specifically, the guidelines are referred to as strengthening the reporting of empirical simulation studies (STRESS), which includes six components: objectives, logic, data, experiments, implementation, and code access (Monks et al., 2019).

3.2 Objectives

The ABM model aims to explore how local people's EI will respond under the current impact of Covid-19, and whether different policies will bring about changes on the supply side of rural tourism. Specifically, the model will try to answer the following questions: (1) whether the epidemic is a catalyst or a deterrent to local EI; (2) whether EI in rural areas is more influenced by community resources and surrounding practitioners; and (3) whether different policy adjustments and support would encourage rural residents to start businesses in the RHT sector. This work-in-progress paper presents the conceptual model of the entrepreneurial implementation intention as a foundation for future empirical work using ABM to test the effectiveness of policy adjustments on the supply side of rural tourism. The model serves as a starting point; further refinement will be needed to better understand the complex dynamics of rural entrepreneurship under Covid-19. Additionally, we demonstrate the potential of ABM as a powerful tool to investigate such dynamics and encourage their further adoption in research on rural entrepreneurship and related areas.

3.3 Data Collection

In ABM, it is essential to capture the behaviour of agents accurately to build an effective model. Both primary and secondary data can be used to achieve this goal. Qualitative and narrative data can help identify the agents' behaviour, such as their rules, goals, and logic, while quantitative data can provide numerical measurements of behaviour, which can be used to validate the model (Baktash et al., 2022). In this study we therefore use questionnaires and interviews, as well as secondary data from different sources, such as local government websites and statistical department reports, to capture the logic and rules behind individuals' behaviour and the goals of their actions.

Between January and April 2022, we conducted a field study in Chongming Island. The field study aimed to understand the challenges and opportunities faced by entrepreneurs in the tourism industry on the island. We interviewed a total of 19 entrepreneurs who were actively involved in the industry, including hotel owners, tour operators, and local attraction providers. During the interviews, we asked the entrepreneurs about their experiences, perspectives, and business practices. The interviews provided valuable insights into the tourism industry in Chongming island. In addition to understanding the experiences of entrepreneurs in the tourism industry on Chongming, we are also interested in exploring the attitudes and willingness of local residents to start their tourism-related business. To achieve this, we have designed a questionnaire that is being distributed online to residents of the island, the questionnaire is provided in the Appendix. This questionnaire aims to explore individuals' attitudes towards entrepreneurship in the tourism industry. It includes questions on basic information, attitudes towards entrepreneurship, perceived self-efficacy and collective efficacy, as well as questions specifically for those who have started their own tourism-related business. Survey data collection is still ongoing.

Based on the interview, secondary data and – once completed – questionnaire responses, we will be able to create a diverse group of agents for our ABM model. These agents will be heterogeneous in terms of their education level, attitudes towards risk perception, financial situation, and motivations for starting a tourism-related business. By incorporating this diversity into the model, we will be able to simulate the behaviour of agents with different characteristics.

3.4 Model Logic

The model is designed based on Esfandiar et al. (2019)'s entrepreneurial implementation intention model (Figure 3). There are three main elements that influence the EI, namely, attitudes towards entrepreneurship (ATE), perceived self-efficacy (PSE) and perceived collective efficacy(PCE).

The equation for ATE is shown in Equation (1). Variable attitudes (AT) is an indicator of whether local people are subjectively interested in or discouraged by entrepreneurship. Variable risk perceptions (RP) is an indicator of whether local people still have confidence in the RHT sector under the influence of epidemic. The two variables combined represent the attitudes of the local people towards entrepreneurship. We use five point scale to measure AT and RP. For example, if an agent has a very positive and optimistic attitude of entrepreneurship, this agent will have an AT score of 5. Similarly, if the agent also feels very positive about future risks, then the RP is also 5. The data for these two variables is collected from interviews and questionnaires.

$$ATE = AT * RP \tag{1}$$

The second part is perceived self-efficacy (PSE), where we use the education level (EL) and available resources (AR) to capture the ability of locals to start a business. These two variables are collected using both primary data and secondary data from the official statistic reports. Equation (2) shows the calculation of PSE. We also use the five point scale to measure EL and PSE. Since each agent may have a different sensitivity to the indicator (Fan et al., 2019), we add a random coefficient β_1 to represent agent's sensitivity to the two capabilities. β_1 is a random number between 0 to 1, where 0 indicates that the agent is completely insensitive to the indicator and 1 indicates that it is extremely sensitive to it.

$$PSE=\beta_1*EL*AR \tag{2}$$

The last component is perceived collective efficacy (PCE). We use the number of people in the agent's community who are already engaged in the RHT sector (CE), and the resources available to the local community (ACR) as representatives. The data for these two variables are collected from government websites. EL and AR are also measured using the five point scale (Equation (3)). Similarly, we use a random coefficient β_2 to demonstrate the different sensitivity levels of the agents. The final intention (IT) is the sum of the three components, and each element has a weight factor W, where the sum of the three weight factors is equal to 1 (Equation (4)).

$$PCE=\beta_2*CE*ACR \tag{3}$$

$$IT=W_1*ATE+W_2*PSE+W_3*PCE \tag{4}$$

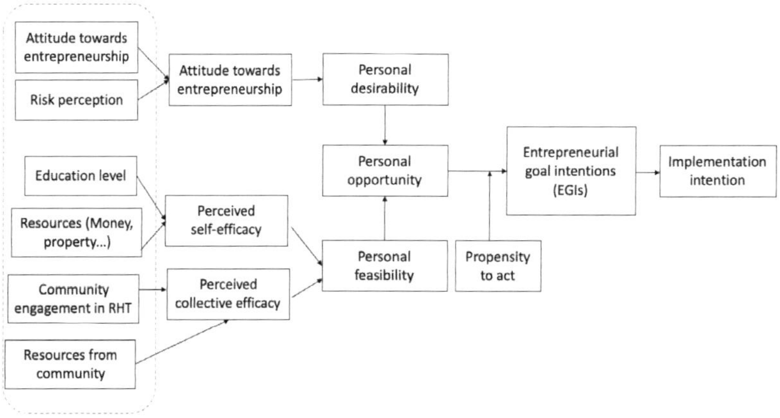

Figure 3 *Extended entrepreneurial implementation intention model (adopted from Esfandiar et al., 2019)*

Figure 4 shows a state chart of the ABM (using AnyLogic software). This conceptual model includes the following elements:

Agents: The agents in the model represent both existing entrepreneurs and potential entrepreneurs in the tourism industry at Chongming Island. They are characterised by various attributes such as their level of education, attitude towards risk and financial situation. The agents' decisions are influenced by their individual attributes and the actions of other agents in the model.

Environment: The model is set in the context of the RHT industry in Chongming Island, with various environmental factors affecting the agents' decisions. These may include government policies, infrastructure, competition, and external events such as pandemics or natural disasters.

Interactions: The agents interact with each other and the environment through various channels, such as collaborations, competition, learning, and information exchange. These interactions can shape the agents' attitudes, behaviour, and decision-making processes.

Agent states: Based on their characterises and interactions, the agents can move into four agent states: 'Locals who are not entrepreneurs', 'Not Interested', 'Become an Entrepreneur', and 'Existing Local Entrepreneur'.

Expected outputs: The model can produce various results such as, the number of local entrepreneurs likely to start up in the future, the level of competitiveness of the industry, the economic and social impact of entrepreneurship, and the factors that promote or hinder entrepreneurial activity.

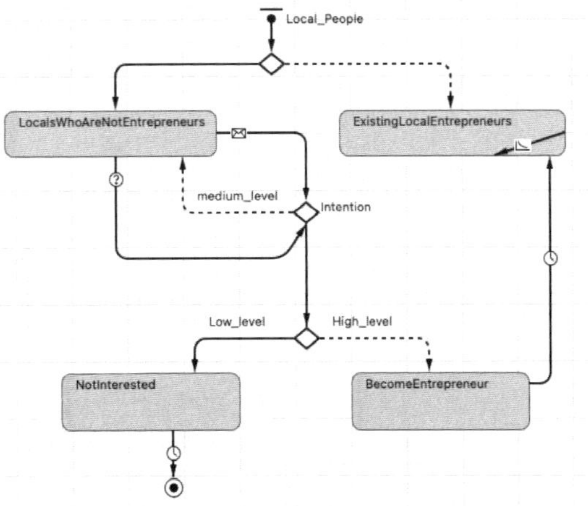

Figure 4 *ABM state chart*

3.5 Experimentation

The entry point is all residents in Chongming, and as each person enters the simulation they will be divided into two categories. The first being locals who are not currently tourism entrepreneurs, and the second being those who are already working in tourism entrepreneurship. Each agent's characteristics, such as education level and perceived risk level, will follow different distributions based on the collected data and will be assigned to each agent when entering the model. The proportion of the two states will be determined based on the entrepreneurs data in Chongming island.

Agents who are at 'Locals who are not entrepreneurs' state will have the chance to generate strong EI and become a local entrepreneur. This transition can be triggered in two paths. The first path is influenced by the existing local tourism entrepreneurs, affecting the agent's perceptions through their friends, relatives or neighbour entrepreneurs. One thing to note is that this influence can be positive, increasing the agent's intention, or it can be negative, de-motivating the agent. The second path is triggered by conditions that calculate the strength of an agent's intention based on variables such as risk perception, education level, and available resources. The EI of each agent is calculated in the 'Intention branch'. The intention will be divided into three different levels: low, medium, and high. Agents with medium-level intention will move back to the 'Locals who are not entrepreneurs' state, where they still have the opportunity to change their minds. Agents with low or high-level intention will move to the 'Not Interested' or 'Become an Entrepreneur' state, respectively. Research suggests that new entrepreneurs can have a significant impact on the EI of people in their social networks (Liñán & Chen, 2009). This is because people tend to be more likely to take advice and follow the example of those they trust and have a close relationship with. As a result, it is possible that these new local entrepreneurs may influence their friends, family members, or neighbours to consider starting a business themselves. This could be particularly impactful in rural areas where social networks are often tightly knit (Huggins & Thompson, 2015). Thus, when individuals become entrepreneurs, their influence can spread randomly throughout their social networks, creating a ripple effect of EI and behaviours (Greve & Salaff, 2003).

4 CONCLUSIONS

This paper demonstrates the potential of ABM in understanding the complex and multifaceted factors that influence the EI of local residents in the RHT sector on Chongming island. Through providing a conceptual ABM model, we aim to simulate the behaviour of heterogeneous agents, including both existing entrepreneurs and potential entrepreneurs. Although the data collection is still ongoing, our preliminary findings suggest that the factors that influence EI are complex and multifaceted, with individual characteristics as well as environmental factors, all playing a significant role. Moving forward, we plan to use the questionnaire results to calibrate and validate our model. We also plan to conduct sensitivity analysis to examine the robustness of the model and explore different scenarios, such as changes in government policies or external economic shocks. We expect to provide valuable insights and recommendations for policymakers and industry stakeholders, as well as academics interested in entrepreneurship, tourism, and agent-based modelling. In summary, this study provides an approach to understanding the EI of residents in a tourism destination. The findings will contribute to the existing literature on tourism entrepreneurship in rural areas, applications of ABM, and can also inform policy and industry decision-making.

A APPENDIX: QUESTIONNAIRE ON THE INTENTION OF RESIDENTS TO START THEIR OWN TOURISM-RELATED BUSINESSES ON CHONGMING.

Section 1: Basic Information
1.1) Please provide your gender? [Male] [Female] [Non-binary] [Prefer not to say]
1.2) Please provide your age range:? [18-24] [25-34] [35-44] [45-54] [55 or older]
1.3) Please provide your education level? [Junior high school degree or below] [High school degree] [Some college or associate degree] [Bachelor's degree] [Graduate degree or above]
1.4) Have you started your own tourism-related business? [Yes] [No]
- If Yes, please skip to Section 5
- If No, please continue to the next question.
1.5) Have you ever considered starting your own tourism-related business? [Yes] [No]

Section 2: Attitude towards entrepreneurship
2.1) How much do you agree with the following statement: "Entrepreneurship is a desirable career choice for me.":
[Strongly disagree] [Disagree] [Neutral] [Agree] [Strongly agree]
2.2) How much do you agree with the following statement: "Entrepreneurship is a risky career choice.":
[Strongly disagree] [Disagree] [Neutral] [Agree] [Strongly agree]
2.3) How much risk do you associate with starting your own tourism-related business?
[Very Low] [Low] [Moderate] [High] [Very High]
2.3) In your opinion, how has the COVID-19 pandemic affected the level of risk associated with starting your own tourism-related business?
[Decreased Risk] [No Change] [Increased Risk] [Unsure/No Opinion]
2.4) What do you consider to be the risks associated with starting a tourism-related business? Please select all that apply.
a) financial risks (e.g., lack of funding, debt)
b) Market risks (e.g., lack of demand, competition)
c) Operational risks (e.g., staffing, supply chain disruptions, lack of knowledge and training)
d) Regulatory risks (e.g., licensing, legal compliance)
e) Force majeure (e.g., natural disasters, other pandemics)
f) Other (please specify): ____________

Section 3: Perceived self-efficacy
3.1) How much do you agree with the following statement: "I feel confident in my ability to start and run a tourism-related business." [Strongly disagree] [Disagree] [Neither agree nor disagree] [Agree] [Strongly agree]
3.2) How much do you think your education/training/specialist knowledge will influence you on starting and running a tourism-related business? [Not at all] [Slightly] [Moderately] [Very much] [Completely]
3.3) Do you feel that your level of education has prepared you to start and run a business? [Yes] [No]
3.4) How much do you think the availability of resources (e.g., money, property) will influence your perceived self-efficacy in starting and running a tourism-related business? [Not at all] [Slightly] [Moderately] [Very much] [Completely]
3.5) Do you have access to financial resources (e.g., savings, loans) to start a business? [Yes] [No]

Section 4: Perceived collective efficacy
4.1) Have any of your friends, family members, or neighbours started their own tourism-related business on the island? [Yes] [No]
4.2) If yes, how much influence did they have on your interest in starting your own tourism-related business? [No influence] [Little influence] [Some influence] [Significant influence]
[Major influence]
4.3) To what extent do you agree or disagree with the following statement: "Seeing family members, friends, or neighbours start their own businesses makes me more confident in my ability to start a tourism-related business."
[Strongly agree] [Somewhat agree] [Neither agree nor disagree] [Somewhat disagree] [Strongly disagree]
4.4) Do you think there are enough resources available in the community to support new tourism-related businesses? [Yes] [No]
4.5) To what extent do you agree or disagree with the following statement: "If there is enough support from the community, I would be more confident in my ability to start a tourism-related business." [Strongly agree] [Somewhat agree] [Neither agree nor disagree]
[Somewhat disagree] [Strongly disagree]
4.6) What types of support do you think the community could provide to help you start a tourism-related business? (select all that apply)
a. Business training and education
b. Mentorship and guidance from experienced entrepreneurs
c. Marketing and promotional assistance
d. Networking and colorations with other entrepreneurs
e. Other (please specify)

Section 5: Experience as an Entrepreneur
5.1) How long have you been running your tourism-related business? [Less than 1 year] [1-3 years] [3-5 years] [More than 5 years]
5.2) How did you finance your business start-up costs? (Multiple answer)
[Personal savings] [Bank loan] [Grants or government funding] [Investors or venture capital] [Crowdfunding] Other (please specify)
5.3) Would you encourage a friend or family member to start their own tourism-related business?
[Definitely not] [Unlikely] [Neutral] [Likely] [Definitely]

REFERENCES

Ajzen, I. (1991). The theory of planned behavior. *Organizational Behavior and Human Decision Processes*, *50*(2), 179–211. https://doi.org/10.1016/0749-5978(91)90020-T

Ali, A., & Yousuf, S. (2019). Social capital and entrepreneurial intention: Empirical evidence from rural community of Pakistan. *Journal of Global Entrepreneurship Research*, *9*(1), 64. https://doi.org/10.1186/s40497-019-0193-z

Alvarez, E., & Brida, J. G. (2019). An agent-based model of tourism destinations choice. *International Journal of Tourism Research*, *21*(2), 145–155. https://doi.org/10.1002/jtr.2248

An, L. (2012). Modeling human decisions in coupled human and natural systems: Review of agent-based models. *Ecological Modelling*, *229*, 25–36. https://doi.org/10.1016/j.ecolmodel.2011.07.010

Badulescu, D., Giurgiu, A., Istudor, N., & Badulescu, A. (2016). Rural tourism development and financing in Romania: A supply-side analysis. *Agricultural Economics (Zemědělská Ekonomika)*, *61*(No. 2), 72–82. https://doi.org/10.17221/94/2014-AGRICECON

Baktash, A., Huang, A., de la Mora Velasco, E., Jahromi, M. F., & Bahja, F. (2022). Agent-based modelling for tourism research. *Current Issues in Tourism*, 1–13. https://doi.org/10.1080/13683500.2022.2080648

Bonabeau, E. (2002). Agent-based modeling: Methods and techniques for simulating human systems. *Proceedings of the National Academy of Sciences*, *99*(suppl_3), 7280–7287. https://doi.org/10.1073/pnas.082080899

Chen, Z. (2019). A Study on the Problems and Countermeasures of the Development of Rural Ecotourism in China. *Journal of Advances in Economics and Finance*, *4*(1). https://doi.org/10.22606/jaef.2019.41006

China Tourism Academy. (2022, July 5). *National cultural consumption data report for the first half of 2022*. http://www.ctaweb.org.cn/cta/gzdt/202208/a07f12fc72b4466483a5302dbfb82495.shtml

Chongming District Bureau of Statistics. (2021). *Chongming Statistical Yearbook 2021*.

Dolhey, S. (2019). A bibliometric analysis of research on entrepreneurial intentions from 2000 to 2018. *Journal of Research in Marketing and Entrepreneurship*, *21*(2), 180–199. https://doi.org/10.1108/JRME-02-2019-0015

Esfandiar, K., Sharifi-Tehrani, M., Pratt, S., & Altinay, L. (2019). Understanding entrepreneurial intentions: A developed integrated structural model approach. *Journal of Business Research*, *94*, 172–182. https://doi.org/10.1016/j.jbusres.2017.10.045

Fan, C., Gong, L., & Li, H. (2019). An agent-based model approach for assessing tourist recovery strategies after an earthquake: A case study of Jiuzhai Valley. *Tourism Management*, *75*, 307–317. https://doi.org/10.1016/j.tourman.2019.05.013

Fortunato, M. W.-P. (2014). Supporting rural entrepreneurship: A review of conceptual developments from research to practice. *Community Development*, *45*(4), 387–408. https://doi.org/10.1080/15575330.2014.935795

Greve, A., & Salaff, J. W. (2003). Social Networks and Entrepreneurship. *Entrepreneurship Theory and Practice*, *28*(1), 1–22. https://doi.org/10.1111/1540-8520.00029

Gurel, E., Altinay, L., & Daniele, R. (2010). Tourism students' entrepreneurial intentions. *Annals of Tourism Research*, *37*(3), 646–669. https://doi.org/10.1016/j.annals.2009.12.003

Huggins, R., & Thompson, P. (2015). Entrepreneurship, innovation and regional growth: A network theory. *Small Business Economics*, *45*(1), 103–128. https://doi.org/10.1007/s11187-015-9643-3

Iorio, M., & Corsale, A. (2010). Rural tourism and livelihood strategies in Romania. *Journal of Rural Studies*, *26*(2), 152–162. https://doi.org/10.1016/j.jrurstud.2009.10.006

Krueger, N. (2009). Entrepreneurial Intentions are Dead: Long Live Entrepreneurial Intentions. In A. L. Carsrud & M. Brännback (Eds.), *Understanding the Entrepreneurial Mind* (pp. 51–72). Springer New York. https://doi.org/10.1007/978-1-4419-0443-0_4

Krueger, N. F., & Carsrud, A. L. (1993). Entrepreneurial intentions: Applying the theory of planned behaviour. *Entrepreneurship & Regional Development*, *5*(4), 315–330. https://doi.org/10.1080/08985629300000020

Kulawiak, A., Suliborski, A., & Rachwał, T. (2022). Research on Rural Entrepreneurship in Terms of the Literature: Definition Problems and Selected Research Issues. *Quaestiones Geographicae*, *41*(2), 7–19. https://doi.org/10.2478/quageo-2022-0020

Liñán, F., & Chen, Y. (2009). Development and Cross–Cultural Application of a Specific Instrument to Measure Entrepreneurial Intentions. *Entrepreneurship Theory and Practice*, *33*(3), 593–617. https://doi.org/10.1111/j.1540-6520.2009.00318.x

Lindkvist, E., Wijermans, N., Daw, T. M., Gonzalez-Mon, B., Giron-Nava, A., Johnson, A. F., van Putten, I., Basurto, X., & Schlüter, M. (2020). Navigating Complexities: Agent-Based Modeling to Support Research, Governance, and Management in Small-Scale Fisheries. *Frontiers in Marine Science*, *6*. https://www.frontiersin.org/articles/10.3389/fmars.2019.00733

López-Fernández, M. C., Serrano-Bedia, A. M., & Pérez-Pérez, M. (2016). Entrepreneurship and Family Firm Research: A Bibliometric Analysis of An Emerging Field. *Journal of Small Business Management*, *54*(2), 622–639. https://doi.org/10.1111/jsbm.12161

McMullen, J. S., & Shepherd, D. A. (2006). Entrepreneurial Action And The Role Of Uncertainty In The Theory Of The Entrepreneur. *Academy of Management Review*. https://doi.org/10.5465/amr.2006.19379628

Monks, T., Currie, C. S. M., Onggo, B. S., Robinson, S., Kunc, M., & Taylor, S. J. E. (2019). Strengthening the reporting of empirical simulation studies: Introducing the STRESS guidelines. *Journal of Simulation*, *13*(1), 55–67. https://doi.org/10.1080/17477778.2018.1442155

Negahban, A., & Yilmaz, L. (2014). Agent-based simulation applications in marketing research: An integrated review. *Journal of Simulation*, *8*(2), 129–142. https://doi.org/10.1057/jos.2013.21

Onggo, B. S., & Foramitti, J. (2021). Agent-Based Modeling and Simulation For Business and Management: A Review and Tutorial. *2021 Winter Simulation Conference (WSC)*, 1–15. https://doi.org/10.1109/WSC52266.2021.9715352

Oppermann, M. (1996). Rural tourism in Southern Germany. *Annals of Tourism Research*, *23*(1), 86–102. https://doi.org/10.1016/0160-7383(95)00021-6

Pato, M. L., & Teixeira, A. A. C. (2016). Twenty Years of Rural Entrepreneurship: A Bibliometric Survey: Twenty years of rural entrepreneurship. *Sociologia Ruralis*, *56*(1), 3–28. https://doi.org/10.1111/soru.12058

Pizzitutti, F., Mena, C. F., & Walsh, S. J. (2014). Modelling Tourism in the Galapagos Islands: An Agent-Based Model Approach. *Journal of Artificial Societies and Social Simulation*, *17*(1), 14. https://doi.org/10.18564/jasss.2389

Reichel, A., Lowengart, O., & Milman, A. (2000). Rural tourism in Israel: Service quality and orientation. *Tourism Management*, *21*(5), 451–459. https://doi.org/10.1016/S0261-5177(99)00099-0

Rosalina, P. D., Dupre, K., & Wang, Y. (2021). Rural tourism: A systematic literature review on definitions and challenges. *Journal of Hospitality and Tourism Management*, *47*, 134–149. https://doi.org/10.1016/j.jhtm.2021.03.001

Schoonhoven, C. B., & Romanelli, E. (2001). *The Entrepreneurship Dynamic: Origins of Entrepreneurship and the Evolution of Industries*. Stanford University Press.

Shapero, A., & Sokol, L. (1982). *The Social Dimensions of Entrepreneurship* (SSRN Scholarly Paper No. 1497759). https://papers.ssrn.com/abstract=1497759

Sharpley, R. (2002). Rural tourism and the challenge of tourism diversification: The case of Cyprus. *Tourism Management*, *23*(3), 233–244. https://doi.org/10.1016/S0261-5177(01)00078-4

Siebers, P. O., Macal, C. M., Garnett, J., Buxton, D., & Pidd, M. (2010). Discrete-event simulation is dead, long live agent-based simulation! *Journal of Simulation*, *4*(3), 204–210. https://doi.org/10.1057/jos.2010.14

Solvoll, S. (2015). Tourism Entrepreneurship – Review and Future Directions. *Scandinavian Journal of Hospitality and Tourism*, *15*, 120–137.

Stathopoulou, S., Psaltopoulos, D., & Skuras, D. (2004). Rural entrepreneurship in Europe: A research framework and agenda. *International Journal of Entrepreneurial Behavior & Research*, *10*(6), 404–425. https://doi.org/10.1108/13552550410564725

Student, J., Kramer, M. R., & Steinmann, P. (2020). Simulating emerging coastal tourism vulnerabilities: An agent-based modelling approach. *Annals of Tourism Research*, *85*, 103034. https://doi.org/10.1016/j.annals.2020.103034

Su, B. (2011). Rural tourism in China. *Tourism Management*, *32*(6), 1438–1441. https://doi.org/10.1016/j.tourman.2010.12.005

Su, M. M., Wall, G., Wang, Y., & Jin, M. (2019). Livelihood sustainability in a rural tourism destination—Hetu Town, Anhui Province, China. *Tourism Management*, *71*, 272–281. https://doi.org/10.1016/j.tourman.2018.10.019

Swinerd, C., & McNaught, K. R. (2012). Design classes for hybrid simulations involving agent-based and system dynamics models. *Simulation Modelling Practice and Theory*, *25*, 118–133. https://doi.org/10.1016/j.simpat.2011.09.002

Thirumalesh Madanaguli, A., Kaur, P., Bresciani, S., & Dhir, A. (2021). Entrepreneurship in rural hospitality and tourism. A systematic literature review of past achievements and future promises. *International Journal of Contemporary Hospitality Management*, *33*(8), 2521–2558. https://doi.org/10.1108/IJCHM-09-2020-1121

Utomo, D. S., Onggo, B. S., & Eldridge, S. (2018). Applications of agent-based modelling and simulation in the agri-food supply chains. *European Journal of Operational Research*, *269*(3), 794–805. https://doi.org/10.1016/j.ejor.2017.10.041

Vaillant, Y., & Lafuente, E. (2007). Do different institutional frameworks condition the influence of local fear of failure and entrepreneurial examples over entrepreneurial activity? *Entrepreneurship & Regional Development*, *19*(4), 313–337. https://doi.org/10.1080/08985620701440007

Vaishar, A., & Šťastná, M. (2022). Impact of the COVID-19 pandemic on rural tourism in Czechia Preliminary considerations. *Current Issues in Tourism*, *25*(2), 187–191. https://doi.org/10.1080/13683500.2020.1839027

Vinogradov, E., Leick, B., & Kivedal, B. K. (2020). An agent-based modelling approach to housing market regulations and Airbnb-induced tourism. *Tourism Management*, *77*, 104004. https://doi.org/10.1016/j.tourman.2019.104004

Wen, J., Kozak, M., Yang, S., & Liu, F. (2020). COVID-19: Potential effects on Chinese citizens' lifestyle and travel. *Tourism Review*, *76*(1), 74–87. https://doi.org/10.1108/TR-03-2020-0110

Wortman Jr., M. S. (1990). Rural entrepreneurship research: An integration into the entrepreneurship field. *Agribusiness*, *6*(4), 329–344. https://doi.org/10.1002/1520-6297(199007)6:4<329::AID-AGR2720060405>3.0.CO;2-N

Xie, L., Flynn, A., Tan-Mullins, M., & Cheshmehzangi, A. (2019). The making and remaking of ecological space in China: The political ecology of Chongming Eco-Island. *Political Geography*, *69*, 89–102. https://doi.org/10.1016/j.polgeo.2018.12.012

Zhu, H., & Deng, F. (2020). How to Influence Rural Tourism Intention by Risk Knowledge during COVID-19 Containment in China: Mediating Role of Risk Perception and Attitude. *International Journal of Environmental Research and Public Health*, *17*(10), Article 10. https://doi.org/10.3390/ijerph17103514

AUTHOR BIOGRAPHIES

YUNFEI GU is a PhD candidate at Southampton Business School, the University of Southampton, UK. She obtained a Bachelor of Management degree in Tourism Management from Zhengzhou University in China and a MSc in Business Analysis with distinction from Warwick Business School, the University of Warwick. Her research interests lie in sustainable tourism development, strategic management and different simulation modelling methodology. Her email address is Y.Gu@soton.ac.uk.

STEPHAN ONGGO is a Professor of Business Analytics at Southampton Business School, the University of Southampton. He is a member of the Centre for Operational Research, Management Sciences and Information Systems (CORMSIS). His research interests lie in the areas of simulation modelling methodology (symbiotic simulation/digital twin, hybrid modelling, agent-based simulation, discrete-event simulation) with applications in supply chain, health care and disaster management. He is the associate editor for the Journal of Simulation and Health Systems. His website is https://bsonggo.wordpress.com. His email address is b.s.s.onggo@soton.ac.uk.

MARTIN KUNC is a Professor of Business Analytics/Management Science at Southampton Business School, the University of Southampton. Previously to becoming an academic, he was a consultant at Arthur Andersen. He has also had independent consulting projects in the media, pharmaceutical, financial services, consumer goods, cement and wine industries. He is a member of the Centre for Operational Research, Management Sciences and Information Systems (CORMSIS). He is interested in the intersection of management science, behavioural science and strategic management. He is interim Director of the Centre for Healthcare Analytics and is Editor-in-Chief of the Journal of the Operational Research Society. His email address is M.H.Kunc@soton.ac.uk

STEFFEN BAYER is a Lecturer in Business Analytics within Southampton Business School at the University of Southampton. He is the Programme Leader of the MSc in Business Analytics and Finance. He is a member of the editorial board of Operations Research for Health Care and a past president of the UK Chapter of the International Systems Dynamics Society. Steffen's main research interest is the planning of health services. His past work includes studies on stroke care, renal care, human resource planning and home-based technology supported health delivery. He uses a variety of approaches in his research including qualitative research, system dynamics, agent-based modelling and discrete event simulation. His email address is S.C.Bayer@soton.ac.uk

Proceedings of the Operational Research Society Simulation Workshop 2023 (SW23)
C. Currie and L. Rhodes-Leader, eds.
https://doi.org/10.36819/SW23.009

MODELLING THE IMPACT OF AMBIDEXTROUS LEARNING ON TEAM PERFORMANCE USING AGENT-BASED SIMULATION

Mr. Yongxing Yan

University of Southampton
Southampton SO17 1BJ United Kingdom
yy12n21@soton.ac.uk

Professor Stephan Onggo

University of Southampton
Southampton SO17 1BJ United Kingdom
b.s.s.onggo@soton.ac.uk

ABSTRACT

In an increasingly competitive environment, organizations need to continuously innovate (explorative learning) while making steady improvements to their existing operations (exploitative learning). The capacity to pursue both exploratory and exploitative learning simultaneously is called ambidexterity. Therefore, ambidexterity has become one of the important research topics in the field of organizational study. This paper focuses on ambidexterity learning at the team level. The objective is to propose a generic agent-based simulation model that can be used to examine how ambidextrous learning affect team performance under different levels of task complexity, communication intensity and communication cost. The experiment shows that the model can reproduce what have been reported in the team performance literature.

Keywords: Ambidextrous Learning, Team Performance, Knowledge Exchange, Agent-Based Simulation

1 INTRODUCTION

With the growing competence of dynamic environment, traditional hierarchical structures are being replaced with team-oriented organization due to performance and productivity. To address poorly structured or complex problems and benefit from the collective skills, experience, and expertise of team members, many firms have successfully used team approaches (Michaelsen and Sweet, 2008). Consequently, learning teams, which are organic, flexible and sustainable groups that create learning atmosphere and motivate members' creativity, are crucial in organizations because of the need for active knowledge exchange, behaviour improvement and system optimization (Chen, 2005). The process of evolution in knowledge chain also provides chances for an individual learning to improve the core capability both at members and team level (Sun and Wang, 2009). Over time, organizational ambidexterity has emerged as a crucial field for organisational research. Ambidexterity is the capacity to pursue both exploratory and exploitative learning simultaneously. In exploitative learning, knowledge is gained using local search (e.g. refinement, selection and reuse of existing routines). In exploratory learning, knowledge is gained through experimentation.

Individually, the ambidextrous learning behaviour of employees involves concurrently investigating new abilities while making use of the skills they have already learned from the project at hand (Kar et al. 2021). Higher levels of individual exploration and exploitation lead to higher levels of innovation performance, which impacts aspects of individual, group, and firm performance (Schnellbächer et al 2019). One of the methods used to investigate staff performance and team performance is simulation (Weaver et al. 1995). Increasingly, simulation has been used as a method in management research, including organisational research (Onggo and Foramitti 2021). Simulation models can be used to test theories or hypothesised team mechanisms (e.g. Wong and Burton 2000, Martinez-Miranda and Pavon 2012). Some studies are more practical. For example, Onggo et al. (2012) and Onggo et al. (2010) used microsimulation modelling to design a new performance appraisal system for the European Commission staff that was implemented across the Commission in 2009. In another work, Spasic and Onggo (2012) estimated software project completion using agent-based simulation which models the performance of individual software developers. The model was validated against the real data from AVL who developed powertrain systems with internal combustion engines, as well as instrumentation and test systems.

The objective of this paper is to propose a generic agent-based simulation model to examine how ambidextrous learning affect team performance under different levels of task complexity, communication intensity and communication cost. We model team members who are continuously switching between exploratory learning and exploitative learning states. While acknowledging that this is still an early work, the experimental result confirms what has been reported in team performance literature, such as the relationship between task complexity, communication intensity, communication cost and team performance. The remaining of this paper is organized as follows. In Section 2, we review the literature on ambidextrous learning. We also review articles that study factors that affect team performance which will be considered in our model. The agent-based model is explained in Section 3 followed by the experimental result in Section 4. We end our paper with conclusion in Section 5.

2 LITERATURE REVIEW

In this section, we will review the literature on ambidextrous learning followed by studies that investigate factors that affect team performance and knowledge exchange. Finally, we will discuss the relationship between learning, knowledge exchange and team performance.

2.1 Ambidextrous learning

After March (1991) distinguished between exploratory learning and exploitative learning. Researchers illustrate exploitative learning with refining present skills, certainty, control, efficiency, decreased variance, while describe exploratory learning with trial of new knowledge, flexibility, experimentation, divergent thinking, risk taking (Katila and Ahuja, 2002; O'Reilly and Tushman, 2008; Turner et al 2013). Ambidextrous learning combines both exploratory learning and exploitative learning and it is used to continuously produce and gather knowledge, benefiting lower-level members (Wang and Jiang, 2009), and enhancing competitive advantages (Huang et al 2020).

The "paradox relationship" is occurred between exploitative and exploratory learning (Smith and Lewis, 2011). Traditionally, exploitation and exploration are divided into independent entities. The contingency theory, which states that the designs of dual structural for innovation depend on specific environmental conditions, helps explain this reasoning (Wang and Jiang, 2009). Exploratory learning as well as exploitative learning are in competition for the same organizational resources that are limited, hence they cannot coexist (Huang et al 2020). According to this school of ambidexterity, exploitation and exploration are two poles of a continuum, and businesses should choose the best relative exploratory position on the continuum (Wei et al 2014).

The success trap causes organizational shortsightedness when there is a focus on exploitation, while the failure cycle can happen when concentrates on exploration too much (Levinthal and March, 1993). Consequently the "balance relationship" is studied. According to conception of yin-yang, a holistic strategy to problem-solving that includes a two-stream operation for sustainable success is required for a firm's sustainable development (Smith and Lewis, 2011). Due to the necessity for balance, low (high) exploration must be combined with high (low) exploitation (Atuahene-Gima and Murray, 2007), determined by the degree of resource or coordination flexibility (Wei et al 2014). The overarching aim of our research is to study the balance between the exploitative and exploratory learning and how it affects team performance. In what follows, we will review factors influencing team performance, knowledge exchange and the relationship between learning, knowledge exchange and team performance.

2.2 Factors influencing team performance

At organizational level, different types of culture have different influence on team proactivity (Erkutlu, 2012), job satisfaction (Guerra et al 2005), and then employee performance (Pyöriä, 2007). The team organizational structure is established by the degree of authority centralization, the level of team independence, and leadership conduct, plus these interactions between team members could positively affect performance (Chen, 2007). Kazanjian et al (2000) discovered that organizational structures had an impact on team members' creative tendencies, noting the team's degree of interdependence could influence individual engagement in the process of creation.

At the team level, the team power, due to the team's position in the official organizational hierarchical level (Ragins and Sundstrom, 1989), is built on resource control, which permits a team to have an impact on other employees (Greer et al 2011). On account of competition among high-power team members for leadership positions, intragroup conflict will be more prevalent than low-power teams (Chattopadhyay et al, 2010), having negative relationship on team performance (Wolfe and McGinn, 2005). The network influences the team performance in terms of the constraints, scale, intensity and centralization aspects. Knowledge sharing is encouraged among team members by dense network, increasing accountability and visibility, reducing social loafing, and facilitating the mobilization of potential and resources of team members (Grund, 2012).

With technology evolution, an organization's most valuable asset is no longer machines, but is knowledge (Raducanu, 2012). There are many types and layers of knowledge, including individual and collective knowledge, explicit and tacit knowledge (Kazanjian et al 2000), as well as specialized knowledge and basic knowledge gaining success in comparison to its competitors (Keen and Wu, 2011). Therefore, at the individual level, the ability to gain knowledge from various sources (i.e. learning) has a big impact on group dynamics and performance. Individual contributions, incentivised to work in fields that suit their preferences and abilities (Margerison et al 1995), make up the collective output, and possibly those with great abilities take on the majority of the workload (Rhee et al 2013), achieving a high team performance to some specific situations (José et al 2011). Therefore, in our model, each team member has a set of abilities that are needed for the agent to gain new knowledge.

2.3 Knowledge exchange

While the knowledge of an individual serves as a resource for the team, the personality of the individual affects communication and interaction with other team members (Stewart, 2006). The group with different personalities can provide better project outcomes than the uniform group when the nature of the assignment necessitates a creative approach and problem-solving skills (Kim et al 2008). Enneagram, a personality classification system, is often used to represent the matching between nine personality types. Communication is enhanced by understanding other people's enneagram types in interacting with individuals (Mládková, 2016), due to the attributes like openness, altruism, genuineness, extraversion, and conscientiousness. Different enneagram types make use of different approaches to knowledge sharing (Mládková, 2014). For example, type "expert" said that combining current knowledge with new explicit knowledge to create new knowledge was the greatest strategy. Employees who are competent are more likely to impart knowledge to their peers. It adds to the perception that they use their important and applicable knowledge to assist others in completing specific tasks and resolving issues at work (Raducanu, 2012). The team members in our model are endowed with a personality type and we consider the personality match to model the communication between any two team members.

Knowledge-sharing structures cannot function without communication between team members. Regular contact fosters long-lasting bonds with other members (Burkink, 2002).Team members with more experience can impart difficult-to-find information or specific abilities to less experienced teammates (Song et al 2015). The satisfaction of an organization with its knowledge sharing in this structure is positively correlated with communication frequency (Wagner and Buko, 2005). The social capital theory states that organizations having a strong internal network of connections enjoy the benefits of open communication, close interaction, resource sharing, and improved teamwork. Employees with strong interpersonal connections have easier access to resources (such as vital knowledge) for their jobs, and they are more likely to perform well at work (Lee et al 2015). Therefore, our model consider level of knowledge and communication frequency in the knowledge exchange between team members.

2.4 The relationship between learning, knowledge exchange and team performance

Exchange of information among employees is encouraged by a knowledge-sharing culture, which will lead to more exploration (Wang and Tarn, 2018) and indirectly, exploitation (Caniëls et al 2017). Exploratory learning enhances individuals' present knowledge and provides new skills, which in turn encourages the acquisition of new knowledge (Huang et al 2020). Exploratory learning significantly affects knowledge transmission when people are closely interconnected and the tasks require radical creativity (Miller et al 2007), laying the foundation for exploitative learning that can provide knowledge base for next exchange of exploratory learning. Furthermore, exploitative learning alone can increase the risk of accumulating path dependence personal bias (Hughes et al 2007) which can be reduced by learning from external environment. Successful exploration, overcoming "path dependence", could serve as a foundation for later exploitation to support performance positively (Huang et al 2020).

Ambidextrous learning is not just importance for team performance but also for the competitive advantage of an organization. Although there are initially some advantages to exploitative learning for performance, these advantages quickly disappear as other organizations embrace, copy and use them (Li et al 2013). So an inverse U-shaped correlation is found between performance and exploitative learning (Katila and Ahuja, 2002) as well as exploratory learning (Wei et al 2014).

Researchers interested in knowledge exchange favour concepts, phenomena and connections at macro level, which means there are unclear presumptions in literature regarding individual behaviour and how they interact. It has the same situation for ambidextrous learning on individual level (Caniëls et al 2017). What's more, there are scarce studies concentrating on how ambidextrous learning relates to knowledge exchange and team performance without empirical method. All these research gaps provide opportunity for purpose of our study.

3 AGENT-BASED SIMULATION MODEL

To study the impact of ambidextrous learning on team performance when given a task under various knowledge dimensions ($|L|$) and communication intensity (β) within one single team, we develop an agent-based simulation model. The agents in the model are team members. The complexity of the task is represented by the number of knowledge dimensions (each dimension represents a complementary skill needed to complete the task). Hence, a higher number of knowledge dimensions indicate that the task is more complex. The simulation model and the experiment design are explained below.

3.1 The agent attributes

Knowledge: At time t, each agent $i \in I$ has knowledge $K_{i,t} = (k_{i,1,t}, k_{i,2,t}, \ldots\ldots, k_{i,L,t})$ where $k_{i,l,t} \in [0, +\infty)$ is the mastery level of agent i on knowledge dimension l and L is the number of knowledge dimensions. At the start of the simulation, each agent is allocated with different mastery levels $k_{i,l,0} \sim U(0,1)$ for all dimension $l \in L$.

Ability: Innovation ability, learning ability, collaboration ability (Liu, 2017) and absorptive ability (Wu et al 2022) are used to characterize the ability of a team member to improve its knowledge. Innovation ability illustrates the self-knowledge learning of an agent from its previous experience, expressed by $S_i \in [0,1]$. Cross learning ability represents the ability of an agent to learn from other agents, denoted by $X_i \in [0,1]$. Collaboration ability shows the ability of an agent to complete a task with other agents, represented by $C_i \in [0,1]$. Absorptive ability demonstrate how much knowledge that can be absorbed by an agent after learning from other agents, showed by $A_i \in [0,1]$. At the start of the simulation, these abilities are assigned a value $\sim U(0,1)$ for all agents.

Personality: For simplicity, we assume that each agent has one dominant personality. An agent can learn from another agent only if their personalities match. To find the match, we use the matrix given in the appendix (Table A-1) which is taken from Xiao (2014). A match between two personality types is denoted by "1" in the matrix. For example, when a person who shares knowledge has "Leader" personality (i.e. aggressive, influential, commanding) and the recipient is of type "Expert" (i.e. rational, analytical, intellectual, introvert) then communication can happen (Mládková, 2016). In the simulation, each agent has an equal chance to be assigned to one of the nine personality types.

3.2 The agent behaviours

The agents show ambidextrous learning behaviour by switching between exploitative and exploratory learning as shown in the state chart in Figure 1. The agents start with the exploitative learning. This represents a period of self-learning. During this period, the agents will update their knowledge using their self-learning or innovation ability S_i (see Equation 1). This knowledge updating is based on the relationship between innovation and the accumulation of exploitable knowledge during learning at individual level (Mukundan, 2015).

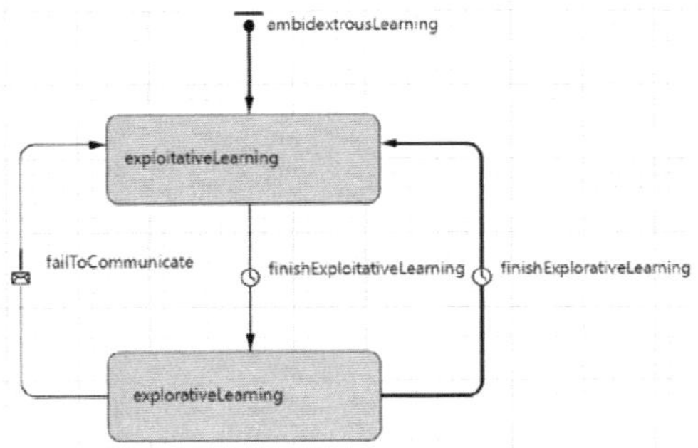

Figure 1 *Agent's state chart diagram*

In the next step, the agents will switch to exploratory learning. For an exploratory learning to happen, each agent must find another agent to learn from. This is when we consider the personality match. If an agent cannot find another agent to learn from then the agent cannot complete the exploratory learning and switch back to exploitative learning via transition `failToCommunicate`. If the agent can find another agent to learn from then the agent will update its knowledge using its cross learning ability X_i and absorptive ability A_i (see Equation 2). The equation shows that agent *i* will update its knowledge dimension *l* from agent *j* from its communication list *J* that has higher knowledge than agent *i* in that dimension. If there are more than one agent with higher knowledge then agent *i* will choose the one with the highest knowledge. The cross learning ability S_i and S_j represents the ability of agent *i* to receive information and agent *j* to give information. The absorptive ability A_i represents the ability of agent *i* to transform the information into knowledge. The number of agents to which an agent can share its knowledge (the size of *J*) is bounded by the communication intensity β as shown in Equation 3. The agent then switches back to exploitative learning via transition `finishExploratoryLearning`.

$$k_{i,l,t+1} = k_{i,l,t} + S_i \quad \forall_{i,l} \tag{1}$$

$$k_{i,l,t+1} = k_{i,l,t} + A_i \max_{j \in J} \max\{X_i X(k_{j,l,t} - k_{i,l,t}), 0\} \quad \forall_{i,l} \tag{2}$$

$$|J| = \beta X_i + 1 \quad \forall_i \tag{3}$$

In this paper, the team performance at time τ, i.e. $P(\tau)$ is defined as the accumulation of the sum of working ability under each knowledge dimension for all agents within the team (Equation 4). In this equation, agent *i* is the team leader that is determined at random. In this paper, we consider a situation in which the team leader allocates tasks to team members and supervise them (which involves a collaboration between the team leader and each of the team members). Hence, we do not consider tasks that require a team member to collaborate with other team members. The parameters C_i and C_j represent the collaborative ability of the team leader and team member *j*, respectively. The unit communication cost $\alpha \in [0,1]$ is used to calculate the communication cost (which increases as the team size increases).

$$P(\tau) = \frac{1}{(1+(|I|-1)\times\alpha)\times|L|} \sum_{t=0}^{\tau} \sum_{l=1}^{L} \left(k_{i,l,t} + \sum_{j\in I\setminus i} C_i C_j k_{j,l,t}\right) \tag{4}$$

4 EXPERIMENTAL RESULTS

Previous research have found that individual and corporate performance will be affected by the level of knowledge complexity (Levinthal, 1997). High levels of knowledge complexity would go against the point of coordination, divert resources away from the endeavour and result in poorer performance. Therefore, in the experiment we will vary $|L|$ with the values of 10, 20 and 30 to represent low, moderate and high levels of knowledge complexity, respectively. Likewise, good level communication affects team performance because high communication frequency may lead to team members exchanging more information, and the increased interchange may increase their knowledge. In the experiment, we vary the communication intensity β with values of 1, 5 and 10, denoting the low, moderate and high levels of intensity of exchange structures, respectively. We arbitrarily set the team size $|I|$ to 10 and α=0.1. The results presented in the next section are based on 35 simulation replications and simulation duration of 56 time steps.

Figure 2 shows the team performance at the end of the simulation $P(\tau)$ for different communication intensity β and number of knowledge dimensions $|L|$. The result shows that the team performance is higher when the task is less complex, i.e. lower number of knowledge dimensions $|L|$. This result is expected and consistent with what we know from the literature. For example, Willem and Buelens (2009) have found that team members may find it challenging to comprehend information from each other when the knowledge complexity is high.

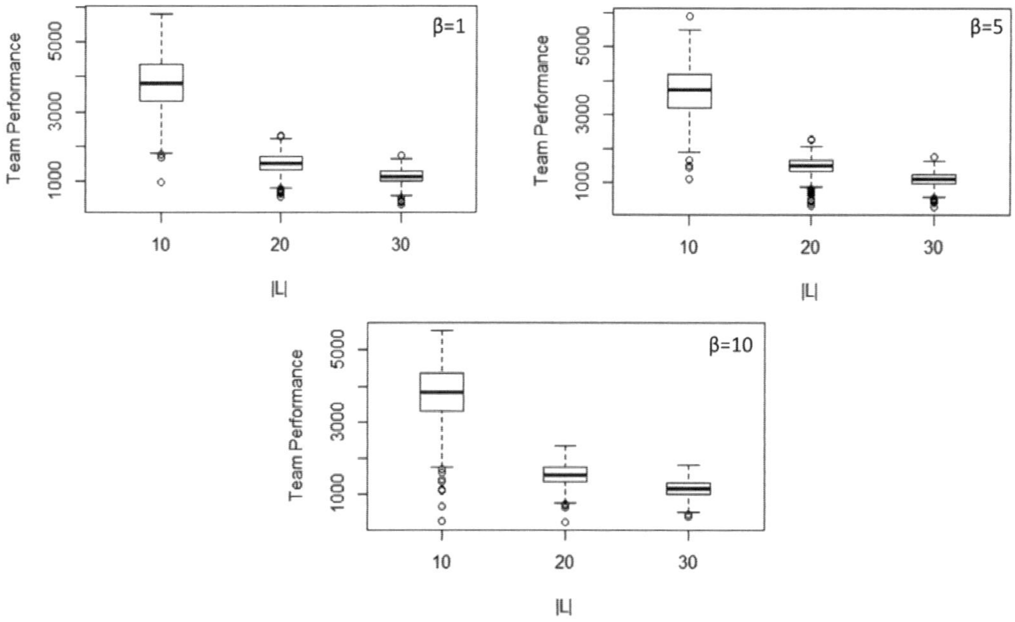

Figure 2 *Team performance for* β={1, 5, 10}, $|L|$={10, 20, 30} *and* α=0.1.

Figure 3 shows the same data but in a different arrangement to show the impact of communication intensity (β) better. The result shows that statistically, β does not have any significant impact on difference team performance. This can be partly explained from the different types of personality in the team. To demonstrate this, we compare the team performance when all team members are perfectionists (can effectively collaborate with the fewest personality types), leaders (can collaborate with the most personality types) and random. The result is shown in Figure 4. It shows that having a team formed by people from different personality types tends to perform better than a team formed by people with the same personality type. The result is particularly bad if all team members are perfectionist.

Despite the impact communication intensity on team performance is statistically insignificant (as shown in Figure 3), we can observe that more frequent communications (β=5 and 10) lead to slightly better average team performance than the bare minimum (β=1). Because we are accumulating the performance, the performance difference will grow bigger in the long term (i.e. if we run the simulation with a longer duration). There the subtle inverse U-shape relationship for $|L|=10$ and 30 where the team performance when $\beta=5$ is better in comparison to $\beta=1$ or 10. The pattern will become more obvious as we run the simulation for longer duration. The inverse U-shape relationship between communication intensity and team performance implies that there exists an optimal level of communication beyond which more frequent communications will decrease the team performance (because less time is used for exploitative learning). The same phenomenon has also been reported in the literature (Li et al. 2013; Katila and Ahuja 2002; Wei et al. 2014). Finally, Figure 5 shows that the unit communication cost (α) has a non-linear impact on team performance. A higher unit communication cost implies a more inefficient communication (more time or effort is needed to achieve the same knowledge transfer).

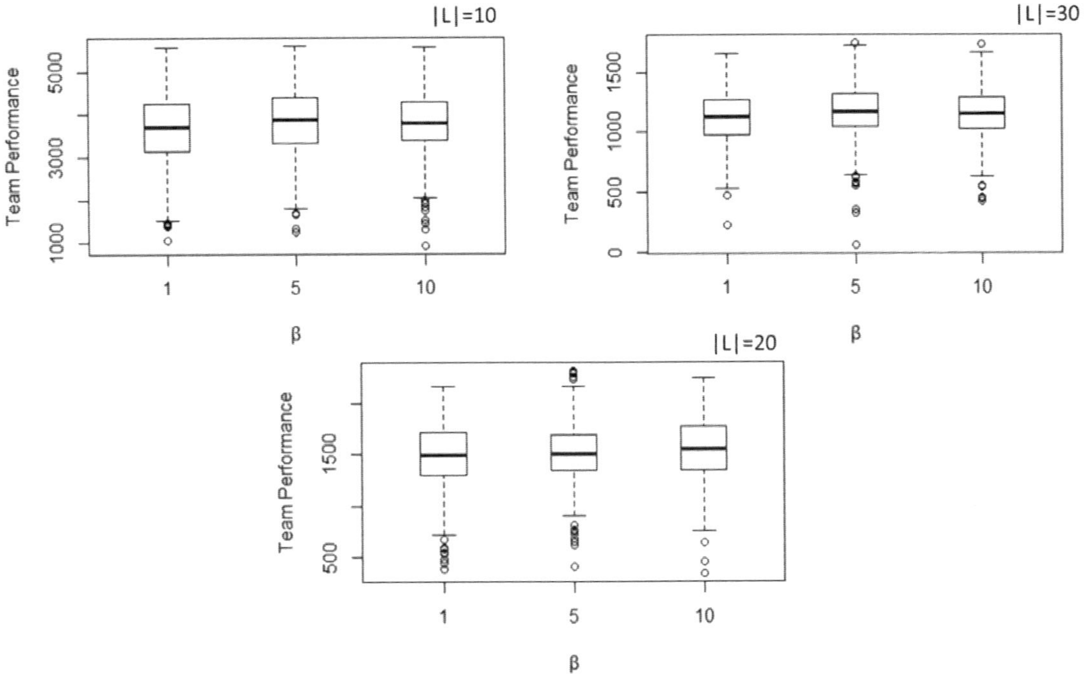

Figure 3 *Team performance for* $|L|=\{10, 20, 30\}$, $\beta=\{1, 5, 10\}$ *and* $\alpha=0.1$.

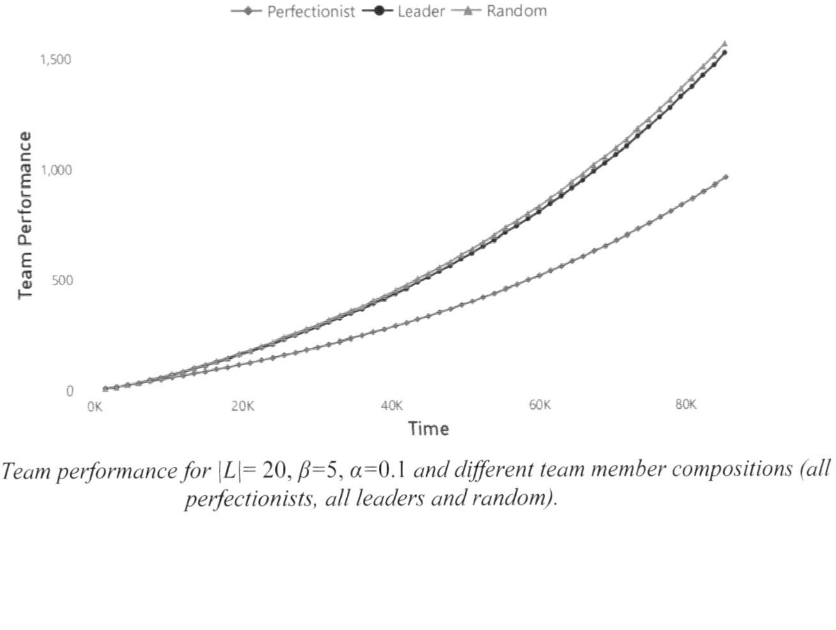

Figure 4 *Team performance for* $|L|= 20$, $\beta=5$, $\alpha=0.1$ *and different team member compositions (all perfectionists, all leaders and random).*

5 CONCLUSIONS

We have presented an agent-based simulation model that can be used to study the impact of ambidextrous learning on team performance under different levels of task complexity (or knowledge dimension) and communication intensity. In this early work, we have focused more on the internal validity of the model by grounding the model structure on the literature on ambidextrous learning and comparing the experimental result to what we know from the literature. For future work, we conduct an empirical study and collect data that can be used to calibrate and validate the model.

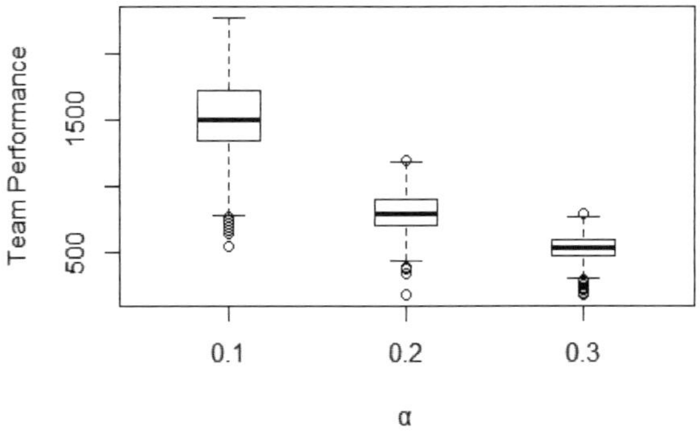

Figure 5 *Team performance for* $|L|= 20, \beta=5,$ *and* $\alpha=\{0.1, 0.2, 0.3\}$.

A APPENDICES

Table A-1 *The Enneagram of personality match*

Giving / Receiving	1 Perfectionist	2 Helper	3 Winner	4 Romantic	5 Expert	6 Realist	7 Adventurer	8 Leader	9 Peace Seeker
Perfectionist	0	0	1	0	0	0	0	1	0
Helper	0	1	1	1	0	1	1	1	1
Winner	1	0	1	0	1	1	1	1	0
Romantic	0	1	0	1	0	0	1	0	1
Expert	1	0	1	0	1	0	0	1	0
Realist	0	1	0	1	0	0	1	0	1
Adventurer	0	1	1	1	0	0	0	1	1
Leader	1	1	1	0	0	1	1	1	0
Peace Seeker	0	1	0	1	1	0	1	0	1

REFERENCES

Atuahene-Gima K and Murray J Y (2007) Exploratory and exploitative learning in new product development: A social capital perspective on new technology ventures in China. *Journal of International Marketing* 15(2): 1–29.

Burkink T (2002) Cooperative and voluntary wholesale groups: channel coordination and interfirm knowledge transfer. *Supply Chain Management: An International Journal* 7(2): 60-70.

Caniëls M C, Neghina C and Schaetsaert N (2017) Ambidexterity of employees: the role of empowerment and knowledge sharing. *Journal of Knowledge Management* 21(5):1098-1119.

Chattopadhyay P, Finn C and Ashkanasy N M (2010) Affective responses to professional dissimilarity: A matter of status. *Academy of Management Journal* 53(4): 808-826.

Chen C J (2007) Information technology, organizational structure, and new product development--the mediating effect of cross-functional team interaction. *IEEE Transactions on Engineering Management* 54(4): 687-698.

Chen S (2005) Task partitioning in new product development teams: A knowledge and learning perspective. *Journal of Engineering and Technology Management* 22(4): 291-314.

Erkutlu H (2012) The impact of organizational culture on the relationship between shared leadership and team proactivity. *Team Performance Management: An International Journal* 18(1/2): 102-119.

Greer L L, Caruso H M and Jehn K A (2011) The bigger they are, the harder they fall: Linking team power, team conflict, and performance. *Organizational Behavior and Human Decision Processes* 116(1): 116-128.

Grund T U (2012) Network structure and team performance: The case of English Premier League soccer teams. *Social Networks* 34(4): 682-690.

Guerra J M, Martínez I, Munduate L and Medina F J (2005) A contingency perspective on the study of the consequences of conflict types: The role of organizational culture. *European Journal of Work and Organizational Psychology* 14(2): 157-176.

Huang S Z, Lu J Y, Chau K Y and Zeng H L (2020) Influence of ambidextrous learning on eco-innovation performance of startups: moderating effect of top management's environmental awareness. *Frontiers in Psychology* 11: 1976–1976.

Hughes M, Hughes P and Morgan R E (2007) Exploitative learning and entrepreneurial orientation alignment in emerging young firms: Implications for market and response performance. *British Journal of Management* 18(4): 359-375.

José M, Aragón-Correa J A and Ferrón-Vílchez V (2011) Job-related skill heterogeneity and action team performance. *Management Decision* 49(7): 1061-1079.

Kar S, Kar A K and Gupta M P (2021) Understanding the S-curve of ambidextrous behavior in learning emerging digital technologies. *IEEE Engineering Management Review* 49(4): 76-98.

Katila R and Ahuja G (2002) Something old, something new: A longitudinal study of search behavior and new product introduction. *Academy of management journal* 45(6): 1183-1194.

Kazanjian R K, Drazin R and Glynn M A (2000) Creativity and technological learning: the roles of organization architecture and crisis in large-scale projects. *Journal of Engineering and Technology Management* 17(3): 273–298.

Keen C and Wu Y (2011) An ambidextrous learning model for the internationalization of firms from emerging economies. *Journal of International Entrepreneurship* 9(4): 316-339.

Kim D, Jang J and Shin S J (2008) The effect of personality type on team performance in engineering materials term projects. *ASEE Annual Conference and Exposition, Conference Proceedings*, pp 13-1221.

Lee S, Park J G and Lee J (2015) Explaining knowledge sharing with social capital theory in information systems development projects. *Industrial Management & Data Systems* 115(5): 883-900.

Levinthal D A (1997) Adaptation on rugged landscapes. *Management Science* 43(7): 934–950.

Levinthal D A and March J G (1993) The myopia of learning. *Strategic Management Journal* 14(S2):95-112.

Li Y, Wei Z, Zhao J, Zhang C and Liu Y (2013) Ambidextrous organizational learning, environmental munificence and new product performance: Moderating effect of managerial ties in China. *International Journal of Production Economics* 146(1): 95-105.

Liu H Z (2017) *Research on the impact of multi-team cooperation and knowledge communication on team performance based on multi-agent system*. Master Thesis, Hunan Normal University.

March J G (1991) Exploration and exploitation in organizational learning. *Organization Science* 2(1): 71-87.

Margerison C, McCann D and Davies R (1995) Focus on team appraisal. *Team performance management: an international journal* 1(4): 13-18.

Martínez-Miranda J and Pavón J (2012). Modeling the influence of trust on work team performance. *Simulation* 88(4):408-436.

Michaelsen L K and Sweet M (2008) The essential elements of team-based learning. *New directions for teaching and learning* 2008(116): 7-27.

Miller B K, Bierly III P E and Daly P S (2007) The knowledge strategy orientation scale: individual perceptions of firm-level phenomena. *Journal of Managerial Issues* 19(3): 414-435.

Mládková L (2014) Impact of personality on work with knowledge. *Proceedings of the International Conference on Intellectual Capital, Knowledge Management & Organizational Learning.* Academic Conferences Limited., pp 298.

Mládková L (2016) The enneagram as a tool of management of knowledge workers. *Proceedings of the European Conference on Knowledge Management.* ECKM, 2016-january, pp 614–621.

Mukundan R (2015) *Product and process innovation: Antecedents and performance outcomes in small IT firms in India.* PhD Thesis, Cochin University of Science and Technology.

Onggo B S, Pidd M, Soopramanien D and Worthington D (2012) Behavioural Modelling of Career Progression in the European Commission. *European Journal of Operational Research* 222(3):632-641.

Onggo B S, Pidd M, Soopramanien D and Worthington D (2010) Simulation of Career Development in the European Commission. *Interfaces* 40:184-195.

Onggo B S and Foramitti J (2021) Agent-Based Modeling and Simulation for Management Decisions: A Review and Tutorial. In *Proceedings of the 2021 Winter Simulation Conference*, pp. 1-15.

O'Reilly C A and Tushman M L (2008) Ambidexterity as a dynamic capability: Resolving the innovator's dilemma. *Research in Organizational Behavior* 28: 185-206.

Pyöriä P (2007) Informal organizational culture: the foundation of knowledge workers' performance. *Journal of Knowledge Management* 11(3): 16-30.

Raducanu R R (2012) *Assessment of employees' attitudes and intentions to share knowledge based on their individual characteristics.* Unpublished Master Thesis, Copenhagen Business School.

Ragins B R and Sundstrom E (1989) Gender and power in organizations: A longitudinal perspective. *Psychological bulletin* 105(1): 51-88.

Rhee J, Parent D and Basu A (2013) The influence of personality and ability on undergraduate teamwork and team performance. *SpringerPlus* 2(1): 1-14.

Schnellbächer B, Heidenreich S and Wald A (2019) Antecedents and effects of individual ambidexterity—A cross-level investigation of exploration and exploitation activities at the employee level. *European Management Journal* 37(4): 442–454.

Smith W K and Lewis M W (2011) Toward a theory of paradox: A dynamic equilibrium model of organizing. *Academy of Management Review 36* (2): 381–403.

Song C, Park K R and Kang S W (2015) Servant leadership and team performance: The mediating role of knowledge-sharing climate. *Social Behavior and Personality: an international journal* 43(10): 1749-1760.

Spasic B and Onggo B S (2012) Agent-Based Simulation of Software Development Process: A Case Study at AVL. In *Proceedings of the 2012 Winter Simulation Conference*, pp. 3646-3656.

Stewart G L (2006) A meta-analytic review of relationships between team design features and team performance. *Journal of management* 32(1): 29-55.

Sun R and Wang T (2009) Study on knowledge exchange and learning in knowledge teams based on knowledge chain. *In 2009 Fourth International Conference on Cooperation and Promotion of Information Resources in Science and Technology*, pp 95-100.

Turner N, Swart J and Maylor H (2013) Mechanisms for managing ambidexterity: A review and research agenda. *International Journal of Management Reviews* 15(3): 317-332.

Wagner S M and Buko C (2005) An empirical investigation of knowledge-sharing in networks. *Journal of Supply Chain Management* 41(4): 17-31.

Wang F and Jiang H (2009) Innovation paradox and ambidextrous organization: A case study on development teams of air conditioner in Haier. *Frontiers of Business Research in China : Selected Publications from Chinese Universities* 3(2): 271–300.

Wang J F J and Tarn D D (2018) Are two heads better than one? Intellectual capital, learning and knowledge sharing in a dyadic interdisciplinary relationship. *Journal of Knowledge Management* 22(6): 1379-1407.

Weaver J L, Bowers C A, Salas E and Cannon-Bowers J A (1995). Networked simulations: New paradigms for team performance research. *Behavior Research Methods, Instruments, & Computers* 27:12–24.

Wei Z, Yi Y and Guo H (2014) Organizational learning ambidexterity, strategic flexibility, and new product development. *Journal of Product Innovation Management* 31(4): 832-847.

Willem A and Buelens M (2009) Knowledge sharing in inter-unit cooperative episodes: The impact of organizational structure dimensions. *International Journal of Information Management* 29(2): 151-160.

Wolfe R J and McGinn K L (2005) Perceived relative power and its influence on negotiations. *Group Decision and Negotiation* 14(1): 3-20.

Wong S S and Burton R M (2000) Virtual Teams: What are their Characteristics, and Impact on Team Performance? *Computational & Mathematical Organization Theory* 6:339–360.

Wu J, Yuan Y and Guo B (2022) Cognitive proximity, technological regime and knowledge accumulation: An agent-based model of interfirm knowledge exchange. *Asian Journal of Technology Innovation* pp 1–20.

Xiao X X (2014) *A knowledge team's behaviour simulation system design and implementation based on CAS theory*. Master Thesis, Dalian University of Technology.

AUTHOR BIOGRAPHIES

YONGXING YAN received a Bachelor of Management in Tourism Management from the Tianjin Normal University Jingu College in China in 2014 and a Master of Management in Corporate Management in the Ningbo University in China in 2017. He completed his MSc in Business Analytics and Management Science at the University of Southampton in 2022. He had worked in Ningbo Polytechnic as a student counselor between 2017 and 2020 in China.

STEPHAN ONGGO is a Professor of Business Analytics at Southampton Business School, the University of Southampton. He is a member of the Centre for Operational Research, Management Sciences and Information Systems (CORMSIS). His research interests lie in the areas of simulation modelling methodology (symbiotic simulation/digital twin, hybrid modelling, agent-based simulation, discrete-event simulation) with applications in supply chain, health care and disaster management. He is the associate editor for the Journal of Simulation and Health Systems. His website is https://bsonggo.wordpress.com. His email address is b.s.s.onggo@soton.ac.uk.

Proceedings of the Operational Research Society Simulation Workshop 2023 (SW23)
C. Currie and L. Rhodes-Leader, eds.
https://doi.org/10.36819/SW23.010

AGENT BASED SIMULATION OF WORKERS' BEHAVIOURS AROUND HAZARD AREAS IN MANUFACTURING SITES

Hanane El Raoui

University of Strathclyde
G1 1XQ, UK
Hanane.el-raoui@strath.ac.uk

John Quigley

University of Strathclyde
G1 1XQ, UK
J.quigley@strath.ac.uk

Ayse Aslan

Edinburgh Napier University
EH10 5DT, UK
A.Aslan@napier.ac.uk

Gokula Vasantha

Edinburgh Napier University
EH10 5DT, UK
G.Vasantha@napier.ac.uk

Jack Hanson

The University of Edinburgh
EH8 9YL, UK
Jack.hanson@ed.ac.uk

Jonathan Corney

The University of Edinburgh
EH8 9YL, UK
J.R.Corney@ed.ac.uk

Andrew Sherlock

National Manufacturing Institute Scotland
PA4 9LJ, UK
A.sherlock@strath.ac.uk

ABSTRACT

Rewards for risk taking behaviour by workers (if accidents do not occur) can be realised in the form of increased productivity or worker idle time. However, frequent unsafe behaviours of workers inevitably results in accidents and an associated loss in productivity. Workers' behaviour towards safety is influenced by management, who can encourage or discourage risk taking behaviour. In this paper, we explore the relationship between the perceived reward by individual workers who expose themselves to hazards and a management response in the form of inspections to monitor and address inappropriate behaviours. We conduct this study by developing an Agent Based Simulation Model, where workers are required to learn paths within a factory exposed to hazardous areas, with inspectors randomly moving around the factory to correct inappropriate behaviour if noticed. We assume workers are maximising their anticipated reward as they learn routes through the factory. This agent based model is used to explore the impact of inspection frequency and reward perception (i.e. parameters which can be influenced by management) on the number of workplace accident. The results demonstrated that the proposed model is a valuable tool to assist the management in predicting the potential safety improvement from safety management practices focusing on safety inspections, and changing workers perceptions.

Keywords:

Safety, behaviour, agent based model, accidents, hazardous areas, rewards, safety inspections

1 INTRODUCTION

Hazards can exist in almost every workplace and despite continuous efforts to mitigate risks and improve the safety, the rate of accidents and injuries still very high in many industries. Indeed, occupational safety organization around the world report alarming trends in workplace safety, for example: in the UK, the Health and Safety Executive (HSE) is the government organization responsible for regulating the development and implementation of safety rules. The HSE collects information on workplace accidents through the Reporting of Injuries, Diseases and Dangerous Occurrences Regulations, 2013 (RIDDOR). The manufacturing sector's non-fatal injury data provided by RIDDOR between 2016/17 and 2020/21 show that 32.6% of injuries have been caused by slips, trips and falls that can occur during movement. Safety violations have been widely recognized as the central cause of injuries and accidents. Hence, in order to improve safety in workplaces, there is ongoing interest in reducing the number of safety violations.

Safety violations can be present in different forms, such as the failure to wear protecting equipment (PPE), or taking shortcuts by travelling paths in the workplace where they are exposed to hazards as opposed to less hazardous paths in compliance with company guidance. In this study, we focus on the latter, by evaluating how the engagement of workers in shortcuts can impact the safety and productivity. Taking shortcuts is an intentional act, but non-malevolent. Indeed, workers expose themselves to hazards to save time, but they do not intend to cause damage to the organization.

Workers are often exposed to contradictory pressures, when for example, they have to balance production pressure with the need for safety. Such conflict may result in behaviour, where the workers start "cutting corners" to boost output. However, aggressive attitudes in the workplace can co-exist with risk avoidance behaviour when for example safety inspectors are around.

In this study, we investigate the impact of safety inspections on workers' accidents through changing behaviour. To model the safety-related behaviour, we introduce an Agent Based Simulation approach to: 1) Model how the workers learn about their environment, to determine their autonomous movement paths to save time, 2) simulate the aggressive behaviour of workers around hazardous areas, and study the consequent potential injuries, 3) Simulate the changes in workers behaviour from being aggressive to avoiding cutting corners when the safety inspector is around, 4) evaluate the impact of safety inspections, and rewards perceptions on the safety behaviour, to help the organization setting an efficient safety management policy.

The rest of the paper is organized as follows: Section 2 describes the academic background and related works, section 3 explains the research methodology adopted, section 4 describes the Q-learning algorithm, the proposed Agent Based Simulation framework is presented in section 5, the simulation results are detailed in section 6, and we end up with concluding remarks and some future works.

2 BACKGROUND AND LITERATURE REVIEW

2.1 Agent Based Modelling and Simulation

Agent Based Modelling and Simulation (ABMS) can be defined as a simulation system with agents that repeatedly interact with each other and with their environment in an autonomous way (Parker (2019); El Raoui, Oudani, and Alaoui (2018)). The agents in ABMS have certain properties and attributes (Wooldridge and Jennings (1994); El Raoui, Oudani, and Alaoui (2020)): autonomous, proactive, interacting, and unique. These properties enable the agent to communicate with other agents, interact with the environment, and make decisions in response.

2.2 Agent Based Modelling for safety behaviour

A wide range of simulation techniques have been used in previous studies to understand and address issues related to safety. Agent based modelling has gained a lot of interest in modelling safety-related behaviours. It was found that agent based modelling surpass the discrete event simulation at micro-level details of modelling, such as the behavioural aspects. Owing to the ability of ABM to capture the interactions between agents.

ABM have been mostly used to study safety behaviours in the construction industry. Lu, Cheung, Li, and Hsu (2016) proposed an ABM to investigate the impact of the different interactions between workers, and the various safety investment on improving the safety on site. In order to investigate ways to limit the risky behaviours of construction workers, Choi and Lee (2018) proposed a socio-cognitive approach based on ABM to simulate the social influence. Taillandier and Taillandier (2014) developed an ABM to access the impact of potential risks and work accidents on the costs and quality of work.

ABM was also combined with System Dynamic (SD) in several recent studies, for a more comprehensive analysis of safety behaviour aspects. A SD-ABM simulation model was developed by Nasirzadeh, Khanzadi, and Mir (2018) to examine the impact of social contagion on the violation of safety rules.

Manufacturing is different from construction, and to the best of the authors knowledge, no existent study has modelled the worker's aggressive behaviour around hazardous areas in manufacturing sites, its impact on the accidents rate and productivity loss, and how the safety inspection can reduce rate of injuries by pushing workers to avoid risk taking behaviour.

3 RESEARCH METHODOLOGY

This paper aims to assess the manufacturing site's safety by considering the workers safety violations around hazardous areas. Introducing safety inspections, and changing worker's valuation of risk taking can play an essential role in reducing the violations, which we try to demonstrate in this paper. The methodology used is as follows:

- Design the learning process of agents, to represent how the workers determine their movement paths on site.
- Define the behavioural rules of agents, and integrate it with the learning step to build the agent based simulation model.
- Define the scenarios to be simulated, and run the model to analyse the potential consequences of workers' behaviour on productivity and safety.
- Analyse the results and identify potential interventions by the organization to improve the safety.

4 Q-LEARNING BASED PATH DETERMINATION

When training an agent, different types of learning can be used such as supervised, and unsupervised learning, and reinforcement learning. The latter is used in this study. Reinforcement learning, can be defined according to Gatti (2015), as a machine learning technique that describes how a set of subsequent decisions will lead to the accomplishment of a goal, which is considered as a trial and error process to learn patterns.

Q-learning algorithm is a model free reinforcement learning algorithms used to train agents to find the optimal action in a Markovian Decision Process Dayan and Watkins (1992).The fundamental idea of Q-learning is that the agent learns an action value function to maximize the total rewards received from the environment. Considering that S is the set of possible states of an agent within an environment, and A the set of possible actions that the agent can choose from. The state, action, and reward of agent i at time t can be represented as:

$$s_t^i \in S, a_t^i \in A, r_t^i \in S \times A \rightarrow \mathbb{R} \tag{1}$$

The Q-Value can be updated as follows:

$$Q(s_t, a_t) = (1-\alpha)Q(s_t, a_t) + \alpha\left[r_{t+1} + \gamma \max Q(s_{t+1}, a_t)\right], \tag{2}$$

where α is the learning rate, and γ is the discount factor. We put, $\alpha = 1$ which assumes that the new value does not take into account the previous values of Q. We consider $\gamma = 0.8$ as it's usually set to in previous studies Samma, Lim, and Saleh (2016) which will make the agent strive for a long-term high reward.

The Q-learning algorithm is integrated in the proposed simulation model, to simulate how the agents learn about the environment to determine the best path to reach storage sites. The states of each agent are the possible neighbourhoods. In this paper, we use the Moore neighbourhood topology (i.e. a two-dimensional square lattice that contains a central cell surrounded by eight adjacent cells) Moore (1964). Changing from a neighbourhood (i.e. cell) to another is the action, therefore, 8 basic actions are available to each agent as shown in Figure 1a.

The goal of our agents is to reach the destination as fast as possible, therefore the designed reward is a function of the distance to the target point. Let $d(o,T)$ be the distance between the neighbour O and the target point T. The reward is computed as follows:

$$r = \begin{cases} 1, o = T \\ \frac{1}{d(o,T)}, \quad o \neq T \end{cases} \tag{3}$$

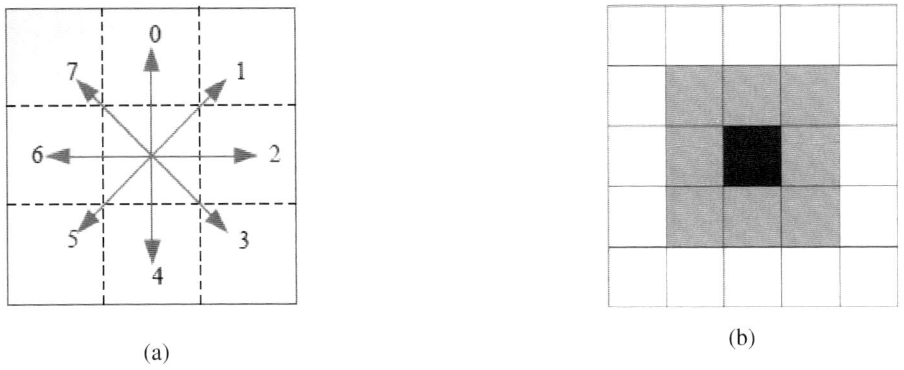

Figure 1: (a) The agent's actions, (b) Moore neighbourhood

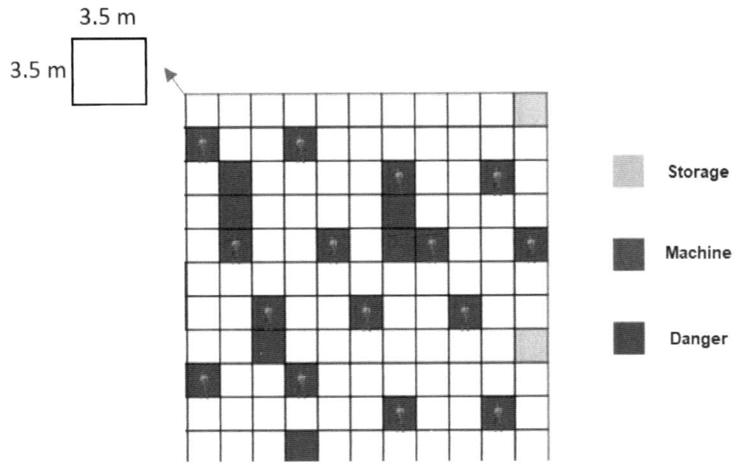

Figure 2: The virtual working environment.

5 ABMS FRAMEWORK

5.1 The virtual working environment

In agent based models, an agent can interact with other agents, and the environment. These interactions are represented in the model space where the agents can move. Different topologies can be used to model the space Macal and North (2009). In this study, we use a discrete spatial topology to represent the workplace. As shown is Figure 2 , the virtual manufacturing site consists of a grid layout of 400 cells characterized by their location (x,y) coordinates, and an attribute to characterize the type of each cell. This attribute is represented in the model by a variable that can assume the following values:

- Normal: cells with no manufacturing activities. At the beginning of the model, all agents cells are generated as normal one.
- Machine: cell that represent the work station.
- Danger: cells with hazardous conditions, placed randomly on the site.
- Storage: cells to represent the storage sites.

The agents move from cell to cell on the grid, and the cells immediately surrounding an agent are its neighbourhood. Two kinds of cell neighbourhood are usually considered: i) the von Neumanns and ii) the Moores (Figure 1b). The latter is considered in this work.

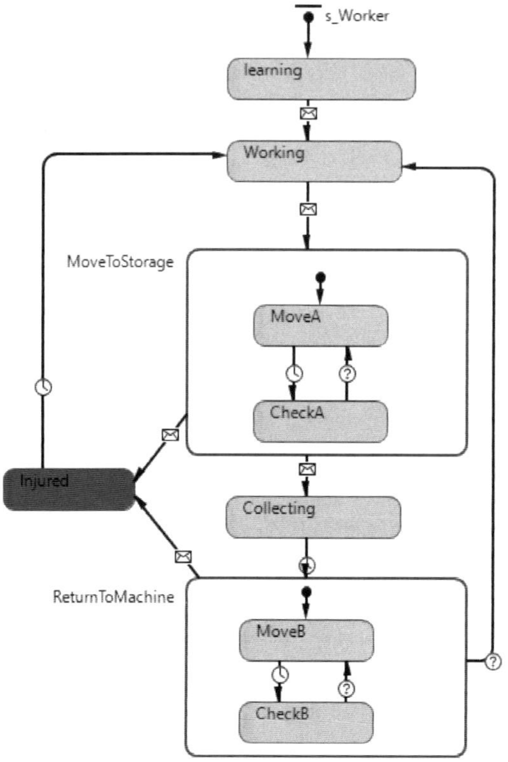

Figure 3: Workers' state chart.

5.2 The agent's behavioural rules

5.2.1 Workers

Workers are the main agents in our model. They are required to move from work stations to the storage sites to pick items, or to store semi-finished products. Accidents and injuries can occur with a certain probability during movements if a worker is in a danger zone. To describe their behaviour, we use the state chart in Figure 3. The first step in our model, before the agents start working is the learning phase. In the learning process, we replicate how the workers learn about the environment, to find the best path between two workplaces using a path determination mechanism, based on the Q-learning algorithm. We assume that the workers have an aggressive behaviour, so they choose shorter, riskier paths to reach the storage sites. However, worker's aggressive behaviour can change if the safety inspector is around. Indeed, workers will tend to avoid stepping on a danger zone to prevent being penalized by the inspector.

5.2.2 Safety Inspector

The inspector roams the site to make safety checkup, moving from one cell to another with a given frequency per day. The inspection aims to discourage undesirable behaviours and increase safety compliance by making the agent avoid danger zones.

6 SIMULATION AND ANALYSIS

The proposed ABM was implemented using the Anylogic software. This study is based on a hypothetical case. We consider a manufacturing site of, 4900 m, with 50 work stations, 50 workers, and 3 storage sites. The site involve 20 danger zones, where each zone occupies 12.25 m.

The model simulates the movement of workers from working to storage stations with a certain frequency. The inputs to our model are the following:

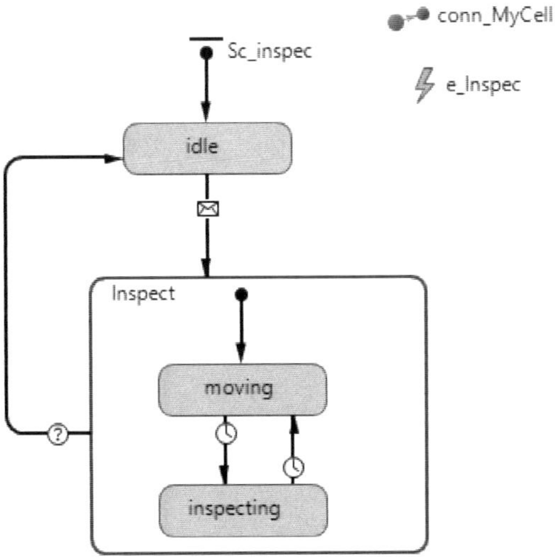

Figure 4: Safety Inspectors' state chart.

- The frequency of movements: controlled through an event generated following a uniform distribution, uniform (2h,3h).The event create a collection order from a random storage site.
- The probability of accidents in danger zones is set to 0.2
- The average walking speed of workers is 1.8 km/h.
- The frequency of inspection was set based on each particular scenario simulated to represent the different management policies.
- The duration of injuries is set to 1 day.

The user interface is shown in Figure 5 that gives control to the steps of the simulation, and provide a visualization of the statistics about the accidents. At the start of the simulation, the agents are trained to find their paths to each of the storages by performing 100 learning-passes. Once the learning process is complete, the user can launch the work process.

The purpose of the model is to assess the relationship between the rate of injuries, perceived rewards for risk taking by workers and managerial inspection policy.

6.1 Scenario 1: Avoider vs Aggressive

In this scenario, we simulate the aggressive and avoider behaviour of agents while moving around the site during 8 hours of work. The purpose is to quantify the potential savings in time from taking shortcuts.

The results in figure 6 shows that aggressive workers can effectively save time while moving to any of the storage sites. The amount of time that can be saved depend on the worker's placement, the destination, and the number of danger zones included in the best path of the worker.

The mean time that can be saved by the workers to storage 1,2,3 is respectively 7.4, 9 and 7.8 minutes. The lower whiskers represent the workers with slight time saving, that can be explained by their proximity to the storage sites, or either their safe fast path, means that they don't need to walk through danger zones to reach the storage. The upper whiskers represent the workers who can gain interesting saving by being aggressive, that could range between 12 and 22 min for storage 1, 12 to 21 min for storage 2, and 11 to 19 min for storage 3. We observe some outliers in storage 2 and 3 which represent the workers who have greater interest in making shortcuts. The provided results about the potential savings can give insight on the possible gain in productivity, and also helps in identifying the workers that are more likely to take shortcuts.

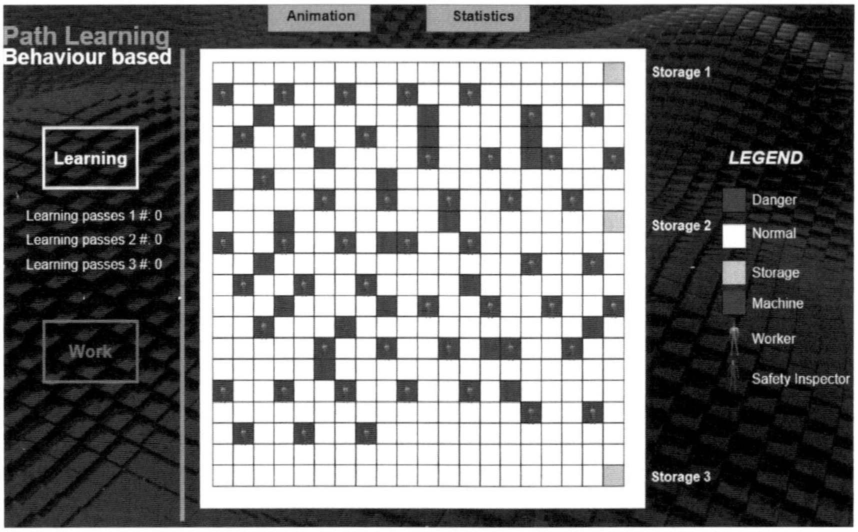

Figure 5: The customized user interface.

6.2 Scenario 2: Safety inspection

This scenario examined the impact of different safety inspection policies on the worker's safety behaviour, and the potential improvement in safety. During an inspection, an inspector is simulated walking through the entire workplace so that they cover it twice in 2 hours. We perform 10 simulation runs of 1000 hours and report the number of accidents. Five inspections policies were considered with different frequencies, starting an inspection every 2, 3, 4, 5 or 6 hours. We first simulate the case without inspection to serve as a reference point for comparison.

The results shown in Figure 7 demonstrate the role of safety inspection in reducing the rate of injuries. Unsurprisingly, We see that the mean number of accidents is decreasing as inspections increase. Introducing the most frequent level of inspections lowered the accidents by 23%. In terms of productivity, the duration of injuries is reduced by 96 hours which would improve productivity. However, inspections would require an inspector working all the time and the optimal solution may be with a more moderate inspection policy.

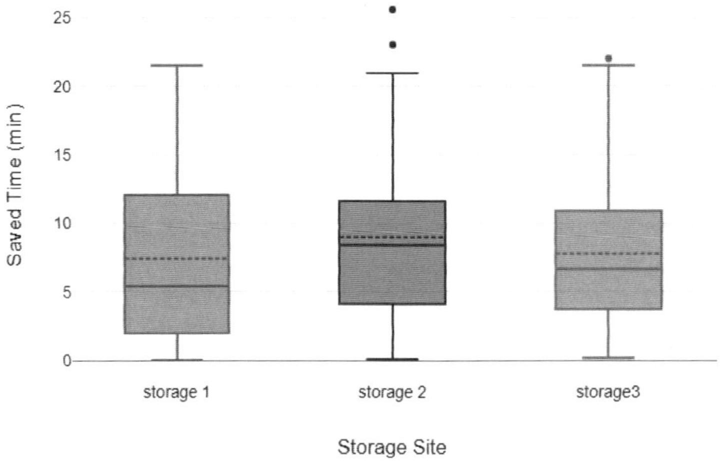

Figure 6: Travel time saved by aggressive workers to each storage site during 1 day work.

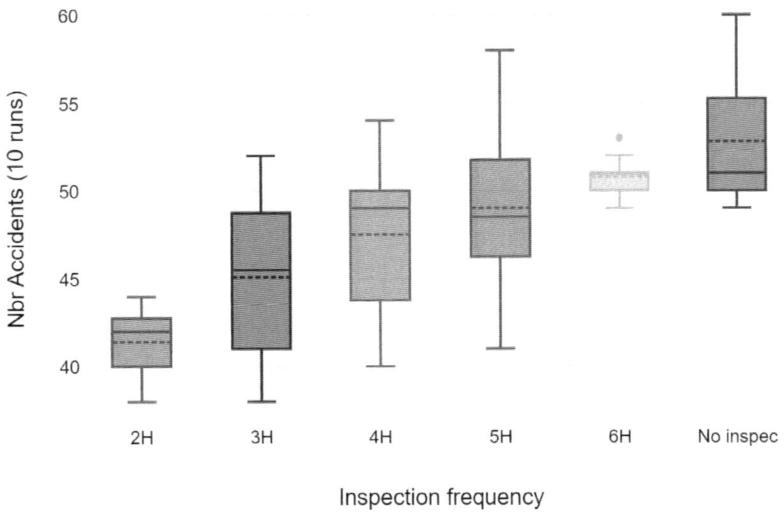

Figure 7: Accidents statistics for each inspection frequency.

6.3 Scenario 3: Perceived risk-taking reward

An alternative influential factor in worker risk taking behaviour is their perceived reward for exposure to hazard. Changing such perceptions provides an alternative route to improving safety in the workplace. The purpose of this scenario, is to examine the impact of diminishing the expected benefits on risk-taking behaviour, by applying a discount factor on the perceived reward.

Let ϕ be the discount factor. The perceived reward from using a hazard zone is:

$$r = (\frac{1}{d(o,T)})\phi \tag{4}$$

We run the experiments for 3 reward profiles corresponding to a discount of 25%, 50%, and 75% to correspond to Low, Moderate and High perceived reward for risk taking. Ten simulation runs of 1000 hours were performed for each combination of risk perception type and inspection policy. The mean number of the potential accidents are shown in Figure 8.

We see under all three risk perception types the higher the frequency of inspection the lower the accident rate. However, the impact of inspections is slight for low and moderate reward types but

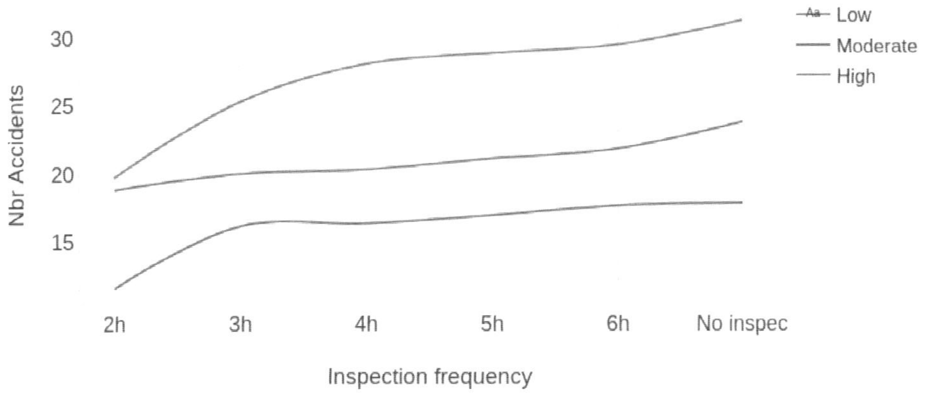

Figure 8: Accidents statistics for each inspection frequency, reward profile.

substantial for high. As such, depending on the characteristics of the work force, inspector may be adding little value. In fact, changing the characteristics of the workforce to move from high reward to moderate or low may be a more laudable goal, as a maximum inspection policy for a high reward workplace would have the same frequency of accidents as a moderate reward workplace with no inspections.

While the inspections can be very costly, changing people's perception can be extremely tricky. Therefore, management need to find the right balance between cost and complexity when implementing safety management practices. The proposed model can assist the management in predicting the potential safety improvement from each strategy profile.

7 CONCLUSIONS AND DISCUSSION

This research proposes an Agent Based Simulation framework to understand the workers' behaviour towards safety when moving around hazardous areas and evaluate the impact of different inspection policies on reducing the accidents rate. The proposed model was developed with three objectives:(1) Simulate the learning process of agents to determine their autonomous movement paths, based on the Q-learning algorithm. (2) model the different behaviours the workers can have towards safety. (3) quantify the potential injuries and productivity loss under different safety management practices.

The proposed model was demonstrated to work effectively through a hypothetical case study. Several scenarios were tested to analyse and asses the impact of the workers'aggressive behaviour on the productivity and safety. The results show that the model is potentially a valuable tool to help organisations understand the impact of risky behaviour and identify the potential ways to improve the safety.

- The potential time saving from taking shortcuts gives insights about the workers that are more likely to violate the safety rules.
- The model can help to predict the potential injuries and productivity loss arising from an aggressive attitude.
- The model can assist the management in determining the effectiveness of the safety inspection practices, focusing on the safety inspections and rewards perceptions, on improving the safety.

Social interactions with the co-workers can influence the safety rules compliance due to the social pressure and the social learning. As future work, the proposed learning process can be improved by integrating the social learning element to the reward design. The integration of the social component can assist the organisations to better understand the emergent behaviours, and design the appropriate strategies to mitigate the risk. The extended version of the model will be applied to a real case study. We are currently collecting the data on workers movement and their environment from a workshop in the University of Edinburgh.

ACKNOWLEDGMENTS

This work was supported by the Engineering and Physical Sciences Research Council, UK [grant number EP/V051113/1-Productivity and Sustainability Management in the Responsive Factory].

REFERENCES

Choi, B., and S. Lee. 2018. "An empirically based agent-based model of the sociocognitive process of construction workers safety behavior". *Journal of Construction Engineering and Management* 144 (2): 04017102.

Dayan, P., and C. Watkins. 1992. "Q-learning". *Machine learning* 8 (3): 279–292.

El Raoui, H., M. Oudani, and A. E. H. Alaoui. 2018. "ABM-GIS simulation for urban freight distribution of perishable food". In *MATEC Web of Conferences*, Volume 200, 00006. EDP Sciences.

El Raoui, H., M. Oudani, and A. E. H. Alaoui. 2020. "Coupling soft computing, simulation and optimization in supply chain applications: review and taxonomy". *IEEE Access* 8:31710–31732.

Gatti, C. 2015. "Reinforcement learning". In *Design of Experiments for Reinforcement Learning*, 7–52. Springer.

Lu, M., C. M. Cheung, H. Li, and S.-C. Hsu. 2016. "Understanding the relationship between safety investment and safety performance of construction projects through agent-based modeling". *Accident Analysis & Prevention* 94:8–17.

Macal, C. M., and M. J. North. 2009. "Agent-based modeling and simulation". In *Proceedings of the 2009 winter simulation conference (WSC)*, 86–98. IEEE.

Moore, E. F. 1964. *Sequential machines: Selected papers*. Addison-Wesley Longman Ltd.

Nasirzadeh, F., M. Khanzadi, and M. Mir. 2018. "A hybrid simulation framework for modelling construction projects using agent-based modelling and system dynamics: an application to model construction workers' safety behavior". *International Journal of Construction Management* 18 (2): 132–143.

Parker, R. A. 2019. "The construction of agent simulations of human behavior". In *International Conference on Intelligent Human Systems Integration*, 435–441. Springer.

Samma, H., C. P. Lim, and J. M. Saleh. 2016. "A new reinforcement learning-based memetic particle swarm optimizer". *Applied Soft Computing* 43:276–297.

Taillandier, F., and P. Taillandier. 2014. "Risk management in construction project using agent-based simulation". In *International Conference on Practical Applications of Agents and Multi-Agent Systems*, 34–43. Springer.

Wooldridge, M., and N. R. Jennings. 1994. "Agent theories, architectures, and languages: a survey". In *International Workshop on Agent Theories, Architectures, and Languages*, 1–39. Springer.

AUTHOR BIOGRAPHIES

HANANE EL RAOUI is a Research Associate in the department of Management Science at the University of Strathclyde. Her current research interests include behaviour modelling and simulation, organizational safety, risk management.

JOHN QUIGLEY is a Professor in the department of Management Science at the University of Strathclyde, and an Industrial Statistician with expertise in developing and applying statistical and stochastic methods to build decision support models.

AYSE ASLAN is a research fellow in the School of Computing, Engineering and the Built Environment at Edinburgh Napier University.

GOKULA VASANTHA is an Associate Professor in Engineering Design in the School of Computing, Engineering and the Built Environment at Edinburgh Napier University. His interests include engineering design, engineering informatics, design methods, system modelling and analysis, crowdsourcing design and Manufacturing.

JACK HANSON completed an MEng in mechanical engineering at Liverpool John Moores University in England in 2017 before completing a PhD in fluid mechanics at the University of Edinburgh. His current research interest is the quantification and characterization of human movement in industrial settings.

JONATHAN CORNEY is a Professor of Digital Manufacturing at the University of Edinburgh. His interests include manufacturing applications of crowdsourcing; cloud interfaces for manufacturing, the interactive search of digital media and, the creation of "predictive CAD systems" by leveraging data analytics.

ANDREW SHERLOCK is a Director of Data-Driven Manufacturing at National Manufacturing Institute Scotland and a Professor of Practice at the University of Strathclyde.His career has focused on the application of AI, data science and search techniques to design and manufacturing.

Proceedings of the Operational Research Society Simulation Workshop 2023 (SW23)
C. Currie and L. Rhodes-Leader, eds.
https://doi.org/10.36819/SW23.011

PROCESS VALIDITY AND FACILITATION IN SIMULATION MODEL PROOF OF CONCEPT DEVELOPMENT – SIMULATION AS A PRODUCT

Dr Daniel JW Arthur
IP Dynamics Ltd
darthur@ipdynamics.org.uk

1 BACKGROUND

This paper presents a methodological and 'reflective practice' perspective on a set of simulation case studies and the elements of successful projects. The issue under consideration is the role of the facilitation processes that are typical and/or required at the early stages of a simulation project, particularly in the Proof-of-Concept (POC) stages, whether they have R&D or more applied commercial focus. The ideas of Test Driven Development indicate that up-front design of testing plans for a project can aid the model process. This short paper is an introduction to thinking about the soft aspects of project design and communication applied to simulation.

2 METHODOLOGICAL OVERVIEW BASED ON CASE STUDIES

This perspective arises from discrete event (DES), Agent Based (ABM) and System Dynamics models (SD) modelling project experience spanning 30 years.

The focus here is on what aspects of project and stakeholder facilitation skills and processes are relevant in diverse settings: from scientific analysis, business and financial analytics, strategy consulting and government policy support based on macroeconomic analysis. Further, how do the facilitation processes need to be changed where a simulation is offered as a product or platform, whether in an interest- or awareness-raising role as opposed to a [management/consultancy function] service?

Soft systems methodology and other Problem Structuring Techniques have been proposed as a way of ensuring the right problem is solved and that a Conceptual Model incorporates or coalesces diverse stakeholder perspectives (Kotiadis, 2007; Kotiadis et al, 2014; Montevechi and Friend, 2014; Pidd, 2007; Vennix, 1996). Forrrester (1994) had put these systems and Soft OR techniques under the 'Describe the system' stage of the modelling process. However, there are client processes needed before this stage – a technical pre-sales stage requiring a sales pipeline and other rapport and issue-identification interactions.

Robinson (2015) refers to 'process quality' which was otherwise called 'process validity' (Arthur and Winch, 1999), where there are 3 components of model quality/validity. The point made is that different types of modelling environments require differing balances of these three dimensions of validity/quality – and indeed that sub-components of these dimensions (ie various validity tests or questions) have a different weight of importance.

Then there is a wide range of potential model purposes: 1) Behavioural forecasting/scenario development 2) Insight generation – current/general or historical 3) Achievement of a specific organisational change, or consensus generation 4) Decision support / clarification of specific issues / remedial policy analysis 5) Hypothesis formulation about complex system behaviour.

3 IMPLICATIONS: PROCESS REQUIREMENTS IN PRODUCT-BASED ADVANCED ANALYTICS METHODS SUCH AS SIMULATION

In cases where the simulation model as a product or platform takes centre-stage, it seems natural that 'content' validity dominates the modelling process, which can become over-technocentric. However, in virtually all cases there is an end-user, stakeholder or 'client'. In cases where the model is being offered as a product, what are the likely necessary project processes to achieve a degree of balance? The suggestion here is that there needs to be a second-round (or further) of process facilitation which might take the form of:

1. a debrief (Winch, 1999) - a business user's reflection on using a simulation
2. a 'discovery workshop': now that certain issues have been surfaced through a process of discovery, can we tweak what questions the model can address?

In what circumstances might a simulation model not need further facilitation? This could imply a short shelf-life for a model if it ceases to address exactly the immediate, extant issue.

The business intelligence move towards self-service analytics is not a suggestion that business users should become fully-capable data scientists but does allow them to explore and discover. A product mindset implies that the product itself must potentially go through several iterations of adaptation and unless the business users become their own creators of adapted architecture, they are still going to rely on specialist modellers and data scientists.

What then are the ongoing facilitation skillsets which would apply in the advanced analytics arena? What are the appropriate processes and facilitation skillsets in advanced analytics provision we might expect to see when applied to a product delivery mindset? I suggest here a simulation product mindset opens the way to subsequent client interactions (ie processes) which might include: 1) education, 2) adaptive discovery, 3) goal modification, 4) shift of the locus of business impact, especially noting the confounding complexity of systems where ripple effects and unintended consequences might arise 5) Implementation opportunities – are model insights actionable? 6) providing metrics on unobserved variables – large datasets in fact do not capture important factors: often, unanticipated relationships across organisational boundaries or soft factor impacts.

The concept of Test Driven Development seems applicable to simulation project design planning – that is, the idea that a suite of tests could be pre-specified as a way of planning the type of project intervention. More broadly, different kinds of modelling project require different validation profiles: different combination sets of tests applied to the planning of different kinds of projects.

REFERENCES

Arthur DJW and Winch GW (1999). Extending model validity concepts and measurements. In: Proceedings of the 17th. International System Dynamics SocietyConference, Wellington, NZ, July

Forrester JW (1994). System dynamics, systems thinking, and soft OR. System Dynamics Review (10),2-3, 245-256.

Kotiadis K (2007) Using soft systems methodology to determine the simulation study objectives, Journal of Simulation, 1:3, 215-222.

Kotiadis K, Tako A and Vasilakis C (2014). A participative and facilitative conceptual modelling framework for discrete event simulation studies in healthcare. *J Oper Res Soc* **65**, 197–213.

Montevechi JAB, Friend J (2012). Using a Soft Systems Methodology framework to guide the conceptual modeling process in discrete event simulation Proceedings of the 2012 Winter Simulation Conference (WSC)

Pidd, M (2007). Making sure you tackle the right problem: Linking hard and soft methods in simulation practice. Winter Simulation Conference

Robinson S (2014). Simulation – the practice of model development and use, 2nd ed. Macmillan International Higher Education / Red Globe Press.

Vennix JAM (1999). Group model-building: Tackling messy problems. System Dynamics Review 15 (4): 379.

Winch GW (1999). Dynamic visioning for dynamic environments. Journal of the Operational Research Society, 4, 354-361.

Winch GW and Arthur DJW (2002). User-parameterised generic models: a solution to the conundrum of modelling access for SMEs? System Dynamics Review, (18) 3, 339-357.

Proceedings of the Operational Research Society Simulation Workshop 2023 (SW23)
C. Currie and L. Rhodes-Leader, eds.
https://doi.org/10.36819/SW23.012

HISTORICAL AND PHILOSOPHICAL PERSPECTIVES ON THE ACTIVITY OF MODELLING

Prof. Stewart Robinson

Newcastle University Business School
5 Barrack Road, Newcastle upon Tyne, NE1 4SE, United Kingdom
stewart.robinson@newcastle.ac.uk

ABSTRACT

The activity of modelling entails the conception, representation and execution of a model. In this paper we explore this activity from a historical perspective, by positioning it within the philosophy of science, and by taking a practice-based perspective. We draw a clear distinction between the internal and implicit conceptual model that exists within the mind of the modeller, and its external and explicit representational and executable forms. As a field we have focused much of our efforts on model representation and execution, while largely skirting around the processes that lie behind how models are conceived. Ultimately, better understanding these processes is important for improving the practice of modelling.

Keywords: History of Modelling, Philosophy of Science, Activity of Modelling

1 INTRODUCTION

At the core of our field of computer simulation is the activity of modelling. On the surface, modelling can be seen as the relatively straightforward task of creating a simplified representation of the real world (that is either in existence or proposed). Dig a little deeper, however, and modelling becomes a much more complex task, not least because of the need to choose an appropriate level of simplification for the problem at hand. It is, therefore, worthwhile exploring the activity of modelling further.

With this in mind, this paper explores the modelling activity, first from a historical perspective, then its positioning within the philosophy of science, and finally from a practice-based perspective. What we identify is that, at its center, modelling is a cognitive process which involves conceiving how to represent a real system. The outcome of this process is an internal and implicit model that exists within the mind of the modeller. This 'hidden' model then needs to be made external and explicit by means of representational and executable forms.

Section 2 provides a brief historical perspective on modelling, dating back nearly 50,000 years. In section 3 we explore how modelling sits within the philosophy of science. Section 4 then takes a practice-based perspective, discussing the activity of modelling with reference to an example. We conclude by discussing the implications for the study of modelling. This work is certainly not the final word on the activity of modelling, but aims to provoke more in-depth consideration of the activity that is at the heart of our discipline.

2 A BRIEF HISTORY OF MODELLING

Modelling is the activity of making a model, that is, a simplified representation of reality. Put more precisely, a "model is an interpretative description of a phenomenon that facilitates access to that phenomenon" (Bailer-Jones 2009). In its broadest sense, modelling is a fundamental part of human activity, the means by which we make sense of the world (Fishwick, 2017). Indeed, what sets Homo sapiens apart is our abilities to understand one another and to imagine different possible futures (Suddendorf, 2014).

At some point in our history, humans started to model, and more importantly they started to make those models 'an external and explicit representation' of the world (Pidd, 2009). This is an important step, since purely mental ('internal and implicit') models have limited benefit, but 'formal' external and explicit models serve as 'transitional' objects from which an individual and others can learn and in so doing make better sense of the world, that is, revise their mental models (Morecroft, 2015).

Dating back 45,000 years or more, possibly the earliest known models are the cave paintings discovered mainly in Europe (Figure 1). These 'models' mostly depict animals in the last ice age, but also represent human figures. Their purpose is not fully known, but explanations include (Lewis-Williams 2004):

- For enjoyment, fun and decoration, although this is largely discredited as a theory
- Magic, for instance, giving hunters power over their prey
- Depictions of conflict and alliances in their social struggle
- Expressions of their understanding of the natural and supernatural organization of the world ('mythograms')

Whichever explanation holds, these paintings were a way of our ancestors etching their thoughts in stone. Perhaps this is the first time that humans expressed their internal and implicit understanding of the world as an external and explicit model of how they understood reality. These paintings may well have acted as transitional objects that enabled both the artist and observers to make better sense of their world. Indeed, today these representations still provide us with a glimpse into our ancestors' minds and how they understood the world.

Figure 1 *Example of a Painting from the Lascaux Caves in France*

Moving on from early artistic representations, analogue models have provided a means for representing and learning about the world in a more tangible, and even dynamic, fashion. Although little is known about the purpose of Stonehenge, built around 5,000 years ago, there is a theory that it is a model of the lunar orbits, sufficient to predict eclipses of the moon (Moir, 2015).

Board games are another example of analogue models. In 6th century India, the board game Chaturanga was developed based on the military forces of the time: elephants, chariots and infantrymen. This game, an early model of military conflict, was a forerunner to modern day chess. In early 20th century climate science, bowls filled with cloudy viscous fluid were used as an analogue for understanding the Earth's climate system (Edwards, 2011).

Modelling has, of course, closely followed the development of mathematics (Schichl, 2004). The counting and writing of numbers dates back to as early as 30,000 BC, but it was not until around 2,000 BC that we see the cultures in Babylon, Egypt and India using mathematics to solve everyday problems. It was then the Greeks that led the way in the development of mathematical theory and modelling across a range of applications, for example, in astronomy Ptolemy devised a model of the solar system (circa 150 AD). Alongside this, developments in mathematical modelling continued in China, India and the Middle East. It is only in the last 1,000 years that we see the emergence of western mathematicians, ultimately leading to the computer age, in which models of massive scale and complexity can be realized.

With the advent of the computer age, we have seen rapid developments in modelling across many fields. Indeed, our own field of operations research/management science owes much of its existence to the rapid increase in computer power over recent decades, enabling algorithms and simulations to be implemented on a scale that would previously have been impossible using only the capacity of humans. Histories of simulation modelling demonstrate the close correspondence of developments in modelling with improvements in computing (Robinson, 2005; Pidd and Carvalho, 2006; Hollocks, 2008, 2017; Nance and Overstreet, 2017).

Despite these developments in modelling through the ages, it is possible that our ability to create internal and implicit models has not changed since the early cave painters. Indeed, some anthropologists and those that have studied cave art argue that Upper Paleolithic people had the same mind as we do (Lévi-Strauss, 1966; Leroi-Gourhan, 1968). However, it is arguable whether human intelligence has remained unchanged, there being a strong body of opinion that intelligence has improved over time as a result of factors such as education, nutrition and technology. Sternberg and Grigorenko (1997) explore the debate around nature versus nurture with respect to intelligence.

Of course, what we can see has changed is the tools available to us for making those models external and explicit, moving from paintings, through analogues and mathematics, to computer models. As a result, the power of our models has increased by orders of magnitude. The possibility for a computer model to represent our internal and implicit understanding, and so to help explain our world and make predictions, is so much greater than a painting etched onto a cave wall. Of course, in so enabling, our minds are now able to conceive of realizing far larger and more complex models than our forebears could; the 'possibility' factor (Chwif et al., 2000). We also have an ever-expanding base of knowledge on which to build our models; the architects of Stonehenge did not have the luxury of information from decades of space exploration to aid their modelling efforts.

What this brief history of modelling shows is the distinction between our mental models and their explicit representation through a variety of every more capable media. Pidd (2009) similarly makes this distinction between our internal and implicit mental models and the external and explicit models that we use in operations research/management science. Morecroft (2015) also distinguishes between the model we carry in our heads of the way something works and a formal model ('a tangible aid to imagination and learning'). So, models are conceived in our minds and made explicit via some media, in our case through documentation and computer code. Having established this context, we now turn our attention to modelling in the philosophy of science.

3 MODELLING IN THE PHILOSOPHY OF SCIENCE

As a starting point it is useful to consider the correspondence between models and theories. According to Caswell (1988), models and theories are not equivalent. Theories can be developed without the need for models and models can be constructed without recourse to theory. Fried (2020) argues that models serve as intermediaries between theory and the real world. Bailer-Jones (2009) also distinguishes models from theory, stating that models are applied to concrete empirical phenomena, while for theories this is not necessarily the case. As such, theories are more fundamental and general than models, while models draw on theories.

Knuuttila (2005) identifies a range of model forms: diagrams, physical three-dimensional things, mathematical equations, computer programs, organisms and laboratory populations. The role of models in science can be seen as for representation, for mediation or as epistemic artefacts.

The structuralist view of models as representations is of a dyadic relationship between the model and the target system. However, more recent discussions have moved to consideration of a triadic relation also involving human agency and interpretation. Both Giere (2004) and Suárez (2004) identify the importance of human agency in modelling and model use.

Giere (2004) understands representation as similarity, arguing that:

$$S \text{ uses } X \text{ to represent } W \text{ for purposes } P$$

where:

S is an individual scientist, a scientific group, or a larger scientific community.
W is an aspect of the real world.
X can be words, equations, diagrams, graphs, photographs, computer-generated images, …

Giere argues that it "is not the model that is doing the representing; it is the scientist using the model who is doing the representing."

Similarly, Suárez (2004) adopts an 'inferential' conception of representation: "A represents B only if (i) the representational force of A points towards B, and (ii) A allows competent and informed agents to draw specific inferences regarding B." So a model's representational 'force' is defined by its capacity to enable a user to consider the target system.

The role of models as mediators is discussed by Morrison and Morgan (1999). They argue that models are partially independent of both theory and the world because they contain 'outside' elements as well as theories and data. Since models sit outside the theory-world axis they are independent, but act as mediators between theories and the world. Models represent some aspects of the world and some aspects of theories, with users learning by building and manipulating the model.

Knuuttila (2005) argues for the role of models as epistemic artefacts; a means for gaining knowledge in diverse ways. She concludes that the representation provided by models is a twofold phenomenon consisting of a material 'sign-vehicle' (historical artefact) and an intentional 'relation of representation' (connecting the sign-vehicle to that which is being represented). Under this interpretation, the sign-vehicle is detached from what it is representing and can be transferred to many different contexts. So, although "it is based on the properties of the sign-vehicle, similarity is nevertheless established in the specific uses of the model that relate it to 'something in the world'." Boon and Knuuttila (2009) suggest that knowledge is gained from models through our interaction with them: building and manipulating them, and using them in different ways.

From the perspective of a theoretical economist, Sugden (2000) attempts to explain how abstract theoretical models explain aspects of the real world. He does so with reference to Akerlof's work on asymmetric information (Akerlof, 1970) and Schelling's modelling of segregation (Schelling, 1971). Such abstract theoretical "models are not abstractions from, or simplifications of, the real world. They describe counterfactual worlds which the modeller has constructed." He again highlights the importance of human agency in filling the gap between the model and the real world through 'inductive inference'.

Frigg and Hartmann (2020) discuss the epistemology of models identifying four primary cognitive functions. Users can learn about models through their construction and manipulation. They can also learn about the target system by translating their knowledge of the model to the target system. Thirdly models can be used for explaining, that is, identifying causal relations; this leads to an interesting discussion of how false models can even provide explanations as a result of their falsity. Finally, models can be used to understand, and this can occur independently of the model's ability to explain.

Discussing specifically the epistemology of modelling and simulation, Tolk (2015) identifies three broad uses for models. First, for providing experience, for instance, through training simulators and games. Second, to gain insights and understanding through experimentation with the simulation. And third, using the model as a repository of knowledge in which we define a concept and bring it to life through the model, especially its animation and visualization.

4 THE ACTIVITY OF MODELLING: A PRACTICE-BASED PERSPECTIVE

Our interest is in models as they are used in operations research/management science (OR/MS), especially simulation models. Of course, models in OR/MS are developed for a range of purposes: understanding the world, changing the world, better managing the world and controlling the world (Pidd, 2009). In what follows we focus on the activity of modelling from conceptualization to creating an executable model. Our discussion is located in a practice-based perspective and we use an example to illustrate how the modelling activity proceeds. For a discussion on the OR/MS modelling process from an epistemological perspective see Eriksson (2003). For a broader discussion on the activity of modelling as it spans many disciplines, see Fishwick (2017).

4.1 An Example of a Model (Based on Actual Events)

I am waiting in a long queue at a self-service restaurant wondering why the service is so slow. I observe the activities being undertaken by the staff and realize that the problem lies in their inefficient working, for which the layout of the facility does not help, and they are almost certainly under resourced; for instance, there are insufficient pay points.

As my expertise is modelling queuing systems, I form a model (simplified representation) of the system in my mind, identifying the elements of the system I would represent and their level of detail in order to find ways of reducing the waiting time of the customers.

Following this experience, I contact the owner of the restaurant to see if she would like me to work with them, using my model, to improve their service system. After some discussion, she agrees that we should work together and we organize to meet. At the meeting I present my ideas for the model, now written down as a simple process flow diagram. This enables me to confirm my understanding of the system, update elements of the model that are incorrect, and to add or takeaway detail as is necessary. It also gives the owner confidence in the proposed model and that it will help identify ways of improving the waiting time. The documented model also makes it possible to identify the data required which the owner will provide.

I then develop a computer model based on the agreed model using a simulation software package. In the process of development, I present the model to the owner on three occasions. Each time, refinements to the model are identified which are reflected back in the process flow diagram. After the third meeting we agree that the model has been developed to the point where the owner is confident to use it as an aid to decision-making.

4.2 Contextualizing the Modelling Activity

In order to make sense of the story above, Figure 2 provides a summary of the modelling activity. The story starts in the problem domain with a real system that could be improved through modelling, that is, reducing waiting times. The modeller forms an understanding of the system (*system description*) by collecting information, in this case by the observing the system in operation. That understanding is not complete and can never be complete; there will always be elements of the system that are not fully known or understood.

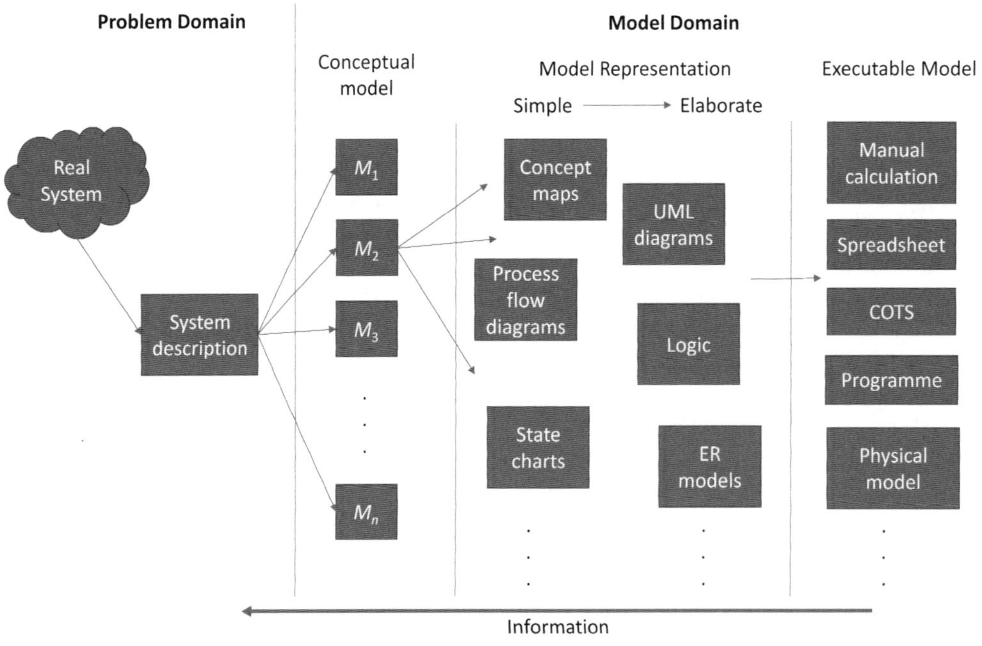

Figure 2 *The Activity of Modelling*

From the *system description* the modeller develops a concept for the model in his/her mind (*conceptual model*). This is an internal and implicit model of the system as it needs to be represented to address the problem at hand. There are a great many (n) potential models, $M_i = M_1, M_2, \ldots, M_n$, where higher values of i imply greater complexity. Indeed, Page et al. (2021) theorize that there is an infinite range of models ($n \to \infty$). Whether the set is finite or not, the modeller is in effect selecting from one of a great many potential models.

In order to share this *conceptual model*, the modeller creates an external and explicit version by creating a *model representation*, in this story, using a simple process flow diagram. Through this process and discussion with the owner the *conceptual model* is refined, that is, $M_i \to M_j$. In general, as the process progresses the model tends to greater complexity, i.e. $j>i$, but this is not necessarily the case.

An *executable model* is developed using a commercial off-the-shelf (COTS) simulation package. As a result of the meetings at which the computer model is presented, refinements are made to the *conceptual model* ($M_j \to M_k$). Again, there is a tendency for the model to become more complex through this process, i.e. $k>j$, but as before, this is not necessarily the case. These refinements are reflected in the *model representation*. This reverse flow of information from the *model representation* and the *executable model* towards the *conceptual model* (as well as the *system description*) is shown by the arrow at the bottom of Figure 2. Model verification and validation activities serve to increase this flow of information back to the *conceptual model*. At the end of this process, the modeller and owner have an agreed model (M_k) which is used for aiding decision-making.

We now look in more detail at each of the three components in the model domain: conceptual model, model representation and executable model.

4.3 The Conceptual Model

There is a very large set of potential models that could be appropriate for representing the real system in order to address the problem at hand. Henriksen (1988), in some measure, demonstrates this multiplicity of models through the example of representing a simple battle. As further examples, the very simplest model (i=1) for a queuing system is a straightforward single-server queue model. For an agent-based model, the simplest model is a single agent that makes random moves on a grid, and for a system dynamics model, a single stock and flow with no information feedback represents the simplest possible model. The most complex model (i=n, where n is very large) would be a full representation of every known facet of the real system, that is, a full representation of the *system description*. Note that by spending more time understanding the real system, the *system description* will become more complex and complete, and as a result there is greater potential to develop a more complex model; n would increase. What we know is that, even given the same *system description*, different modellers will almost certainly come-up with different *conceptual models* ($1 \leq i \leq n$) for the same problem. This is a result of a variety of factors including different levels of experience and modellers' different cognitive processes for forming the *conceptual model*.

As the modelling activity progresses the *conceptual model* is continuously refined as new information becomes available. An initial model (M_i) is formed very quickly (in the story, while the modeller stands in the queue), but through successive meetings and model presentations, the model changes in the light of the new information ($M_i \to M_j \to M_k \to \ldots$).

4.4 The Model Representation

There are many methods for representing *conceptual models*, some of which are listed in Figure 2. These range from quite simple representations to much more elaborate means of documenting the *conceptual model*. The choice of method of representation is primarily a matter of modeller and client preference, but also the way in which the representation will be used. The representation may only need to be quite basic to communicate the key facets of the model. However, it might need to be highly specified to the level that a programmer could create a computer program that faithfully executes the model without the need for further reference to the real world or the modeller.

There could be multiple representations of the model, for example, a simple communicative representation for sharing with the client and a detailed model specification for use by the model developer. However many representations are created, they are all representations of the same *conceptual model*.

It is good practice to create a *model representation*, but the representation is not strictly necessary. The *conceptual model* exists whether it is made explicit or not. Of course, without a *model representation*, there is no means for sharing, validating or improving the *conceptual model*. It is also good practice to continually revise the *model representation* as the *conceptual model* is updated ($M_i \rightarrow M_j \rightarrow M_k \rightarrow \ldots$).

It is unlikely that the *model representation* faithfully captures the *conceptual model* in every aspect. So the model that is communicated to the client is not a perfect representation of the intended model. As a result of this miscommunication, overlaid by the client's interpretation of the *conceptual model*, the client's perception of the model differs from the modeller's conception, i.e. $M_j^C \neq M_j^D$, where C and D are the client's and modeller's perception of the model respectively. Further, if the *model representation* is used directly by a model developer, then errors in the *model representation* will be translated into the final executable model.

4.5 The Executable Model

The *executable model* is normally created on a computer, but it could be a physical model or a manual pen and paper exercise. If the model is programmed on a computer, a range of approaches could be adopted from spreadsheets, through commercial off-the-shelf packages to programming languages. The modeller could choose to create multiple *executable models* of the same *conceptual model*, although it is questionable whether there is any benefit in doing so. If this were to happen, all the models are executable versions of the same *conceptual model*.

As stated above, in the activity of creating the *executable model*, changes may be made to the *conceptual model*. It is good practice to document these changes by updating the *model representation*. Meanwhile, the *executable model* may not faithfully capture the *model representation* (verification errors) and hence the *executable model* is not a perfect representation of the intended model. As such, $M_k^E \neq M_k^D \neq M_k^C$, where E is the *executable model*.

5 CONCLUSION: IMPLICATIONS FOR THE STUDY OF MODELLING

We have briefly explored the history, philosophy and activity of modelling. We see that modelling is a long-standing human activity dating back to our early ancestors. It may even be that our cognitive abilities to conceive our world as a model have not essentially changed. What has changed is the repository of knowledge on which we can call when conceiving a model and the technology available to realize our models.

There is a strong human element not only in the way models are conceived, but also in how they are used and interpreted. This moves us away from a positivist view of modelling, towards a constructivist perspective (Hoffman, 2005). Positivism views the world as something that can be fully known and experiments (models in our case) enable us to determine the true nature of phenomena. Constructivism takes the view that different subjective realities coexist and that the observer (modeller, stakeholder, client, … in our case) is not neutral. As such, the goal of knowledge seeking is not the absolute truth, but viability, that is, knowledge which is helpful for life. In the constructivist paradigm models reflect a subjective reality and they are based on an individual's perception and beliefs.

We should also be cognizant of the distinction between internal and implicit models versus external and explicit models. Fundamentally, the model exists within the mind of the modeller; it is an internal and implicit model. As the model is made external and explicit, that is, information about the model is communicated with the client and other stakeholders, they each form their own, and slightly different, interpretations of the model. *Model representations* and *executable models* are external and explicit expressions of the internal and implicit *conceptual model* that exists in the mind of the modeller. As information flows from the real system and from feedback on the model representation and executable model, the *conceptual model* is subject to continuous refinement. Indeed, the *conceptual model* is a persistent artefact, but one that is subject to constant change. It exists from the beginnings of understanding the real system, through the development and use of a model of that system, and beyond. Adjustments are made to the conceptual model as new information and understanding emerge. Many years after the modelling has been complete, it is still possible for the conceptual model to be adjusted based on revised information about the real system, albeit that the adjusted (internal and implicit) model is unlikely to be realized (made external and explicit) at this point.

The implication of the above is that whilst the study of modelling needs to focus on *model representations* and delivering *executable models*, it also needs to focus on how models are conceived. At present our field tends to focus more on model representation and execution than it does on model conception. We need to consider questions such as: what are the cognitive processes that lead to internal and implicit conceptual models? How can these processes be improved to deliver more effective models? What is the interplay between the *conceptual model* and information feedback from its representational and executable forms? In-depth study of such questions is important for understanding and improving the practice of modelling. Our aim here is to provoke further study of this important topic.

REFERENCES

Akerlof G A (1970). The market for ‘lemons’: quality uncertainty and the market mechanism. *Quarterly Journal of Economics* **84**: 488–500.

Bailer-Jones D M (2009). *Scientific Models in the Philosophy of Science*. University of Pittsburgh Press.

Boon M and Knuuttila T (2009). Models as epistemic tools in engineering sciences: a pragmatic approach. In: Meijers A (ed). *Philosophy of Technology and Engineering Sciences.* Elsevier/North-Holland, pp 687-720.

Caswell H (1988). Theory and modes in ecology: a different perspective. *Ecological Modelling* **43**: 33-44.

Chwif L, Barretto M R P and Paul R J (2000). On simulation model complexity. In: Joines J A, Barton R R, Kang K and Fishwick P A (eds). *Proceedings of the 2000 Winter Simulation Conference.* Institute of Electrical and Electronics Engineers: Piscataway, New Jersey, pp 449–455.

Edwards P N (2011). History of climate modeling. *WIREs Climate Change* **2**:128–139.

Eriksson D M (2003). A framework for the constitution of modelling processes: a proposition. *European Journal of Operational Research* **145**: 202-215.

Fishwick P A (2017). Modeling as the practice of representation. In: Chan W K V, D'Ambrogio A, Zacharewicz G, Mustafee N, Wainer G and Page E (eds). *Proceedings of the 2017 Winter Simulation Conference.* Institute of Electrical and Electronics Engineers: Piscataway, New Jersey, pp 4276-4287.

Fried E I (2020). Theories and models: what they are, what they are for, and what they are about. *Psychological Inquiry* **31(4)**: 336-344.

Frigg R and Hartmann S (2020). Models in science. In: Zalta E N (ed). *Stanford Encyclopedia of Philosophy*, https://plato.stanford.edu/archives/spr2020/entries/models-science (accessed October 2022). Department of Philosophy, Stanford University.

Giere R N (2004). How models are used to represent reality. *Philosophy of Science* **71(5)**: 742-752.

Henriksen J O (1988). One system, several perspectives, many models. In: Abrams M, Haigh P and Comfort J (eds). *Proceedings of the 1988 Winter Simulation Conference.* Institute of Electrical and Electronics Engineers: Piscataway, New Jersey, pp 352-356.

Hofmann M A (2005). Modeling assumptions: how they affect validation and interoperability. In: *Proceedings of the European Simulation Interoperability Workshop.* SISO: Bedford, MA, pp 10-18.

Hollocks B W (2008). Intelligence, innovation and integrity— KD Tocher and the dawn of simulation. *Journal of Simulation* **2 (3)**: 128-137.

Hollocks B W (2017). A history of simulation development in the United Kingdom. In: Chan W K V, D'Ambrogio A, Zacharewicz G, Mustafee N, Wainer G and Page E (eds). *Proceedings of the 2017 Winter Simulation Conference.* Institute of Electrical and Electronics Engineers: Piscataway, New Jersey, pp 60-74.

Knuuttila T (2005). Models, representation, and mediation. *Philosophy of Science* **72 (5)**: 1260-1271.

Leroi-Gourhan A (1968). *The Art of Prehistoric Man in Western Europe.* Thames & Hudson: London.

Lévi-Strauss C (1966). *The Savage Mind.* University of Chicago Press: Chicago.

Lewis-Williams D (2004). *The Mind in the Cave: Consciousness and the Origins of Art.* Thames and Hudson: London.

Moir G (2015). Hoyle on Stonehenge. *Antiquity* **53(208)**: 124-129.

Morecroft J D W (2015). *Strategic Modelling and Business Dynamics: A Feedback Systems Approach.* 2nd ed. Wiley: Chichester, UK.

Morrison M and Morgan M S (1999). Models as mediating instruments. In: Morrison M and Morgan M S (eds). *Models as Mediators.* Cambridge University Press: Cambridge, UK, pp 10-37.

Nance R E and Overstreet C M (2017). History of computer simulation software: an initial perspective. In: Chan W K V, D'Ambrogio A, Zacharewicz G, Mustafee N, Wainer G and Page E (eds). *Proceedings of the 2017 Winter Simulation Conference.* Institute of Electrical and Electronics Engineers: Piscataway, New Jersey, pp 243-261.

Page E H, Thompson J R and Koehler M (2021). Sic semper simulation - balancing simplicity and complexity in modeling and analysis. In: Kim S, Feng B, Smith K, Masoud S, Zheng Z, Szabo C and Loperds M (eds). *Proceedings of the 2017 Winter Simulation Conference.* Institute of Electrical and Electronics Engineers: Piscataway, New Jersey.

Pidd M (2009). *Tools for Thinking: Modelling in Management Science*. 3rd ed. Wiley: Chichester, UK.

Pidd M and Carvalho A (2006). Simulation software: not the same yesterday, today or forever. *Journal of Simulation* **1(1)**: 7-20.

Robinson S (2005). Discrete-event simulation: from the pioneers to the present, what next? *Journal of the Operational Research Society* **56(6)**: 619-629.

Schelling T C (1971). Dynamic models of segregation. *Journal of Mathematical Sociology* **1**: 143-186.

Schichl H (2004). Models and the history of modeling. In: Kallrath J (ed). *Modeling Languages in Mathematical Optimization.* Boston, Springer: MA, pp 25-36.

Sternberg R J and Grigorenko E (1997). *Intelligence, Heredity, and Environment.* Cambridge University Press: Cambridge, UK.

Suárez M (2004). An inferential conception of scientific representation. *Philosophy of Science* **71**: 767–779.

Suddendorf T (2014). *The Gap: The Science of What Separates Us from Other Animals*. Basic Books: New York.

Sugden R (2000). Credible worlds: the status of theoretical models in economics. *Journal of Economic Methodology* **7(1)**: 1-31.

Tolk A (2015). Learning something right from models that are wrong: epistemology of simulation. In: Yilmaz L (ed). *Concepts and Methodologies for Modeling and Simulation: A Tribute to Tuncer Ören*. Springer International Publishing: Cham, Switzerland, pp 87-106.

AUTHOR BIOGRAPHY

STEWART ROBINSON is Dean and Professor of Operational Research at Newcastle University Business School, UK. His research focuses on the practice of simulation model development and use. Key areas of interest are conceptual modelling, model validation, output analysis and alternative simulation methods (discrete-event, system dynamics and agent based). Professor Robinson is author/co-author of six books on simulation, co-founder of the Journal of Simulation and co-founder of the UK Simulation Workshop conference series. He was President of the Operational Research Society (2014-2015). www.stewartrobinson.co.uk

Proceedings of the Operational Research Society Simulation Workshop 2023 (SW23)
C. Currie and L. Rhodes-Leader, eds.
https://doi.org/10.36819/SW23.013

IMPROVING HOSPITAL DISCHARGE FLOW THROUGH SCALABLE USE OF DISCRETE TIME SIMULATION AND SCENARIO ANALYSIS

Dr. Zehra Önen-Dumlu
University of Bath
School of Management
Claverton Down
BA2 7AY
zod23@bath.ac.uk

Dr. Paul Forte
NHS BNSSG ICB
Modelling and Analytics
360 Bristol
BS1 3NX
paul.forte@nhs.net

Dr. Alison Harper
University of Exeter
Medical School
Magdalen Road
EX1 2HZ
a.l.harper@exeter.ac.uk

Prof. Martin Pitt
University of Exeter
Medical School
Magdalen Road
EX1 2HZ
m.pitt@exeter.ac.uk

Prof. Christos Vasilakis
University of Bath
School of Management
Claverton Down
BA2 7AY
cv280@bath.ac.uk

Dr. Richard Wood
NHS BNSSG ICB
Modelling and Analytics
360 Bristol
BS1 3NX
richard.wood16@nhs.net

ABSTRACT

Inadequate patient flow from hospitals into community care is often blamed for bed blockages in the acute setting. This is bad for patient experience and outcomes and has an upstream knock-on effect for Accident and Emergency performance and, in turn, ambulance offload delays and response times. Despite the large numbers of acute bed-days lost to delayed discharges and the ambition to expand home-based community care, there has been a deficit of modelling studies investigating the dynamics of this pathway and providing the relevant insights to service planners. Working closely with healthcare managers, this paper reports on the development and deployment of versatile simulation tools for modelling both the home-based and bedded community step-down pathways, known as 'Discharge to Assess' or D2A in England's NHS. Developed in open source 'R', these tools offer scalable solutions for exploring different scenarios relating to demand, capacity and patient length of stay.

Keywords: Healthcare management, Resource allocation, Community services, Discharge planning.

1 INTRODUCTION

Intermediate care describes services typically provided in the community that support the integration and continuity of care between different health and care settings, such as bridging the transfer from hospital-based care to patients' normal residence. Key purposes are to facilitate timely discharge from the acute hospital setting for mainly older and frail patients (Levin and Crighton, 2019; McGilton et al, 2021) and to promote faster recovery, maximise independence, and prevent readmission (Sezgin et al, 2020). With rising life-expectancy globally, health and care systems must support a growing number of frail and elderly patients in a cost-effective and technically productive way. Yet, there remain various challenges to achieving this, not least a lack of fluidity in the movement of patients along the acute discharge pathway, leading to the propagation of discharge delays in the more costly upstream settings (Cadel et al, 2021).

Health and care systems have responded to the growing proportion of elderly people in their populations with strategies that attempt to distribute resources more optimally between acute and community healthcare, and between health and social care. In many countries, the provision of integrated care models has been a focus for decades, aiming to improve system efficiency and quality of care by working across multiple services, providers, and settings (WHO, 2016). In England, recent reforms have seen the establishment of statutory Integrated Care Systems (ICS) which mandate collaboration within and between National Health Service (NHS) organisations and Local Authority social care, with intermediate care and acute discharge planning a key focus (Department of Health and Social Care, 2021).

In the English NHS, the Discharge to Assess (D2A) service (NHS, 2021) encompasses three time-limited pathways which provide intermediate 'step-down' care for patients discharged from hospital for a period of up to six weeks (Figure 1). The underlying rationale is to reduce unnecessary use of acute hospital resources beyond the point a patient is deemed medically fit for discharge by transferring them to a setting in which their longer term health and social care needs can be properly assessed. Patients discharged on D2A pathway 1 (P1) return to their usual place of residence and receive domiciliary visits from community health services. If more intensive post-acute rehabilitation is required, then patients – who are expected to return to their usual place of residence eventually – may be discharged on D2A pathway 2 (P2), which involves transfer to a non-acute bedded facility for up to six weeks. D2A pathway 3 (P3) is also non-acute bed-based care, but is reserved for those requiring the most complex health and social care need assessments. Many of the patients in this pathway will subsequently go on to a long-term care home placement. In England, of those that enter a D2A pathway following an acute admission, it is expected that at least 90% will require P1, with a maximum of 8% and 2% requiring P2 and P3 respectively (NHS, 2021).

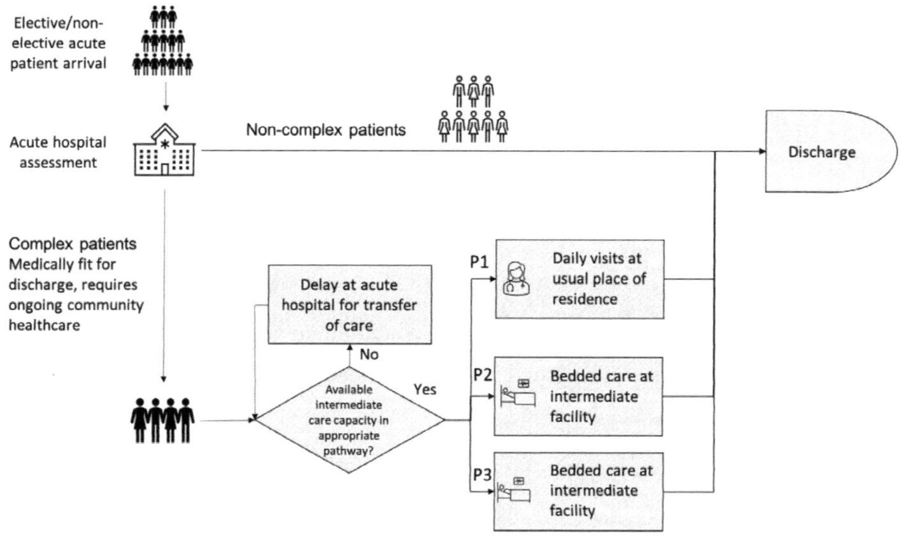

Figure 1 *Organisation of intermediate care services in the English National Health Service (NHS).*

For modelling community care, investigators have considered bedded facilities (Patrick et al, 2015) and home visits (Demirbilek et al, 2019). However, a review of modelling studies of community services (Palmer et al, 2017) found that multiple care settings are rarely considered and also that time-varying demand is not captured. In those studies that do consider the interaction between acute and community care, this is often limited to a specific type of disease, such as stroke care, childhood asthma, or other chronic diseases (Monks et al, 2016; Wood and Murch, 2020). Nonetheless, these accounts do demonstrate the value in modelling the wider context of the pathway. Without considering these interactions, it is difficult to meaningfully assess the full extent of various capacity considerations in the performance of upstream services, and to derive a more thorough understanding of the wider health and care system operation.

The objective of this study is to report on the development and practical use of a purpose-built computer simulation tool for modelling the D2A P1-3 complex discharge pathways, from the point a patient is deemed medically fit for discharge from acute hospital, to admission and discharge from the intermediate care provider. The simulation method employed for modelling bed-based intermediate care services (P2 and P3) is detailed in Section 2.1. This is extended in considering visits-based intermediate care services (P1) in Section 2.2. Examples of real-life use in a major health and care system in and around Bristol is presented in Section 3, through a selection of scenarios requested by service managers and modelled via the simulation tool. Finally, Section 4 contains a discussion regarding practical implementation, strengths and limitations, and further research opportunities.

2 METHODS

2.1 The Bed-based Care Pathway Model

After discharge from the acute hospital, patients assigned to the D2A P2 or P3 pathways are transferred to a step-down bed-based care facility for a given length of stay. Daily referrals and lengths of stay are sampled from appropriate statistical distributions using empirical data or other sources of information. The capacity in this setting is represented in terms of the number of intermediate care beds available with the patient occupying such a bed for the entirety of their length of stay. If a referred patient is unable to be transferred to a community care bed due to capacity being fully occupied, then he or she will continue to occupy the acute hospital bed until capacity in the community is available. The queue discipline is first-in first-out (FIFO). By setting the capacity to a sufficiently large number, the model allows any patient to enter the assigned pathway without any delay in the acute hospital. The resulting estimate of the number of beds occupied illustrates the maximum capacity needed to have zero delayed discharges from the acute hospital. The above events are simulated over discrete time-steps, with each bed-based care pathway (in this case P2 and P3) treated independently.

2.2 The Visits-based Care Pathway Model

Patients assigned to a visits-based D2A P1 pathway will reside in their usual place of residence after being discharged from the acute hospital and will be provided care by one or more care workers during regular visits. As in the bed-based model, patients referred but unable to be discharged into the pathway due to unavailable capacity, remain in an acute hospital bed and represent a delayed discharge. As with the bedded-care pathway model, daily referral rates and duration of service can be generated from statistical distributions estimated from the user-defined data. Likewise, the initial and final daily visit requirements are sampled from user defined distributions. Discussions with the intermediate care providers revealed that the number of visits tapers over the duration of the service, which is a particularly interesting dynamical property.

For this pathway model, the capacity is defined as the number of available visits in the system, i.e. the number of patients that can be admitted into the pathway (also defined as 'slots'), multiplied by the average number of visits required per day. Hence, at full capacity a patient with a high number of required visits and/or long service time may prevent patients with lower service requirements from accessing the community care and this will create delays with respect to discharge from the acute hospital. Figure 2 shows an illustration of how patients waiting in the queue are admitted when capacity becomes available. On the left-hand side there are sampled daily visit demands of the queueing patients over their sampled lengths of stay, and on the right-hand side the remaining visit requirements of the patients currently admitted in the service. At each discrete time step, the queue is searched for the longest waiting patient(s) whose demand can be accommodated, based upon service capacity (dashed line). Then, the corresponding patient is admitted to the P1 pathway. There are many different rules that can be assumed while deciding on which patients in the queue should be admitted first (e.g., FIFO, highest load first etc.). Our decision rule is based on current practice in the healthcare system for which this model is applied. However, as these patients are especially characterized with complex needs, there is no "one rule fits all" and alternative admission rules can be interchangeably used in real practice. In the scope of this work, we have controlled that with the assumed admission rule, patients with higher initial visit requirements are not over-harmed. Comparison of different queuing admission rules can be conducted as part of future work.

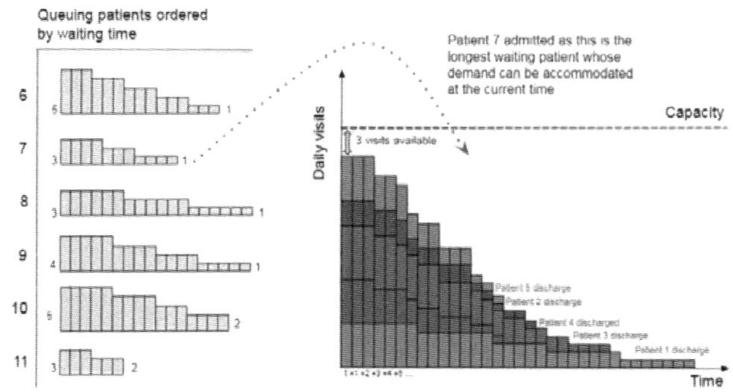

Figure 2 *Illustrative example of state of the system for P1 at a particular time in the simulation.*

2.3 Model Implementation

The simulation tool is designed to be scalable such that it enables flexibility in designing the modelled system in terms of numbers and types of pathways, localities, and parameters, as well as parameter combinations relating to various what-if scenarios. The tool uses a Microsoft Excel spreadsheet into which the user enters the required input information including:

- Daily arrival rates into each pathway.
- Initial conditions of each pathway and locality, including both the initial occupancy of each pathway (how many patients are currently in the service at the beginning of simulation period, by pathway and locality) and the initial queue in each pathway (how many patients are ready to be discharged into a pathway waiting in the acute at the beginning of simulation period).
- Capacity in each pathway, in terms of either visits/slots (P1) or beds (P2 and P3).
- Distribution of length of stay in each D2A pathway with related parameters (e.g. mean, standard deviation, median).
- For P1 pathway, the initial number of visits, final number of visits and corresponding distribution assumed for an average patient over their service time.
- Average costs of intermediate care service of each pathway and acute hospital bed.

The Excel spreadsheet can then be read in by the simulation model, which employs the routines described in Sections 2.1 and 2.2, and is coded in 'R'. On completion, csv files containing the full simulation output data are automatically generated and deposited in the user's working directory. Outputs are also provided in Microsoft Word reports, produced automatically through the 'R Markdown' package. These reports include the inputs used by the model for each scenario. Plots are used to illustrate each referral scenario, and tables for length of stay and capacity under each scenario. For each day in the pre-determined horizon period, the model estimates the mean number of patients in each pathway, the mean number of acute patients delayed, and the mean number of days patients are delayed under given capacity constraints. For each pathway, three plots show the mean daily number of patients in D2A service, mean daily numbers of acute patients delayed, mean number of acute delayed days and corresponding cumulative costs over the prediction horizon.

3 APPLICATION

3.1 Study Setting and Context within the BNSSG Healthcare System

The Bristol, North Somerset and South Gloucestershire (BNSSG) health and care system serves a resident population of a million across three local authorities and a mixture of large metropolitan, rural, and coastal locations. The city of Bristol contains a higher proportion of younger individuals and has a more culturally and ethnically diverse demographic than either North Somerset or South Gloucestershire. Rural and coastal areas contain a greater proportion of older individuals and pockets of severe deprivation. Overall, however, the age profile of BNSSG is similar to the England average.

Acute care is provided by two hospital trusts with intermediate care provided by a single community services provider. At the time of the study (October 2022), the D2A pathways had high occupancy rates with all 234 available caseload slots in service in the P1 pathway, 92% of the 199 beds and 98% of the 154 beds in use in P2 and P3 respectively. This contributed to significant delays to accessing the three pathways, with 81, 84 and 85 patients in acute care ready and awaiting discharge (henceforth, this is referred to as the number of 'blocked' acute beds). Mean D2A durations of service were 16, 38 and 45 days for the P1-3 pathways, with respective medians of 6, 32 and 39 days illustrating the significant variation.

A particular over-arching complexity of D2A pathways, which makes them difficult to model and plan for, is that responsibility, accountability and control for them are not the remit of any single organisation in the ICS. For example, the effects of mitigating circumstances for one aspect of the pathway or organisation. Reducing delays in discharging patients from acute beds may have significant adverse consequences elsewhere for the pathway or other organisations; in this case on pressures to increase downstream capacity or increase system productivity. This means, in turn, that there is a large number of influential stakeholders whose agendas, priorities and budgets do not always align with each other and need to be taken into account with due recognition afforded to potential trade-offs between them. The creation of the ICS is the latest attempt to support a joint approach to the D2A issue and the acute, community health, and social care interface more generally. However, this, in turn, depends critically on good data flows, access to them, and methodologies for their transformation into usable and effective information which can act as a central focus for exploring planning scenarios in a holistic, system-focused manner. The purpose of this modelling is to support that 'data to information' transformation both through quantification of planning scenarios and a greater overall local awareness of which data are relevant.

3.2 Scenarios

The starting point for the model is the current system state in each local authority: the number of patients occupying each pathway and the number of patients in acute beds who are currently delayed while waiting to be discharged into a D2A pathway. These data are directly pulled from existing, routinely collected datasets. Referrals into D2A settings are based on the actual trend over the previous six months by local authority using anonymised patient-level data. This provides the 'baseline scenario' and enables comparisons to be made between current and target proportions of patients entering each pathway.

Three length of stay (LOS) scenarios (current baseline, target and intermediate – the middle point between current baseline and target), and two arrival rate options (current baseline and target) are automatically generated, and all combinations of these scenarios are outputted in the routine report for each local authority area. This combination provides a total of six scenarios for nine pathways (i.e. for each D2A Pathway P1-3 in each of the three local authorities). Additionally, unrestricted capacity scenarios are considered for each LOS (current, intermediate and target) and pathway split (current and target) in order to assess maximum capacity required if none of simulated patients had to join the queue in the acute hospital. The target mean D2A LOS are 10 days, 21 days and 28 days and the target pathway split is 70%, 10% and 10% for P1-3 respectively. The remaining 10% of the acute referrals correspond to patients with complex mental health diseases and are out of scope of the modelling. Table 1 summarises all twelve scenarios designed based on the combination of different options.

Table 1. *Scenarios considered for the simulation modelling.*

Scenario	Capacity	LOS (mean days)	Arrivals (P1-3 pathway split in percentage)
1	Baseline	Baseline	Baseline
2	Baseline	Baseline	Target (70;10;10)
3	Baseline	Target (10;21;28)	Baseline
4	Baseline	Target (10;21;28)	Target (70;10;10)
5	Baseline	Interim	Baseline
6	Baseline	Interim	Target (70;10;10)
7	Unrestricted	Baseline	Baseline
8	Unrestricted	Baseline	Target (70;10;10)
9	Unrestricted	Target (10;21;28)	Baseline
10	Unrestricted	Target (10;21;28)	Target (70;10;10)
11	Unrestricted	Interim	Baseline
12	Unrestricted	Interim	Target (70;10;10)

3.3 Results

For each of the nine combinations of BNSSG locality (the three local authorities) and D2A P1-3 pathways, modelled parameters and scenario combinations are summarised and plotted (an example is provided in Figure 3). For each of the twelve scenarios, outputs include: the number in service (the number of patients estimated to be utilising the service given service capacity); the number awaiting service (the number of patients estimated to be delayed in the acute sector awaiting D2A); and the mean days delayed (the mean number of days delayed in the acute sector). In the plots, some of the scenarios have similar results as reflected by an overlap of the lines. This happens either because the mean number of patients in service hit capacity (illustrated by a flat horizontal line on fixed capacity input figure), or the mean number of patients in the queue and mean number of days delayed hit 0 when capacity is significantly higher than the demand (e.g., scenarios with unrestricted capacity assumption).

Looking at the baseline scenario, which represents the current status of parameters and assumes that input parameters will remain the same over the simulation horizon, one can see that the queues in the acute will build up extensively in all localities for the P2 and P3 pathways. The combination of scenarios allows the decision maker to see what the impact of different strategies may have on the outputs. For example, achieving the target pathway split of 70%, 10% and 10% for pathways P1-3 respectively has a much higher impact on reducing the number of patients delayed in the acute compared to improving the LOS to intermediate values. However, the target pathway split will increase the arrivals through P1 which will in turn cause delays in the acute. So, a strategy that combines LOS reductions and achieving target pathway split may be a better solution for controlling negative outcomes.

The unrestricted capacity scenarios (Scenarios 7-12) allow to see what capacity should be in order not to have patients delayed in the acute as any patient who is referred to a pathway is immediately admitted as there is always available capacity. Based on these scenarios, the outcomes show that given current LOS in the community and daily demand rates, the capacity needed can go up to three times more than current capacity for certain locality-pathway pairs. But, on the other hand, if target pathway split and target LOS are achieved than current capacity would be sufficient and could even be reduced.

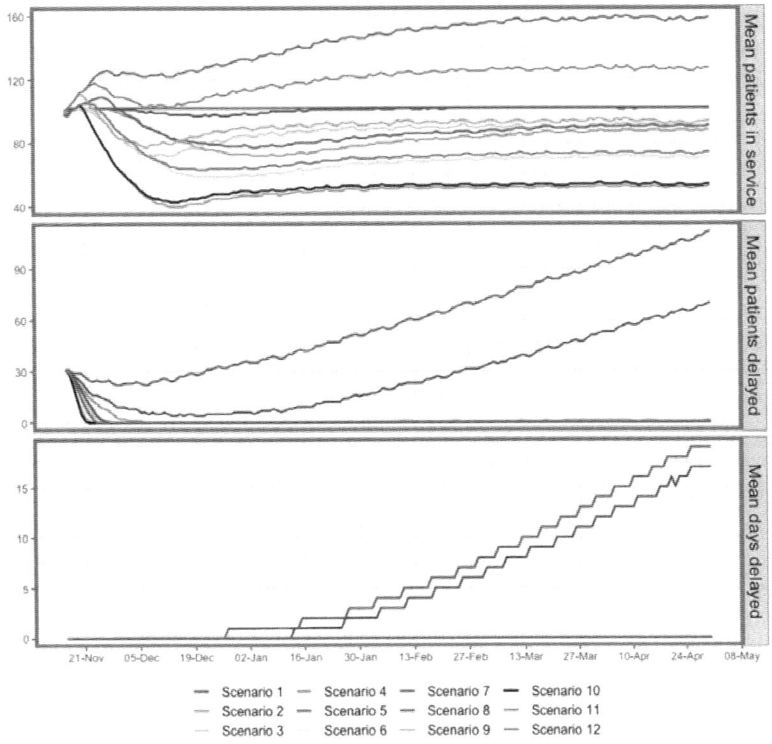

Figure 3 *Example output for Bristol Local Authority D2A P2 pathway for all considered scenarios.*

4 DISCUSSION

4.1 Reflections on Practical Implementation

As a project to develop a model that was both practical and straightforward to understand and implement by non-technical modellers, it was recognised that there was an imperative to engage and work closely with acute and D2A stakeholders from the outset. This was always seen as a two-way process: to understand from them what a 'useful model' looked like and how it might fit with local agendas and decision-making structures, and for the project team to explain and convey what the model was capable of doing for them in supporting a knowledge and skills transfer (not least so modelling could become independent of technical support by the project team). At the time of writing this remains an ongoing process with a good degree of trust between the research team and the stakeholders from all of the principal organisations involved.

The modelling development process has itself generated a greater awareness locally of what constitutes relevant data for D2A planning purposes and increased broader awareness of the whole-system nature of D2A. Also important in this is a recognition that the model has relevance from a care professional as well as strategic planning perspective.

The process of implementation has been iterative from the start. It began with outputs from an initial construct of the model – itself modified through early discussions with local stakeholders – being presented to stakeholders through existing meeting forums focused on D2A issues. Maintaining a standing presence at these forums, and demonstrating an understanding of the D2A system generally, also helped in keeping current an awareness of the model amongst stakeholders and gaining trust that the project team were properly engaged with pertinent D2A issues. A breakthrough in acceptance was the incorporation of modelling estimates into supporting a successful business case proposal of approximately £13 million to further develop and maintain the D2A system locally, and to support its subsequent implementation phase. Using the IPACS model and results from the scenario analysis, stakeholders have now the ability to not only quantify how different strategies (e.g., reduction in LoS, change in spread of the demand) and their combinations can impact key outcomes of their system but also compare different local authorities to characterize best practices. The model is now used as a decision supporting tool which allows better informed decision making for intermediate care service commissioning. There has also been recent interest in using the model to explore scenarios for activity planning for 2023-24.

4.2 Strengths and Limitations

It is crucially important to optimise the complex discharge flow as the efficient and effective movement of patients along the D2A pathway has direct and indirect consequences for the wider acute and community care system. An under-performing discharge pathway results in delayed patients in acute hospitals and reduced available bed capacity. This, in turn, imposes operational delays for Accident and Emergency departments and ambulances and reduced elective capacity (where theatre capacity may be left idle). A large number of patients as a result will not be in the best location for receiving optimal care for their conditions and at greater risk of decompensation and hospital-acquired infection.

Despite these major issues to be addressed, the field has seen a relatively low levels of interest from the academic community so, alongside the above-mentioned practical benefits to the Bristol healthcare system, a strength of this current study is in addressing that research deficit. This has been approached through a flexible modelling tool that can approximate the various pathways, parameters and scenarios currently under consideration. The designed flexibility of the tool will enable the model to be adapted and scaled up to address similar issues in other health and social care systems. While individual ICS areas have unique legacies and starting positions, the underlying issues of D2A are the same right across the UK.

Limitations of this modelling study are, as always, multifaceted. From a technical perspective the current scope of the model does not cover all aspects of the whole discharge pathway system (for example, specific social care inputs, specialised palliative care pathways or longer-term post-D2A pathways such as permanent placement capacity in care homes, or ongoing community health and social care services). These are identifiable gaps, but outside the scope of the current research project. Other limitations lie outside the model and reflect ongoing issues with data completeness and accuracy across organisations. This issue is not confined to Bristol and will be repeated to various degrees across the NHS.

4.3 Further Research Opportunities

With plans to enhance use of home-based care, there exist various opportunities for future simulation and other advanced analytics efforts. These may include developing the pathway model to include social care, which would enable similar analysis of the effect of capacity of this service on upstream services, including D2A as well as acute hospital care. Further work could also examine the effect of acute blockages on Accident and Emergency performance and, in turn, ambulance offload delays and ambulance response times. Investigators may also wish to consider the efficiency of the home-based service from a vehicle routing perspective, in ensuring an optimal schedule of home visits based on a given capacity. Finally, efforts could be directed toward analytical solutions or approximations, thus bypassing the need for relatively computationally-costly simulation models.

ACKNOWLEDGEMENTS

This work was supported by Health Data Research UK, which is funded by the UK Medical Research Council, Engineering and Physical Sciences Research Council, Economic and Social Research Council, National Institute for Health Research, Chief Scientist Office of the Scottish Government Health and Social Care Directorates, Health and Social Care Research and Development Division (Welsh Government), Public Health Agency (South Western Ireland), British Heart Foundation and Wellcome (award number CFC0129). The funders had no role in study design, data collection and analysis, decision to publish, or preparation of the manuscript. Alison Harper and Martin Pitt are funded by the National Institute for Health Research Applied Research Collaboration South West Peninsula. The views expressed in this publication are those of the author(s) and not necessarily those of the National Institute for Health Research or the Department of Health and Social Care.

REFERENCES

Cadel, L., Guilcher, S. J., Kokorelias, K. M., Sutherland, J., Glasby, J., Kiran, T., & Kuluski, K. (2021). Initiatives for improving delayed discharge from a hospital setting: a scoping review. BMJ open, 11(2), e044291. http://dx.doi.org/10.1136/bmjopen-2020-044291.

Demirbilek, M., Branke, J., & Strauss, A. (2019). Dynamically accepting and scheduling patients for home healthcare. Health care management science, 22(1), 140-155. https://doi.org/10.1007/s10729-017-9428-0.

Department of Health and Social Care (2021). Integration and innovation: working together to improve health and social care for all. https://www.gov.uk/government/publications/working-together-to-improve-health-and-social-care-for-all/integration-and-innovation-working-together-to-improve-health-and-social-care-for-all-html-version.

Levin, K. A., & Crighton, E. (2019). Measuring the impact of step down intermediate care on delayed discharge: an interrupted time series analysis. J Epidemiol Community Health, 73(7), 674-679. http://dx.doi.org/10.1136/jech-2018-211628.

McGilton, K. S., Vellani, S., Krassikova, A., Robertson, S., Irwin, C., Cumal, A., ... & Sidani, S. (2021). Understanding transitional care programs for older adults who experience delayed discharge: a scoping review. BMC geriatrics, 21(1), 1-18. https://doi.org/10.1186/s12877-021-02099-9.

Monks, T., Worthington, D., Allen, M., Pitt, M., Stein, K., & James, M. A. (2016). A modelling tool for capacity planning in acute and community stroke services. BMC health services research, 16(1), 1-8. https://doi.org/10.1186/s12913-016-1789-4.

NHS (2021). Hospital Discharge and Community Support: Policy and Operating Model. https://assets.publishing.service.gov.uk/government/uploads/system/uploads/attachment_data/file/999443/hospital-discharge-and-community-support-policy-and-operating-model.pdf.

Palmer, R., Fulop, N. J., & Utley, M. (2017). A systematic literature review of operational research methods for modelling patient flow and outcomes within community healthcare and other settings. Health Systems, 1-21. https://doi.org/10.1057/s41306-017-0024-9.

Patrick, J., Nelson, K., & Lane, D. (2015). A simulation model for capacity planning in community care. Journal of Simulation, 9(2), 111-120. https://doi.org/10.1057/jos.2014.23.

Sezgin, D., O'Caoimh, R., Liew, A., O'Donovan, M. R., Illario, M., Salem, M. A., ... & Hendry, A. (2020). The effectiveness of intermediate care including transitional care interventions on function, healthcare utilisation and costs: a scoping review. European geriatric medicine, 1-14. https://doi.org/10.1007/s41999-020-00365-4.

WHO (2016). Integrated care models: an overview. https://www.euro.who.int/__data/assets/pdf_file/0005/322475/Integrated-care-models-overview.pdf.

Wood, R. M., & Murch, B. J. (2020). Modelling capacity along a patient pathway with delays to transfer and discharge. Journal of the Operational Research Society, 71(10), 1530-1544. https://doi.org/10.1080/01605682.2019.1609885.

AUTHOR BIOGRAPHIES

ZEHRA ÖNEN-DUMLU is a research associate at the University of Bath School of Management in the Information, Decisions and Operations division. She earned her Ph.D. in Industrial Engineering and Operations Management from Koç University. Her research interests focus healthcare operations management. She is mainly interested in developing and investigating models under uncertainty that will help improve healthcare operations and global health systems.

PAUL FORTE is an independent health planning and management consultant and currently holds a part-time role at NHS Bristol, North Somerset and South Gloucestershire Integrated Care Board. He is a Visiting Research Fellow at University of Bath School of Management and holds a part-time teaching role with the London School of Hygiene & Tropical Medicine. He earned his Ph.D. in the School of Geography at the University of Leeds. His interests include whole system planning and management approaches in health and social care and working closely with stakeholders to support and educate the application of models in local decision making.

ALISON HARPER is a research associate at the University of Exeter Medical School, in the Peninsula Collaboration for Health Operational Research and Development. She gained her PhD from University of Exeter Business School. Her research interests are applied health and social care research using data science and quantitative methods to model and improve services, with a particular focus on implementation and impact.

MARTIN PITT is Director of The Peninsula Collaboration for Health Operational Research and Development, Exeter University: Medical School. He is particularly interested in using data science for healthcare improvement. He helped to establish and co-ordinates MASHnet - The UK Network for Modelling and Simulation in Healthcare. He was recently appointed as first President of the Association of Professional Healthcare Analysts (AphA).

CHRISTOS VASILAKIS is Chair and Subject Group Lead in Management Science in the Information Decision and Operations (IDO) Division and Director, Bath Centre for Healthcare Innovation and Improvement (CHI^2) both at the University of Bath School of Management in the UK. In 2020 he was a Visiting Scholar at Stanford University, Department of Management Science and Engineering and SURF Stanford Medicine research centre.

RICHARD WOOD is Head of Modelling and Analytics at NHS Bristol, North Somerset and South Gloucestershire Integrated Care Board and is a Visiting Senior Research Fellow at University of Bath School of Management. His interests are at the interface of academic theory and practical application. His background is in applied mathematics and through his career he has worked in various fields including mathematical biology, quantitative risk assessment, and capital modelling.

Proceedings of the Operational Research Society Simulation Workshop 2023 (SW23)
C. Currie and L. Rhodes-Leader, eds.
https://doi.org/10.36819/SW23.014

USING SIMULATION FOR BED MODELLING IN CRITICAL CARE

Debbie Lentle
NHS Wales Delivery Unit
Debbie.Lentle@Wales.nhs.uk

Victoria Sachser
NHS Wales Delivery Unit
Victoria.Sachser@Wales.nhs.uk

1 BACKGROUND

This project is one of a programme of projects supported by the "Promoting joint analytical problem solving building on the Welsh Modelling Collaborative" programme, which is part of the Health Foundation's Building Capability programme. The Health Foundation is an independent charity committed to bringing about better health and health care for people in the UK.

This work was commissioned by the Critical Care (CC) department in the University Hospital of Wales (UHW). The overarching aim was to determine the future physical combined number of Level 2 and Level 3 CC beds, split by Emergency Intensive Care Unit (ICU) and Elective Post-Anaesthesia Care Unit (PACU). Providing permanent and sustained increases in CC provision has many potential limiting factors as the service is expensive and resource intensive. As a result of this need to plan, emphasis is placed on two key time points, 2030 (expected end of life of the current building) and 2040, to ensure infrastructure is adequate to support the service. Previous models in this area have been developed but the scope of that work did not meet the needs of the stakeholder with this specific question.

Critical Care provides organ support and monitoring to patients who have potentially reversible disease, and who are at a high risk of dying or sustaining long term morbidity. A full Critical Care Unit with no capacity to admit leads to delayed, or denied, patient access to Critical Care. This leads to patient harms and downstream system disruption providing sub-optimal Critical Care in sub-optimal locations.

2 APPROACH

The objective of this project is to model capacity demand in the CC department at UHW, a discrete event simulation (DES) was built in Simul8 covering the flow of patients in a one-year period through ICU and PACU at UHW. The model has been populated using forecasted demand profiles and historic data to model realised patient length of stay (LOS) and typical discharge behaviour. The model captures delays accessing CC, delays exiting CC and occupancy within ICU and PACU.

The simulation model focuses on the movement of patients within the CC department with wider ward capacity modelled at a high level to ensure the key relationship between the CC unit and the rest of the hospital is captured. Due to the variability of patients seen within CC and the potential service developments, the model splits patients into workstreams. This means a simple structure has been implemented to capture a patient's journey, which is then replicated for each workstream in each scenario.

In order to populate the model, it was key to understand the demand profiles for the two years in question, 2030 and 2040. Conventional time series methods were explored but due to the length of the time series data available, the inclusion of the Covid-19 pandemic in the data set and the known service changes for CC, a subject matter expert (Delphi) approach was undertaken to agree the forecast parametrisation.

An iterative approach was taken to generate the scenarios modelled. Initially the model was run for the different years concerned with unlimited bed resources in CC and the general pool of ward beds. From there, the number of CC beds was limited; in 2030 this was based on a physical constraint of the current ward space and for 2040 based on the performance seen across the unconstrained trial already undertaken. From there, scenarios were run with both CC and ward beds constrained to highlight that flow within CC can be hindered by a lack of capacity to discharge into.

3 IMPACT

This project has recently concluded, and the full impact of this work is still being assessed.

UHW has recognised that the current physical capacity in their existing building is not fit for purpose for changes in demand and the latest advancements in healthcare technology. This work has enabled the Critical Care department to use robust evidence based modelling to describe what their bed capacity requirements may need to be up to and including 2040. The outputs of this model will be used to facilitate conversations on the future needs of the department considering the proposed service developments and expected changes to the population.

This project incorporated the wider hospital setting as availability of ward beds has an influence on the ability to discharge patients from CC. This modelling work has been able to demonstrate the importance of the CC to ward relationship and the requirement for decision makers to account for the interdependencies when making decisions.

ACKNOWLEDGEMENTS

The project team: Antuela Tako, Bahman Rostami-Tabar, Craig Spencer, Debbie Lentle, Eduard Incze, Jennifer Morgan, Michael Ware, Victoria Sachser.

REFERENCES

UHW: First minister says state of cardiff A&E 'unacceptable' (2022) *BBC News*. BBC. Available at: https://www.bbc.co.uk/news/uk-wales-politics-63214751 (Accessed: January 12, 2023).

Williams E, Szakmany T, Spernaes I, Muthuswamy B, Holborn P. Discrete-Event Simulation Modeling of Critical Care Flow: New Hospital, Old Challenges. Crit Care Explor. 2020 Sep 14;2(9):e0174. doi: 10.1097/CCE.0000000000000174. PMID: 32984824; PMCID: PMC7491890.

Hyndman, R.J., & Athanasopoulos, G. (2021) Forecasting: principles and practice, 3rd edition, OTexts: Melbourne, Australia. OTexts.com/fpp3. Accessed on 12/01/2023

Proceedings of the Operational Research Society Simulation Workshop 2023 (SW23)
C. Currie and L. Rhodes-Leader, eds.
https://doi.org/10.36819/SW23.015

CONCEPTUAL MODELLING OF EMERGING TECHNOLOGIES - THE USE OF NOVEL ELECTRIC AIRCRAFT FOR EMERGENCY MEDICAL SERVICES

Theresa Degel MSc,
Dr. Durk-Jouke van der Zee

University of Groningen

P.O. Box 800, 9700 AV
Groningen, The Netherlands
theresa.degel@gmail.com,
d.j.van.der.zee@rug.nl

Jannik Krivohlavek MSc,
Jaap Hatenboer MSc

Emergency Medical Services,
University Medical Center Groningen
Vriezerweg 10 9482 TB,
Tynaarlo, The Netherlands
j.krivohlavek@rav.nl,
j.hatenboer@rav.nl

ABSTRACT

Implementation of new technologies in operations systems sets specific requirements on simulation conceptual modelling, relating to uncertainties with respect to their specifications, changes implied for operations systems and regulatory frameworks restricting their operation. Furthermore, modelling objectives may have to be tailored to innovation agendas made by potential adopters of technologies. In this paper, we explore the needs for extending current frameworks to facilitate conceptual modelling of new technologies, using a case study on the introduction of novel electric aircraft (eVTOLs) for Emergency Medical Services. Extensions proposed concern the choice of modelling objectives – which should be aligned with an innovation agenda, technology representation as model content, inputs and outputs – accounting for various uncertainties, and the modelling process – requiring a careful concerting with engineering efforts.

Keywords: Conceptual Modelling, Emerging Technologies, Emergency Medical Services

1 INTRODUCTION

The aviation industry is on the cusp of a significant system change, where novel electric aircraft are enabling a wider adoption of air operations (Schwab et al., 2021). The main enablers for electric aircraft are advancements in key areas such as electric motors, batteries, sensors, connectivity, lightweight materials and advanced manufacturing processes. Perceived benefits of electric aircraft are in their sustainability (no direct emissions), operational costs, lack of noise and higher levels of automation. Two categories of novel aircraft may be distinguished resembling airplanes and helicopters, i.e., electrically-powered Conventional-TakeOff-and-Landing aircraft (eCTOLs) and electrically-powered Vertical-TakeOff-and-Landing aircraft (eVTOLs). Their market availability is expected around 2025, depending on the timely introduction of new regulatory frameworks (European Union Aviation Safety Agency, 2021).

The potential of electric aircraft for enhancing operations may be significant throughout various domains, including Emergency Medical Services (EMS) (ADAC, 2020). Their high speed, cost level, independence of ground infrastructure, and sustainability make them an interesting alternative for ground ambulances and helicopters. At the same time, implications of their use for EMS may be significant, in terms of large investments over a longer period of time, and considerable changes in EMS network and operations. Hence, a timely, and careful assessment of gains implied by their use, and changes required in EMS organization is required. Simulation, being much applied in optimizing EMS network design and use, may offer support in doing so. However, such simulation use would go along with specific modelling challenges. This is due to the emergent status of electric aircraft, possibly impacting on model validity and credibility. Some main uncertainties to address concern their specifications, like speed and range, changes implied for EMS operations and regulatory frameworks within which they have to operate – being not fully clear yet. In addition, also managerial context may matter, requiring a tailoring of modelling objectives to technology development stages as perceived by the EMS provider. Such tailoring may be reflected in, for example, model accuracies required, and alternative system configurations studied in experimenting.

Figure 1 *eVTOLs for EMS – Exploring the Near Future (Source: AXIRA)*

Motivated the observed lack of support offered by simulation conceptual modelling frameworks in addressing emerging technologies, in this paper, we explore the needs for extending guidance for the analyst. To do so we perform a case study concerning a simulation project on the introduction of eVTOLs for a Northern Netherlands subregion (Degel, 2022). Essentially, we compare modelling network extensions using an emerging technology (i.e. eVTOLs) vs. a known technology (i.e. ground ambulances). By scrutinizing differences in formulating modelling objectives, choice of model content, inputs and outputs and set-up of the model process we seek to identify extensions required for modelling frameworks in addressing emerging technologies.

Many EMS systems worldwide face similar challenges. Firstly, fewer people are available or willing to work as professionals in the EMS, causing staff shortages (Uppal and Gondi, 2019). Secondly, budget cuts by local governments lead to further restrictions and centralization of these resources. Especially in rural areas, this results in longer transportation times (Matinrad, 2019). In addition to declining supply, EMS faces increasing demand for service (Hegenberg et al., 2019), aging populations being among the main reasons (Veser et al., 2015). Being "caught" in these developments, the EMS transportation system is under pressure to increase productivity. EVTOLs may contribute EMS productivity by reducing transportation times in an efficient and sustainable way vs. ground ambulances and helicopters (Figure 1). Scarce evidence suggests that eVTOLs may be a well performing alternative for current transport modalities in bringing doctors to a scene (ADAC, 2020; Nakamoto, 2022). At the same time, being an emergent technology, its gains implied for EMS and requirements set on their operations are still largely unclear.

This paper is structured as follows. In Sections 2, and 3 we consider perceived benefits of eVTOLs for EMS and introduce the case setting. Next, in Section 4 we discuss simulation study set-up and key findings. In Section 5 we identify, and scrutinize decisions made in modelling eVTOLs – representing emerging technologies. In Section 6 we discuss how findings in Section 5 may underpin the need for extensions of modelling frameworks. Finally, in Section 7, we summarize main conclusions.

2 GUIDANCE ON SIMULATION CONCEPTUAL MODELLING

In recent years considerable research efforts have been put in improving guidance for the analyst in specifying simulation conceptual models, meant as a precursor for model coding and experimenting. Robinson (2008a) distinguishes three basic approaches on simulation model development: principles of modelling, methods of simplification, and modelling frameworks. Most research efforts, however, concentrated on the development of modelling frameworks (Robinson, 2020). Modelling frameworks go beyond other approaches by specifying what to model. To do so, they provide a procedural approach for detailing a model in terms of its elements, their attributes, and their relationships. Several modelling frameworks have been developed, differing in intended field of application (for example, health, operations systems or the military), scope (including or excluding model inputs, outputs and modelling objectives) and process support (help in exploring the problem situation). For overviews of modelling frameworks, see Robinson (2008a, 2020), Karagoz and Demirors (2011), Van der Zee et al. (2011) and Furian et al. (2015). So far, however, frameworks proposed lack guidance on how to address specific challenges in modelling emerging technologies being implemented in operations systems.

3 CASE: USING EVTOLS FOR EMS IN THE FRISIAN LAKE AREA

The Frisian Lake area in the northern part of The Netherlands is a sparsely populated (130 inhabitants per 1000 km^2) (Cybo Company, 2022) rural area. The area is part of a larger region, the province of Friesland, being served by an EMS provider. The area has a complicated road infrastructure due to many lakes and water canals, delaying ground transportation (Turcanu, 2012), see Figure 2. EMS serves the area by five stations, located in Bolsward, Sneek, Koudum, Joure, and Lemmer. Each station operates one ambulance 24/7. In addition, the station located in Sneek, operates one ambulance during a late shift (3-11PM) and a night shift (11PM-7AM). Each ambulance is staffed with a nurse and an ambulance driver. Dutch EMS providers employ nurses to staff ambulances, while internationally many ambulance services employ paramedics.

Figure 2 *Map of EMS system in the Frisian Lake Area (Source: Google Maps)*

On a province level, 95% of urgent (A1) calls must be served (i.e. reached by a medical professional) within 15 min, according to Dutch norms. In the Frisian Lake area, the performance has been below this threshold for the past 4 years, see Figure 3. Moreover, the performance has become significantly worse in the previous 2 years, partly due to extra safety measures in relation to the Covid19 pandemic but also due to a lower average speed of ambulances. It must be noted that the Frisian Lake area is a challenging area due to its geography implying that a performance of 95% is harder to reach than in other areas of the same province. At the same time, being part of a larger region some tolerances were allowed with respect to norm settings, as worse performance may be "compensated" in other subregions of the Friesland region, especially urban regions for which adherence to norms may be easier to realize. Nevertheless, the observed deviation from the norms for the area underpins the need to explore possible interventions.

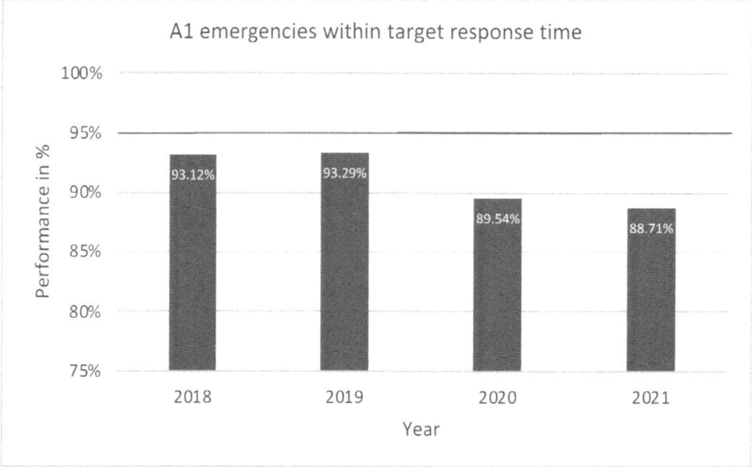

Figure 3 *EMS Performance in the Frisian Lake Area 2018 - 2021 (Source: UMCG Ambulancezorg)*

Additional ground units and eVTOLs have been considered as solutions to improve the performance in the Frisian Lake area. Extending the EMS system with additional ground unit(s) is perceived as too labor-intensive, due to existing staff shortage, and as inefficient due to an expected underutilization of the units. On the other hand, an eVTOL unit can cover a far greater area than a ground unit (within the same time) because of travelling in a straight line bypassing the limited road infrastructure. Therefore, eVTOLs were deemed to be more resource efficient and operationally viable. However, more insights on system performance gains and the parameters influencing these were desired.

4 SIMULATION STUDY

4.1 Extending the EMS System Using eVTOLs

The simulation study did go together with an engineering effort in EMS systems (re)design, given its extension with eVTOLs. Essentially, the engineering effort resulted in clarity with respect to changes required to current EMS operations when implementing eVTOLs and choice of EMS system configurations with eVTOLs to be studied (Tables 1,2). Table 1 captures required changes as assumptions – to be reflected in model content. Table 2 specifies alternative system configurations through design parameters, acting as simulation model inputs. Together with various sensitivity analyses on eVTOLs specifications these were underlying experiments.

Adoption of eVTOLs would imply an extension of the existing ground-based system (base line, see Section 3) towards a hybrid system operating both ambulances and eVTOLs. While various uses of eVTOLs for EMS could be distinguished, it was chosen to restrict use of eVTOLs to a role as so-called rapid responders. As such, eVTOLs would be used to bring nurses to the scene, offering a fast alternative for ground transport. Hence, EMS performance on responsiveness would be expected to improve. However, patient transport from scene to a hospital, if deemed necessary, was still to be executed by a ground ambulance, called by the nurse that arrived by eVTOL on the scene. Further refinements of uses eVTOLs as rapid responders were considered by studying alternative dispatching rules. For example, "what if the eVTOL was primarily used as a backup for ground ambulances if these cannot arrive in time". The choice of restricting the role of an eVTOL to a rapid responder was motivated by the state of art in eVTOL technology, suggesting patient transport not to be feasible before 2030 (Mihara et al., 2021).

4.2 Study Results

Using simulation the various configurations (Table 2) were put to the test. See Figure 4 for an overview of main study outcomes relating to experimental factors. It shows how implementation of eVTOLs may have a relevant impact on system performance. For further details, including sensitivity analyses on

Table 1 *Assumptions on eVTOL Operation in a Hybrid EMS system*

Subject	Assumption
Regulatory Framework	The framework for eVTOL operation is in place (European Union Aviation Safety Agency, 2022).
Infrastructure	Infrastructure elements such as charging facilities and vertiports are in place and corresponding regulations (for building and operating these) exist
eVTOL specifications	eVTOLs can carry a pilot and one medical specialist.
eVTOL specifications	Take-off time is constant and is 2 minutes (UMCG Ambulancezorg, 2022).
eVTOL specifications	Landing time is constant and is 3 minutes. This includes finding a landing spot and the landing procedure (UMCG Ambulancezorg, 2022).
Personnel	Pilot licensing for eVTOLs is possible (European Union Aviation Safety Agency, 2022).
Personnel	An eVTOL is staffed by a licensed pilot and a nurse with the same qualifications as nurses in ground ambulances.
eVTOL operation	The approximate response time for each vehicle in the system (ambulance and eVTOLs) can be estimated by the dispatch center and it is the basis for assigning a vehicle.
eVTOL operation	An eVTOL must be charged after every ride. The charging time is constant and is 15 minutes (Liu et al., 2021)
eVTOL operation	eVTOLs will first operate during daytime. This means they can be dispatched after sunrise and until 45 minutes before sunset (UMCG Ambulancezorg, 2022).
eVTOL operation	At any given time, there is a 1% chance that an eVTOL is not able to launch due to poor weather conditions (UMCG Ambulancezorg, 2022).

Table 2 *Design Parameters for a Hybrid EMS System with eVTOLs*

Design Parameter	Values
Dispatch rule	Various dispatch rules, for example: "An eVTOL is dispatched if the estimated response time of a ground ambulance would exceed 12 minutes".
Location of eVTOL Station	Current locations of stations, additional eVTOL station in the center of the region.
Number of eVTOLs in system	1 eVTOL
Type of eVTOL	Average cruise speed of 180km/h, minimum range of 150 km.
Operation mode	24h, day time only

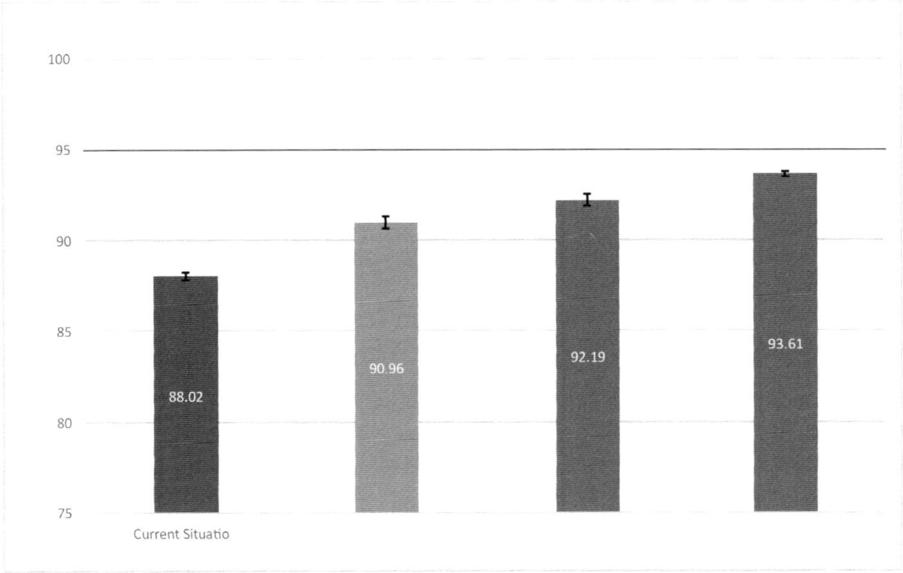

Figure 4 *EMS Performance in the Frisian Lake Area for Selected Configurations*

eVTOLs' speed, take-off and landing time, charging times, and availability due to weather conditions see Degel (2022). Clearly, progress made with respect to the latter aspects may further enhance system performance.

5 EXPLORING REQUIREMENTS ON CONCEPTUAL MODELLING OF EVTOLS

In this section we explore specific requirements on conceptual modelling of eVTOLs for EMS use (Table 3). We do so in three steps:

1. Modelling of new technology as a part of the conceptual model: determine the way eVTOLs have been represented in the conceptual model. In doing so, we adopt the Robinson (2008b) modelling framework, implying modelling objectives, model inputs, outputs, and content to be components of a conceptual model. See Table 3; only those model elements being specific for eVTOLs are included, i.e., being different from straightforward resource extensions using ground ambulances.
2. Scrutinize modelling decisions: assess choices with respect to which components to include or exclude, assumptions and simplifications, paying attention to the (many) uncertainties on eVTOLs' specifications, their effective use and performance. Which trade-offs does this imply with respect to quality criteria for a conceptual model (i.e. validity, credibility, utility, feasibility)?
3. Assess implications for organizing the modelling process: who is involved (expertise, roles), characteristics of modelling activities (efforts, nature, team set-up) , and information sources.

6 DISCUSSION – EXTENSIONS OF MODELLING FRAMEWORKS

6.1 Relating Simulation Modelling to Technology Development and Use

Strategic decisions on emerging technologies to be adopted in business set a specific context for a simulation study, bringing technology adopters and developers together. Ideally, context specifics are reflected in organisational aims motivating the study, as well as modelling objectives that should contribute to these. Next to establishing potential of a new technology in terms of performance improvement or even optimization, the simulation study may contribute to (realizing) innovation agendas by answering "what does it take", i.e., exploring changes to current operations to be realized to successfully embed the new technology. Moreover, modellers and models may act as linking pins between technology developers and potential adopters. Starting from this role they may facilitate discovery of promising use cases by linking model-based estimates on systems' operational performance to technology specifications – thereby highlighting enablers and barriers yet to overcome.

6.2 Modelling & Engineering

As technology is emerging, simulation modelling efforts go together with engineering efforts or may sometimes even trigger these, compare the development of dispatching rules for allocating eVTOLs, see Section 5. Note how the latter example also illustrates how engineering efforts may result from existing and new technology working together. Furthermore, modelling requires assessing future engineering contributions to operational performance. For example, eVTOLs speed and range may be expected to improve considerably over time.

6.3 You Cannot be Certain: Need for Sensitivity Analysis

New technologies go with a clear need for sensitivity analysis. Key uncertainties may be in infrastructure, resource specifications, availability and use, staffing, and regulatory frameworks. Clearly, scope and detail of sensitivity analysis has to be weighed against needs suggested by technology development stage and innovation agendas.

Table 3 *Exploring Requirements on Conceptual Modelling of eVTOLs*

Step 1: Elements of conceptual model (excerpt, focus on use of eVTOLs) Organisational aims - Assess potential of eVTOLs as rapid responders for EMS in terms of their responsiveness to calls for help, and costs and sustainability of their use (gains to be expected). - Identify key decisions to make for adapting current operations for eVTOL based rapid responders and changes implied for its system configuration and operation (what does it take). - Set a time path for eVTOLs implementation as rapid responders for EMS (how to plan to make it happen). Modelling objectives - Explore potential of eVTOLs as rapid responders for contributing to EMS responsiveness, measured as the percentage of calls for help met within 15 minutes, taking into account existing eVTOL technology and their developments in the near future (5 years horizon). Dutch EMS standards require at least 95% of urgent calls to be addressed within 15 minutes at a regional level. However, deviations to this norm may be acceptable at a subregional level, if compensated by other subregions. Model inputs - Dispatch rule - Location of stations - Operation mode - Number of eVTOLs - Type of eVTOLs - Sensitivity analysis: speed, take-off and landing time, weather conditions, charging times Model outputs - % of calls served within 15 minutes - Average response time to call for help - Number of yearly missions per eVTOL Model content (assumptions, see Table 1) - Activities o Dispatch: dispatch rule, availability overnight, weather conditions o Travel to patient: flight time model o Treatment of patient: determine need for ambulance transport to hospital o Handover of patient: time required to hand over patient to ground ambulance o Return to station: flight time model + 15 minutes (charging time) - Resources o eVTOLs: location
Step 2: Scrutinize modelling decisions Organisational aims Apart from insights in EMS performance the study has a relevant role in strategic resource planning in (1) underpinning decisions to make and (2) actions to undertake and setting a timeline for these. The inclusion of new performance criteria on sustainability adds to the strategic nature of decision making. Modelling objectives The many uncertainties going together with emergent technologies classify the study as explorative. Both significance of technology developments expected and the investment horizon make peaking ahead to future eVTOL technology relevant as an input for the study. Model accuracy (validity, credibility) has to be interpreted according to the technology development stage – allowing for wider tolerances in earlier stages. Model inputs Next to common decisions to be made on transportation vehicles (location of stations, number) specific decisions concern the "operation mode" and "choice of dispatch rule". The operation mode is determining eVTOLs availability (daytime, 24h). Day time use is assumed in the first years of eVTOLs operation. After building experience on eVTOL use and securing its safe operation, 24h use would likely be allowed. As eVTOLs are part of a hybrid solution, including ground ambulances, dispatching rules have to be adjusted, clarifying call priorities in allocating eVTOLs. Such adjustments imply an engineering effort – as such rules are non-existing. Key operational uncertainties concerning eVTOL specifications are its speed, time required for take-off and landing, vulnerability to weather conditions, and battery charging times. Hence, a need is acknowledged for sensitivity analysis on respective factors. Model outputs No specific model outputs relating to use of eVTOLs vs. ground ambulances have been defined. One may wonder about outputs related to eVTOLs use, like percentage of calls being allocated to eVTOLs, in explaining its contributions to specific system configurations. Model content Set-up of operations has changed due to use of eVTOLs (compare Table 1). Main differences are explained by the facts that (1) eVTOLs travel by air (vs. ground transport) requiring a flight time model and setting restrictions to their availability (weather, overnight), and (2) solutions are hybrid (entailing eVTOLs and ground ambulances) requiring more advanced control logic (dispatching rules).

Table 3 (continued) *Exploring Requirements on Conceptual Modelling of eVTOLs*

Step 3: Implications for organizing the modelling process
Modelling & engineering The modelling of the system did go together with engineering efforts concerning changes of EMS operations and their control (dispatching rules), compare Tables 1,2. Concerting activities was required. Information sources on system set-up and operation - Information was found also outside the business context, i.e., aircraft manufacturers, researchers, and literature. - Roles of domain experts were extended in exploring future settings. For example, helicopter pilots were asked to assess eVTOLs' operations. Business context The project had to be linked to a development stage in eVTOL implementation. Choice of interventions and precision (model accuracy) were to be related to this stage (maturity level). In principle, the earlier the stage, the less accuracy is accepted. Project team composition An analyst did both the modelling and engineering activities – safeguarding their concerting in this way. Inclusion of the manager of the EMS helicopter team and EMS innovation manager guaranteed easy access to relevant domain experts (especially pilots, and ambulance nurses) and alignment with project objectives and EMS agenda on innovation. Finally, information on state-of-the art of eVTOLs was easily accessible through the EMS project manager being involved in research projects concerning new transport modalities.

6.4 The Process Matters!

Success of modelling efforts cannot be considered loose from the modelling process. Modelling emerging technologies sets additional requirements to this process with respect to concerting modelling and engineering efforts, collecting data on system specifications, consultation of domain experts and – overarching – relating simulation project to the companies innovation agenda. Ideally, the project team and set-up reflects these requirements. As far as project set-up is concerned, facilitated use of simulation may be considered in according to which analysts work jointly with the client and stakeholders in model development, adding to model accuracy and identification of feasible solutions (Robinson et al. 2014; Tako and Kotiadis 2015; Tako et al. 2021). We found how facets of such facilitated use – seeking cooperation with both adopters and developers of technology did benefit the simulation study.

6.5 What's a Good Quality Model?

Clearly, this is usually not an easy question to answer. It may be better to ask: "which quality would be acceptable". In answering this question one might consider aspects like technology development stage, including uncertainties on its specifications, timing and size of investments, and estimated performance gains relative to existing system performance. Note how the question on model accuracy (validity, credibility) may be linked to model feasibility: setting extensive demands on model accuracy while modelling uncertainties are high, likely puts a (too) high burden on the shoulders of the analyst due to extensive data collection and/or efforts to be put in experimenting – if possible at all.

6.6 Extending Modelling Frameworks for Emerging Technologies

Issues addressed in 6.1-6.5 suggest a clear need to extend guidance offered by modelling frameworks. Starting from the Robinson (2008b) modelling framework more support is required for "understanding the problem situation" by linking it to a client company's innovation agenda. Moreover, specific attention for "understanding solutions" with respect to their specifications and implications for company's operations may also be of importance. Among others this may be reflected in model inputs, including sensitivity analysis. Finally, to make things work, organizing the process in a collaborative manner and facilitating effective communication among parties involved (Haveman and Bonnema, 2015) in utilizing cross-disciplinary knowledge is paramount – which is scarcely being addressed in modelling frameworks so far.

6.7 Limitations

Our findings are based on a single case study. Clearly, further studies are required to explore issues raised in greater depth. However, the study clarifies how issues are very much present in practice.

7 CONCLUSIONS

In this study we explored specific requirements on simulation conceptual modelling of emerging technologies. Starting from a case study on a simulation project exploring use of eVTOLs for EMS we established needs for extending modelling frameworks to benefit the practice of modelling.

Case study findings suggest that in conceptual modelling of emerging technologies a good understanding of the problem at hand and solutions provided requires the identification of an innovation agenda for the potential adopter. This agenda should be reflected in the modelling objectives and choice of model inputs. Organizing the process of modelling such that (1) reliable information on technology specifications is obtained – to be reflected in choice of model content and inputs, and (2) concerting of modeling and engineering activities is realized, is considered paramount.

Future research is directed towards developing modelling frameworks facilitating interfacing with the engineering scene, while taking due notion of requirements set to the modelling process.

REFERENCES

ADAC Luftrettung gGmbH (2020) Multicopter in the rescue service - Feasibility study on the application potential of multicopters as emergency doctor shuttles. Available at: https://luftrettung.adac.de/app/uploads/2021/02/LRG_Machbarkeitsstudie_engl.pdf (Accessed: October 26, 2022).

Cybo Company (2022) Postal Codes in Friesland. Available at: https://postal-codes.cybo.com/netherlands/friesland/ (Accessed: June 5, 2022).

Degel, T (2022) The Use of Electric Vertical Take-Off and Landing Aircraft as rapid responders in Emergency Medical Services in Rural Areas. MSc Thesis. University of Groningen.

European Union Aviation Safety Agency (2021) Terms of reference for rulemaking task RMT.0230: Introduction of a regulatory framework for the operation of unmanned aircraft systems and for urban air mobility in the European Union aviation system – Issue 3. Available at: https://www.easa.europa.eu/en/downloads/126656/en (Accessed: November 4, 2022)

European Union Aviation Safety Agency (2022) Urban Air Mobility (UAM). Available at: https://www.easa.europa.eu/en/domains/urban-air-mobility-uam (Accessed: November 1, 2022).

Furian, N, O'Sullivan, M, Walker, C, Vössner, S and Neubacher, D (2015) A conceptual modeling framework for discrete event simulation using hierarchical control structures. *Simulation Modelling Practice & Theory* **56**: 82–96.

Haveman SP, Bonnema GM (2015) Communication of simulation and modelling activities in early systems engineering. *Procedia Computer Science* **44**: 305-314.

Hegenberg K, Trentzsch H, Gross S, Prueckner S (2019) Use of pre-hospital emergency medical services in urban and rural municipalities over a 10 year period: an observational study based on routinely collected dispatch data. *Scandinavian Journal of Trauma, Resuscitation and Emergency Medicine*, 27(1), p. 35

Karagoz N A and Demirors O (2010) Conceptual Modelling Notations and Techniques. In: Robinson S, Brooks R J, Kotiadis K. and Van der Zee D J (eds). *Conceptual Modelling for Discrete-Event Simulation*. CRC/Taylor & Francis: Boca Raton, pp 179-210.

Liu T, Yang X G, Ge S, Leng Y and Wang C Y (2021) Ultrafast charging of energy-dense lithium-ion batteries for urban air mobility. *eTransportation*, 7, 100103.

Matinrad N (2019) An Operations Research Approach for Daily Emergency Management. MSc Thesis. Linköping University.

Mihara Y , Nakamura T, Payuhavorakulchai P, Nakano M.(2021) Cost Analysis of eVTOL Configuration Design for an Air Ambulances System in Japan. doi:10.5703/1288284317285

Nakamoto A, Mihara Y, Motomura T, Matsumoto H and Nakano M. (2022) The applicability of eVTOLs in emergency medical care in japan. Keio University, Japan (preprint, available as: https://doi.org/10.21203/rs.3.rs-1218578/v1).

Robinson S (2008a). Conceptual modelling for simulation Part I: definition and requirements. *Journal of the Operational Research Society* **59(3)**: 278-290.

Robinson S (2008b). Conceptual modelling for simulation Part II: a framework for conceptual modelling. *Journal of the Operational Research Society* **59(3)**: 291-304.

Robinson S, Worthington C, Burgess N and Radnor ZJ. 2014. Facilitated Modelling with Discrete event Simulation: Reality or Myth? *European Journal of Operational Research,* 234 (1): 231–240.

Robinson S (2020). Conceptual modelling for simulation: Progress and grand challenges. Journal of Simulation **14(1)**: 1-20.

Schwab A, Thomas A, Bennett J, Robertson E and Cary S (2021). Electrification of Aircraft: Challenges, Barriers, and Potential Impacts. Golden, CO: National Renewable Energy Laboratory. NREL/TP-6A20-80220. https://www.nrel.gov/docs/fy22osti/80220.pdf.

Turcanu, L. (2012) Rural Vitality in the Netherlands. Available at: https://spinlab.vu.nl/wp-content/uploads/2016/09/Rural_Vitality_in_the_Netherlands.pdf (Accessed: November 1, 2022).

UMCG Ambulancezorg (2022) "Personal Interview with Domain Expert: Helicopter pilot, Data Scientist, Innovation Manager, Medical Personnel ."

Uppal B N, Gondi B S (2019) Addressing the EMS workforce shortage: How medical students can help bridge the gap. Journal of Emergency Management, 17(5), pp. 380–384.

Tako A.A. and K. Kotiadis (2015) PartiSim: A Framework for Participative Simulation Modelling. *European Journal of Operational Research,* 244 (2): 555–564.

Tako AA, Robinson S, Gogi A, Radnor Z, Davenport C (2021) Using facilitated simulation to evaluate integrated community-based health and social services. *Proceedings of the Operational Research Society Simulation Workshop 2021 (SW21)*: 97-106.

Van der Zee D J, Brooks R J, Robinson S and Kotiadis K (2011). Conceptual Modelling: Past, Present and Future. In: Robinson S, Brooks R J, Kotiadis K. and Van der Zee D J (eds). *Conceptual Modelling for Discrete-Event Simulation.* CRC/Taylor & Francis: Boca Raton, pp 473-490.

Veser A, Sieber , Gross S, Prueckner S (2015) The demographic impact on the demand for emergency medical services in the urban and rural regions of Bavaria, 2012–2032. *Journal of Public Health*, 23(4), pp. 181–188.

AUTHOR BIOGRAPHIES

THERESA DEGEL is a junior researcher at the Faculty of Economics and Business, University of Groningen, The Netherlands. Currently, she is working as a consultant in supply chain and operations management in the private sector. Her research interests include simulation methodology and applications in health care logistics.

DURK-JOUKE VAN DER ZEE is associate professor of Operations at the Faculty of Economics and Business, University of Groningen, The Netherlands. He holds a MSc and Phd in Industrial Engineering. His research interests include health care logistics engineering, simulation methodology and applications, simulation & serious gaming, and manufacturing planning & control. He is a member of the INFORMS-SIM. His webpage is http://www.rug.nl/staff/d.j.van.der.zee.

JANNIK KRIVOHLAVEK is a junior project manager at the Emergency Medical Service of the University Medical Center Groningen (UMCG), The Netherlands. He holds a MSc in Mechanical Engineering and Operational Management. His research interests include the electric aviation sector and its implications on the EMS and wider healthcare sector. He is managing the participation of the UMCG in the European research project AiRMOUR which researches and demonstrates medical applications of electric aircraft (cargo and passenger drones) in the urban space.

JAAP HATENBOER is innovation advisor at the Emergency Medical Service of the University Medical Center Groningen, The Netherlands. He holds a MSc in Industrial Engineering and is at present mainly working on the energy and transport transitions in the EMS system.

Proceedings of the Operational Research Society Simulation Workshop 2023 (SW23)
C. Currie and L. Rhodes-Leader, eds.
https://doi.org/10.36819/SW23.016

PLASTICITY SCENARIOS AND MODELLING: PLASTIC WASTE COLLECTION IN URBAN ENVIRONMENTS

Dr Regina Frei

University of Surrey
Stag Hill, Guildford, UK
r.frei@surrey.ac.uk

Prof. Diego Vazquez-Brust

University of Portsmouth
Portland Street, Portsmouth, UK
diego.vazquez-brust@port.ac.uk

Dr Mengfeng Gong

University of Sussex
Falmer, Brighton, UK
mengfeng.gong@sussex.ac.uk

Dr Virginie Litaudon

University of Portsmouth
Portland Street, Portsmouth, UK
virginie.litaudon@myport.ac.uk

ABSTRACT

Within the scope of PlastiCity, an Interreg2Seas project (2019-2022), a set of scenarios were developed for the collection of plastic waste in the urban environments of Ghent (Belgium), Douai (France), The Hague (The Netherlands), and Southend-on-Sea (UK). This included the exploration of alternative vehicles like CargoBikes and electro-vans, in comparison to conventional diesel-powered refuse collection vehicles. For each city, we developed an individual scenario and executed optimisations to compare different collection strategies and frequencies in terms of distance travelled, time used, as well as costs and emissions generated. We used OptiFlow, a logistics optimisation software made available by Conundra, a startup from Ghent University. The main challenge was the unavailability of realistic data on plastic waste volumes for different types of small and medium businesses and organisations in these urban environments, which was at least partially due to pandemic restrictions. Thus, our modelling is mostly to be understood as ways to explore different scenarios and constraints, such as a very limited loading capacity on CargoBikes.

Keywords: Plastic recycling, Waste collection, Sustainability

1 INTRODUCTION

The PlastiCity project[1] aimed to capture plastic waste lost in urban environments, focusing on the cities of Ghent (Belgium), Douai (France), The Hague (The Netherlands), and Southend-on-Sea (UK). This included the development of a mobile processing unit to demonstrate newly developed recycling technology for dealing with multi-material films, the creation of products made from the recycled materials, and plastic waste collection scenarios.

The initial scenario development (PlastiCity Consortium, 2020) looked at reverse logistic solutions that might fit the four cities from a more abstract perspective; what might be possible in the future? The subsequent modelling was more concrete, applying sophisticated logistics optimization software to test scenarios looking at numbers and costs for doing the collections of plastic waste with different vehicles.

A limitation of this work is that calculations are only as good as the data they are based on. When input data or assumptions change, the outputs change. Therefore, these scenarios should be seen as a tool to support thought experiments rather than as evidence-based solutions. The optimisations are not capable of providing exact answers or reliable numbers, given that they are based on many assumptions - some of them more realistic, others more hypothetical, depending on the data available. When the input data changes, e.g. vehicle emissions, waste owner addresses or waste quantities, the results

[1] https://www.plasticityproject.eu

change. Even when all inputs stay the same, the outputs may vary as the optimisations are not entirely deterministic. This is reflected in reality: for instance, traffic conditions change continuously. However, the key aspect of the optimisations is the like-to-like quantitative comparison between alternative scenarios, and - all other things being equal - the optimisations allow to identify the best (e.g. less cost and less pollution) scenarios developed from a set of assumptions. These best scenarios provide the starting point for developing real-world strategies for reverse plastic logistics in each city.

A lot of the data required to build realistic scenarios were unavailable at the time of execution, due to a variety of reasons including the pandemic. Therefore, the modelling must be seen as thought experiments, reflecting the ideas present in each PlastiCity city. They are all different; different types of stakeholders are involved, and different types of data to reduce emissions and costs are available. The overall goal is always collecting plastic waste in an efficient and sensible way, adapted to the local situation.

We developed individual scenarios for each city, depending on how the involved stakeholders (waste collection companies, city councils, etc.) see the situation. For Douai, the optimisations are based on a client database provided by a waste collection company, and we investigated the lorry trips required to collect Ampliroll containers from client locations. For Southend, three scenario optimisations were conducted: two milk-run collection tours executed by Ford E-Transit vehicles and the lorry trips required to serve the 10 mini-hubs distributed across town. For The Hague, we considered three circular bands around the city centre: up to 1km from the city centre, CargoBikes would collect waste and drop it off at the hub; between 1 and 3km a Nissan e-nv200 would collect plastics and drop them off at the hub; and between 3 and 5km from the city centre a diesel lorry would be used, dropping off the waste at a local waste processing company. Finally, for Ghent, we explored the use of alternative vehicles and checked the necessary collection radius for gathering a certain amount of plastic waste, e.g. 5 tonnes.

The overall goal was always to determine a sensible approach to plastic waste collection that minimises emissions, impact on pedestrian zones, and costs, whilst keeping plastics in the best condition possible (e.g. compaction only, no shredding before quality-assured sorting).

2 RELATED WORK

The environmental consequences of plastic solid waste are visible in the ever-increasing levels of global plastic pollution both on land and in the oceans. Although there are important economic, social and environmental incentives for plastics recycling, recycling rates remain low. Within the urban environment, a lot of 'lost plastics' is available that would be eligible for recycling but is not effectively validated, partly because the economic opportunities are not fully known and understood, collection logistics not fully developed, and the sorting facilities not well equipped or stakeholders not fully engaged. These are all barriers to realise the full potential of plastic in the circular economy and need to be overcome in delivering the EU strategy for plastics in the Circular Economy.

When searching for alternatives to traditional waste collection vehicles, one of the possible solutions are CargoBikes (Sheth et al., 2019). Besides significantly reducing congestion (Cairns and Sloman, 2019), they are also able to access narrow roads in town centres more easily and may be allowed to enter pedestrian zones. CargoBikes are often powered by human muscles assisted by batteries which can be exchanged easily. Many models have small boxes or crates, e.g. to hold food for delivery. Those used for waste collection require larger loading containers and are ideally equipped with a compactor.

The CargoBike used by The Hague in the PlastiCity project is powered by a replaceable battery and has a loading volume of 2.2m3. About 7m3 of plastics in their original state can be compressed down to fit into the CargoBike. The electrical van is assumed to be a Ford e-Transit[2] with a loading volume of 15m3, with a press for compacting plastics on-board.

Research by Eunomia (2020) concluded that it is favourable to replace diesel waste collection vehicles by electrical ones: " (...) switching the UK's fleet of diesel powered refuse collection vehicles (RCVs) for electric trucks would have multiple benefits. These include reducing UK greenhouse gas emissions by 290 kilotonnes of CO2 each year – the equivalent of recycling almost 16 billion plastic bottles – eliminating associated exhaust fumes, and saving local authorities money in the long run." An

[2] https://www.autocar.co.uk/car-news/new-cars/new-electric-ford-e-transit-revealed-217-mile-range

additional benefit is that electrical vehicles are far less noisy than diesel powered ones, contributing to a better life quality especially in cities.

In 2018, Veolia trialled the conversion of old diesel RCVs (26t) into electrical vehicles, powering them with energy gained from the incineration of non-recyclable waste. In 2021, they announced the launch of a new fleet of electric RCVs in the City of London and in Westminster, where also electric sweeping vehicles and trikes for collecting recycling. The trikes are similar to the CargoBike used in PlastiCity but aimed at collecting street recycling.

There are challenges with electrical vehicles. Their high purchasing prices are high due to the cost of batteries. Assuming a lithium-ion battery capital cost equal to 90 €/kWh, acceptable pay-back periods (about 6 years) were obtained (Calise et al., 2019).

Most importantly, end-of-life solutions for the batteries are urgently required. Battery recycling is complicated and expensive, and there is a risk of batteries being sent to developing countries where health and safety regulations are less stringent and less enforced. Car manufacturers are legally obliged to recycle the batteries and first pilot recycling plants are operational, some going beyond the (very low) obligation of recycling 50% of the metals. It is also possible to give car batteries a second life before recycling, using them for energy storage which is required for many renewable energy sources.

The advantages and downsides of small and large vehicles with alternative power sources, such as LNG, LPG, biogas, biodiesel and hydrogen were discussed. We concluded that for the purpose of the modelling, the fuel did not matter, only the type of vehicle and its loading volume. Once the optimisations were concluded, the emissions generated can be calculated for various fuel types.

The literature, discussed in PlastiCity Consortium (2022), provides a strong case for environmental gains to be achieved by reducing emissions when switching to electric vehicles (large and small). However, the actual impact on wheel-to-wheel emissions depends on the country's energy mix. In countries where the generation of energy is heavily dependent on fossil fuels, the use of electric vehicles does not reduce the country's total emissions (Woo, Chong, Ahn, 2017). In the UK this could be the case, while in France electric vehicles will deliver substantial gains. Natural Gas vehicles (CNG and LNG) do not bring significant improvements. Hydrogen vehicles are potentially the most environmentally beneficial but there is not yet evidence to support the most optimistic claims.

3 SCENARIO SETUP AND FINDINGS FOR EACH CITY

Each city adopted a very different concept for its plastic waste collection, and hence the results are not directly comparable. The software used to explore various logistic scenarios is OptiFlow[3], a commercial logistics optimisation platform that is mostly used by companies organising deliveries. OptiFlow can process a maximum of 2000 deliveries or collections at a time, which meant that we had to split the bigger cities (The Hague and Ghent) into strategic areas. The software allows for analysing transport time and distance, presenting the most efficient routes and schedules. It is non-deterministic, meaning that it can produce different outputs for the same inputs. However, the outputs are not generated through a random search process. Below is a summary and some highlights of each set of scenarios based on three key strategies: the fetch, milk-run and concentric strategies, as well as an improved combination. Details are available in PlastiCity Consortium (2020, 2022).

3.1 The Fetch Strategy in Douai

The PlastiCity partner in Douai is Theys Recyclage, and hence our modelling was designed from their perspective. Theys provided access to databases with actual volumes of waste collected yearly from their clients. As a result, the strategies produced have a business perspective and will provide particularly useful insights for plastic waste business models. Theys is currently building an innovative recycling sorting plant at their base, expanding their existing capacity. Theys has a distinctive waste collection strategy as part of their business model. We call this the "fetch strategy". Their clients notify Theys when the ampliroll containers are (almost) full, and according to contract, Theys needs to collect them within two days. Our modelling uses Theys' client database in anonymised form. However, only the collection frequency per client per year was recorded, not the dates when the collections happened.

[3] https://www.conundra.eu/optiflow-route-optimization-software

Hence it was not possible to run optimisations taking into account collection distribution over time. Instead, we assumed all collections needed to be done subsequently, in random order, and calculated the time required, distance travelled, emissions generated and costs incurred.

Due to the nature of ampliroll containers, each requires an individual trip, and a milk-run scenario is not possible. However, the literature suggested potential cost and emission efficiencies when lorries with trailers are used. Lorries with trailers are not suitable for collecting waste in city center areas with narrow and twisty streets but Theys clients do not operate in old central areas. This suggested that while Theys is using lorries without trailers, substantial gains could be obtained if they changed this. Thus we compared the calculations on their current basis (lorries without trailers) to a situation where all lorries pull trailers, and can hence transport two ampliroll containers at the same time.

Theys' database contains 108 clients with a varying number of collections. Collectively, they generate 2700 orders. As OptiFlow allows for a maximum of 2000 orders to be calculated simultaneously, not all orders could be included in the modelling. However, this should not influence the validity of the results; e.g. if using trailers could save 50% of distance travelled, this applies whether 80% or 100% of the orders are considered. The base scenario includes 5 lorries without trailers. We compare this against scenario 1, where the 5 trailers travel with trailers. In scenario 2, we use only 4 lorries with trailers.

3.1.1 Fetch Strategy Findings

Figure 1 shows an example of a scenario calculated for Douai, in this case using 5 lorries with trailers. Our results agree with common sense: adding a trailer, the distance travelled and operational costs incurred are roughly half. However, the time is only reduced by 40%. This is mainly due to the time it takes to load and unload, which remains the same, whilst travelling is reduced and the time added to manoeuvres by handling the trailer. An additional factor to consider is that trailers require a capital investment as well as further training for drivers, as they require a special driving license for operating lorries with trailers.

Our conclusion is that if the goal is to reduce costs, emissions and distance travelled, scenario 2 is the best choice. However, scenario 1 requires fewer days to complete the waste collection, given that there is one more lorry, and still reduces costs, distance and emissions compared to 1. When the business model is based on speed of collection, scenario 1 would be the best choice.

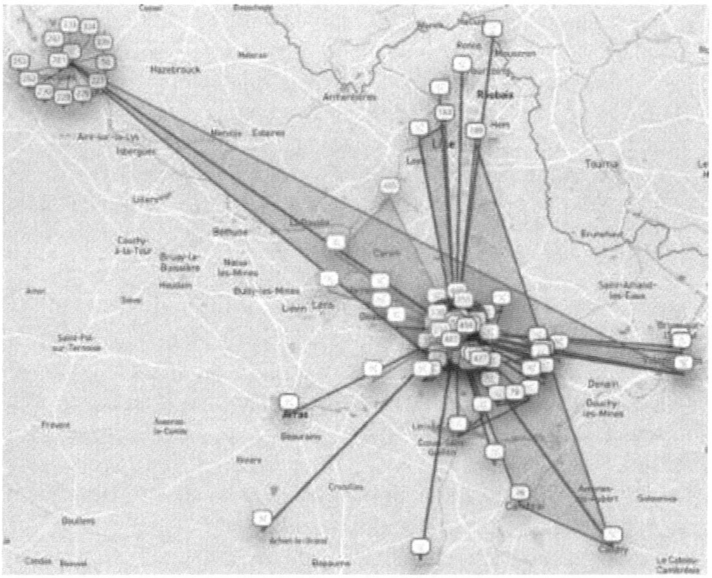

Figure 1 *Possible routes for Douai with 5 lorries with trailers*

3.2 The Fetch Strategy for Mini-hubs and the Milk-runs Strategy in Southend

Southend-on-Sea Borough Council (SBC) is the local PlastiCity partner in the UK and interested in improving plastics waste logistics in various ways. Three scenarios were built to explore a concept with 10 mini-hubs distributed across the town, each with a roll-on/roll-off container requiring individual lorry trips similar to the scenario in Douai, based on a "fetch strategy". We compared the use of 3 different waste management companies to service the mini-hubs; one company is located out of town, the second is within the outskirts of town, and the third is in town.

However, the scenarios are more complex because Southend wanted to explore the use of electric vehicles, which present substantial reductions in emissions compared to fuel-based vehicles. Accordingly, the other two scenarios in Southend are milk-runs executed by electrical vans: one route in an industrial area called Temple Farm and one route in the town centre, where a CargoBike is used additionally. Both locations also host a mini-hub, where the collected plastics can be accumulated. Southend's town centre includes a high street with lots of small and some bigger shops, restaurants and cafes as well as a shopping centre. The shopping centre also hosts a mini-hub. The high street is reserved for pedestrians during the day, hence a CargoBike is used for serving this location. The town centre milk-runs are executed once a week or once a fortnight. SBC provided us with a list of businesses in their town centre. As there is no data on waste volumes, random numbers were used. We assume the collections are executed by a two-person crew.

3.2.1 Combined Strategy Findings

Unsurprisingly, the mini-hub service could save considerable time and reduce emissions by operating via the waste management company located in town. This may need to be weighed against prices charged by each company as well as whether it is desirable to use (and hence support) a waste treatment plant in town, with its space requirement and the noise, smell and traffic generated.

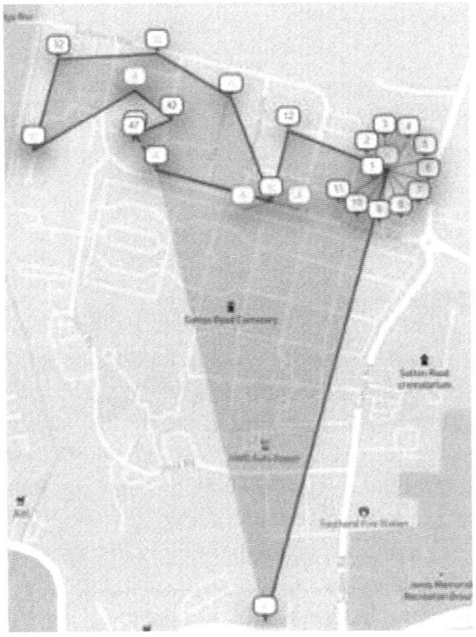

Figure 2 *Southend Temple Farm, possible route for monthly collections with triple waste volume and one container at the mini-hub*

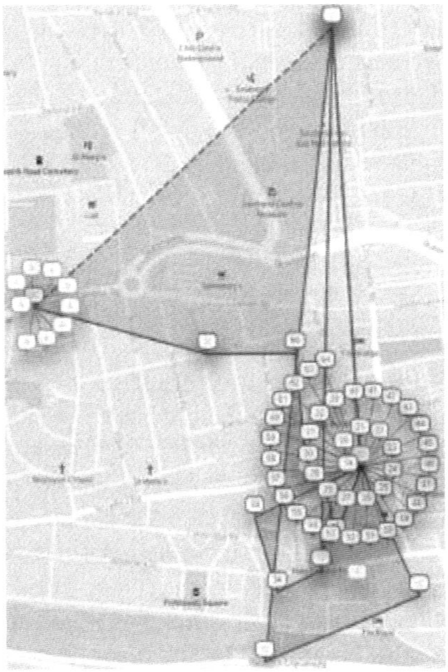

Figure 3 *Southend town centre, possible route for one vehicle*

For the Temple Farm milk-run (Figure 2), we explored weekly, fortnightly and monthly collections with assumed single, double and triple waste volumes and two different local drop-off locations. We found that with higher waste volumes, it is necessary to have more than one container available at the mini-hub to receive the collected waste. Also, for certain situations, a number of collection orders remained unfulfilled, meaning that the vehicle had run out of time.

For the town centre, shown in Figure 3, we found that it is necessary to define a strategy for how to deal with cases where a shop has waste volumes that are above the CargoBike's loading capacity: Will the CargoBike take as much as possible and return, or will a larger vehicle be called instead?

3.3 The Concentric Strategy in The Hague

The Hague conducted a waste collection trial to get an idea of plastic waste quantities. The targeted business types were based on a survey conducted in Ghent at the beginning of the project to find out how much plastic waste can be found at different types of NACE[4] codes. The results suggested focusing on retail and offices. However, due to the pandemic lock-down, offices were not occupied and therefore did not respond to our query. Therefore, waste could only be collected from retail businesses, and the number of participants was small. The trial was conducted over a period of 6 weeks, with weekly collections, but some companies were closed during some weeks and hence did not have any plastics to collect. Whilst this trial gives a glimpse of possible waste numbers, there is not enough data to provide typical waste volumes per retail business type, or even for retail as a whole.

The core idea of the scenario for The Hague was to explore the use of different vehicles for different circular zones around the city centre. We call this strategy: "the concentric" strategy. The local partner is the city council. They supplied a database with businesses in the city, which was filtered for relevant potential waste owners per zone.

Within a radius of 5km (Zone 1) from the assumed centre point, a CargoBike would be used to collect plastic waste only. The drop-off point is the Hub. Within a radius of 5-10km (Zone 2), an electrical van would be used, and in Zone 3, from 10 to 15km from the center point, a diesel lorry would be used. In Zones 2 and 3, mixed recyclables would be collected. However, it turned out that these radii were too big for The Hague. To focus on the city and its immediate agglomeration only, the zones were defined as up to 1km, 1-3km, and 3-5km, instead. Two waste treatment facilities in the outskirts of The Hague were considered as drop-off points. The electrical van assumed to be used is a Nissan E-NV200, with one van available for collecting plastics. The loading space is 4.2m3. We established three scenarios to be able to make comparisons between the vehicles, with weekly collections:

- Scenario 1 (base line): all by diesel lorry
- Scenario 2: zone 1 collection of plastics by CargoBike, zone 2 and 3 by diesel lorry
- Scenario 3: zone 1 collection of plastics by CargoBike, zone 2 by electro-truck, zone 3 by diesel lorry
- Scenario 4: zone 1 collection of plastics by CargoBike, zone 2 and 3 by electro-truck

3.3.1 Concentric Strategy Findings

A trial-and-error approach was used to determine a suitable number of vehicles in each scenario, as shown in the example of Figure 4. The calculations showed us that the intuitively nice idea of concentric circular zones is not very practical in reality, as streets and activity zones rarely follow this pattern.

Very unsurprisingly, emissions are lowest for electrical vehicles. But of course, the wheel to wheel (WTW) emissions are only lowest for electricity-powered vehicles if we assume that the electricity is fully generated by renewable energy sources. Otherwise, the emissions are just moved out of the city. As noted before, a country-specific analysis of the energy generation mix is needed to assert the extent to which electric vehicles produce less emissions than diesel vehicles. We do not have quantitative information about WTW emissions of cargo bikes vehicles in Woo et al. (2017) for the Netherlands, but we know that in 2021 The Netherlands was ranked 27th out of 27 European countries in share of renewables, and has the highest rate of emissions per KWH generated. 46% of its energy mix came from oil (38%) and coal (11%) with a further 38th from natural gas and only 11% from nuclear, wind, solar, hydropower and geothermal.[5] Such an energy mix is unlikely to result in gains in emissions from

---

[4] https://ec.europa.eu/eurostat/statistics-explained/index.php?title=NACE_background

[5] US International Trade Administration 2021: Netherlands, Country Commercial Guide, Energy.

the use of electric vehicles compared to diesel, and very likely to result in electric vehicles contributing more than diesel vehicles to emissions.

Figure 4 *Optimisation result data for The Hague*

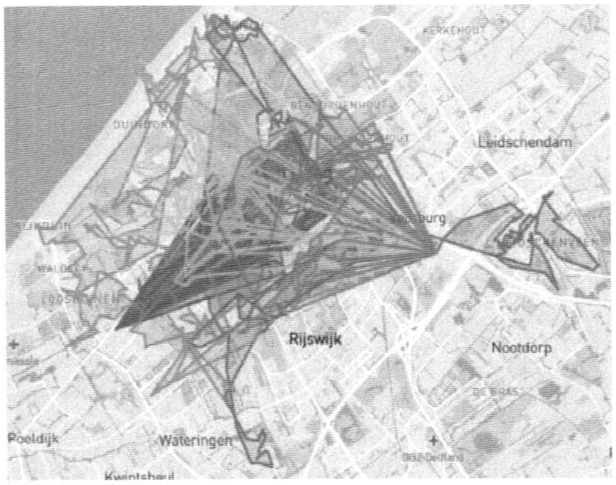

Figure 5 *Possible route for The Hague scenario 3.2, using six vehicles: 2 CargoBike (z1), 2 electro-lorries (z2) and 2 diesel lorries (z3)*

A scenario where only cargo bikes are used could result in less emissions but our modelling approach shows that such a scenario is unfeasible. The cheapest scenario is that with all electric vehicles. However, the savings compared to the second cheapest scenario - all diesel vehicles- amount to only € 1000/year, suggesting savings are not significant from the operational cost of view. However, there is, as noted during the interviews carried out for sensitization, a significant difference in acquisition costs between electric vans and diesel lorries, with the former being more pricey. This suggests that policy intervention subsidizing costs of electric vehicles may be needed to scale up its use. However, our preliminary WTW emissions analysis based on Woo et al. (2017) suggests that such interventions may be counterproductive if the country does not radically change its energy mix to reduce dependence on oil, coal and gas. Further calculations are required to determine the actual WTW emissions produced by using electric vehicles, taking into account the actual WTW emission factor for cargo bikes and electric vehicles vans in the Netherlands.

3.4 Leveraging Previous Knowledge About Fetch, Milk-run and Concentric Strategies in Ghent

The knowledge acquired from the previous scenarios based on the fetch, milk-run and concentric strategies was applied to a more complex scenario in Ghent. The optimisations conducted for Ghent include 7 different scenarios including weekly, fortnightly and monthly collection and alternative transportation modes and alternative vehicles like the CargoBike. To determine plastic waste quantities, a waste collection trial was conducted in Herentals, Belgium, amongst retail businesses. Two collections were executed: the first collection in the period of July-August 2020 and the second one in the period from Sept 2020 to January 2021. It is not known for how long the businesses were accumulating the waste collected in the first round - potentially 1-2 months, but the waste collected in the second round was likely accumulated during 1-5 months. This wide variation makes it impossible to create typical values for waste volumes. For the sake of the exercise, we shall assume 2 months on average, and hence the volume for a monthly collection would be half.

The database used for the scenario calculations contained 10136 entries in total. Those without plastic waste volume data and those with 0.000 tons per year (which might be zero because of rounding) were removed. 9930 addresses remained in the database. Out of these, three zones were defined for the modelling, based on the shopping areas. The PlastiCity Hub is situated at Farmanstraat 40, a site in the southern part of the North Sea Port district, situated in the middle between the historical city centre with a lot of retail and gastronomic venues and the major companies along the harbor.

We assume that there is space for a 40m3 container to receive the collected plastics at the mini-hub locations. These containers then get taken to the hub at Farmanstraat in scenarios 5/6/7. The building at Farmanstraat can be reached by road, tramway (there are rails next to the road) and boat.

3.4.1 Improved Combined Strategy Findings

The very high density of businesses in Ghent city led to visualisations as shown in Figure 6. The fact that the (assumed) waste quantities cannot fit into the smaller vehicles at once needs to be discussed when deploying them. Possible solutions are:

- Call for a larger vehicle, especially in case of large items
- Revisit this business
- Collect more frequently
- Increase the loading volume of the CargoBike, potentially by adding a trailer
- Encourage the business to be a mini-hub where a larger container can be placed, hence also serving the neighbouring businesses

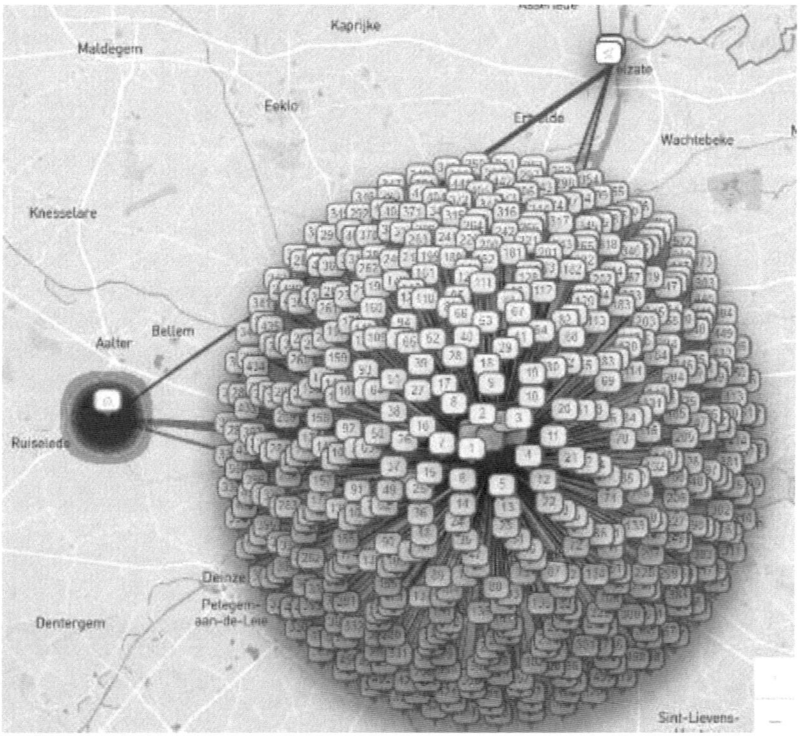

Figure 6 *Possible routes for a Ghent scenario in 'Zone South' with fortnightly collections*

4 DISCUSSION, CONCLUSION AND RECOMMENDATIONS

Each city needs to make individual decisions. How companies handle their waste depends on many factors, including whether they are individual locations or part of a chain. For instance, one chain of electronics equipment with many shops in the Netherlands, Belgium, Germany and Luxembourg stated that they separate their recyclables (paper, cardboard and plastic) locally and then take them to the main location for proper disposal. It is possible that the reason behind this strategy is that they need to pay for their recyclables to be collected, and it is cheaper to do this in one location, only, even if it requires company-internal transportation. It may make sense from an environmental perspective as well if the company can benefit from reverse logistics, that is, for the forward distribution to take the recyclables back.

Large companies often use reverse logistics in this way. Some even include the customers in the chain, by asking them to return certain items or materials to the store. For instance, the British supermarket chain Tesco collects post-consumer soft plastics (plastic films of any type) for chemical recycling, as councils in the UK currently do not collect these materials and they would end up in landfill / incineration otherwise.

The scenario considered for Douai is very simple, and the conclusion from it is equally clear: to reduce emissions and costs in the long run, lorries picking up large containers should run with trailers. It would be interesting to explore further possibilities for this town, considering also companies that do not host their own 40m3 container, and using a milk-run approach for collecting their waste. This could be an interesting project for an MSc student, for instance. In addition, policy makers in Douai should actively encourage the use of electric vehicles. The positive impact of EV in emissions reductions depends on each country's energy mix. France's energy mix is heavily dependent on nuclear and renewable energies, and results in the lowest emissions per KW of electricity generated in Europe. In such conditions, the use of EV could result in 60-70% less emissions of reverse logistics than diesel lorries and 80% less than gasoline lorries. Our cost analysis in the other cities suggest that the costs of using electric vehicles in Douai will not be substantially higher than current alternatives. However, acquisition costs are steep and would need policy intervention, for instance with subsidies.

For Southend, it is difficult to make a qualified recommendation due to the complete lack of waste quantity data when the optimisations were run. However, their idea of creating mini-hubs for accumulating plastics locally certainly makes sense, especially in areas with many small businesses like in high streets, shopping malls and industrial areas with many small and medium sized companies. This idea has been adopted by Ghent. However, the use of a CargoBike with a very small loading capacity is not useful. For the collection service to reach a sensible level of efficiency, it is essential to have the largest loading capacity possible and to have a press on-board. In terms of emissions, the UK energy generation mix is still heavily dependent on fossil fuels, resulting in some of Europe's higher emissions per KW. As a result, the environmental advantages of using EV are minimal.

The modelling conducted for The Hague taught us that "distance from the centre" is not necessarily a useful criterion to organise logistics as streets and neighbourhoods are usually not organised in a concentric way. Most importantly, the waste drop-off location is essential especially when using vehicles with small loading capacities. While the use of cargo bikes seems intuitively appropriate for old areas of the city, cargo bikes will always need to be combined with other means of transport with larger carrying capacity, in what could be called a two step strategy. A combination of bikes and electric vans will result in lower running costs, but increase acquisition costs. Environmentally, since The Netherlands has one of the most polluting energy mix of Europe, the use of electric vans is not advisable. The use of cargo bikes may still have benefits but positive effects are more likely to derive from reductions in congestion. A detailed well to wheel analysis of cargo-bikes emissions in The Netherlands is recommended.

For the Ghent modelling, many of the above-mentioned lessons were taken into account. Collections were arranged by shopping areas, with milk-run drop-off at local mini-hubs which were optimised for the further transportation mode (next to a tramway or river/canal if this modus is considered to empty the mini-hub). The loading volume of the CargoBike used here is larger and it is equipped with a press. However, given that some small companies have waste quantities that still exceed the loading volume, the use of a trailer should be considered. As in the case of France, Belgium has a relatively clean energy mix, where low carbon emitting sources dominate. As a result, the use of electric cargo bikes and larger electric vehicles is definitely advisable from an environmental point of view. Policy incentives should be provided to drive the electrification of waste and reverse logistic companies afloat.

It would make sense for all cities (and rural areas as well) to move away from each waste management company to organise their own logistics. Currently, some streets get visited by 5 recyclables collection vehicles, plus several more for other types of waste. Whilst potentially difficult to negotiate, the use of a joint collection service would reduce emissions, noise and traffic, improving quality of life.

ACKNOWLEDGMENTS

We would like to thank Conundra[6], and in particular Luis D'hondt, for their generous support in allowing us to use the OptiFlow software for free. PlastiCity has received funding from the Interreg 2 Seas programme 2014-2020 co-funded by the European Regional Development Fund under subsidy contract No 1/12/02/2019 16:43:38.

REFERENCES

Cairns, S., & Sloman, L. (2019). Potential for e-cargo bikes to reduce congestion and pollution from vans in cities. Transport for Quality of Life Ltd. Available online: https://www.bicycleassociation.org.uk/wp-content/uploads/2019/07/Potential-for-e-cargo-bikes-to-reduce-congestion-and-pollution-from-vans-FINAL.pdf

Calise, F., Cappiello, F. L., Cartenì, A., d'Accadia, M. D., & Vicidomini, M. (2019). A novel paradigm for a sustainable mobility based on electric vehicles, photovoltaic panels and electric energy storage systems: Case studies for Naples and Salerno (Italy). *Renewable and Sustainable Energy Reviews*, 111, 97-114.

Eunomia (2020). Ditching diesel - a cost-benefit analysis of electric refuse collection vehicles. Available online: https://www.eunomia.co.uk/reports-tools/ditching-diesel-analysis-electric-refuse-collection-vehicles

PlastiCity Consortium (2022). Logistics Simulations Report. Available online: https://www.plasticityproject.eu/downloads

PlastiCity Consortium (2020). Logistics Scenario Report. Available online: https://www.plasticityproject.eu/downloads

Sheth, M., Butrina, P., Goodchild, A., & McCormack, E. (2019). Measuring delivery route cost trade-offs between electric-assist cargo bicycles and delivery trucks in dense urban areas. *European transport research review*, 11(1), 1-12.

Woo, J., Choi, H., & Ahn, J. (2017). Well-to-wheel analysis of greenhouse gas emissions for electric vehicles based on electricity generation mix: A global perspective. Transportation Research Part D: Transport and Environment, 51, 340-350.

[6] https://www.conundra.eu

Proceedings of the Operational Research Society Simulation Workshop 2023 (SW23)
C. Currie and L. Rhodes-Leader, eds.
https://doi.org/10.36819/SW23.017

USES OF THE SKEW-LOGISTIC FUNCTION FOR MULTI-WAVE FUNCTIONS

Dr. Russell Cheng

University of Southampton
Highfield, SO17 5BJ
United Kingdom
cheng@btinternet.com

Dr. Brian Williams

SACEMA
Stellenbosch University,
South Africa
williamsbg@me.com

ABSTRACT

The Skew-Logistic (SL) function has been proposed to model a real-life dynamic process which rises monotonically to a peak followed by a monotonic falling back. It was introduced to model the first stage of the Covid-19 pandemic to forecast its behaviour. Then, with different controls and variants, Covid-19 - rose and fell in what might be called a Multi-Wave (MW) behaviour; with waves not necessarily the same size. This paper shows how using the SL function for one wave can be easily modified to model the MW situation. We apply it to two examples. One is to Covid-19, to examine its most recent behaviour. We also apply it to climate change, the most serious issue of our time. Ensuring that the world simply achieves carbon-equality is not enough. We have to rapidly achieve carbon-negativity to prevent bringing an end to the known world.

Keywords: Covid-19 waves, climate change, carbon negative

1 INTRODUCTION

The Skew-Logistic (SL) function was introduced in Dye et al (2020) to compare the first wave of the Covid-19 epidemics in different European countries. The mathematical form is described in the Supplemental Materials of the paper, which also describes how the model can be fitted to real data using standard statistical techniques. This paper shows how the method can be used to firstly fit to the individual waves of a sample of Multi-Wave (MW) data and how all these individual waves can be then combined to create an overall simultaneous fit to the entire data sample. It will be convenient to summarise the single wave fitting first, which is done in Section 2. Then, in Section 3, we discuss the MW fitting method. The present paper is based on Cheng and Williams (2022).

The SL function has already been used by Cheng et al (2020) who introduce SEPIR, an extension of the well-known SEIR compartmental model, to analyse the Covid-19 epidemic. The SL function was fitted to a single wave to assist which was then used to study the behaviour of the epidemic. In our application here, we fit the most recent Covid-19 UK Active Cases that occurred from March 2022 until November 2022 covering not one but the most recent 3 waves. We then use this to predict the likely case level in the next few weeks.

The purpose of this paper is to discuss how to fit the MW version of the SL function but, as discussed in Cheng et al (2020), the usefulness of estimating the SL parameters is that the values of the infectiousness and transmission parameters in the SEPIR model can be expressed as functions of the SL function parameters so that estimating these latter parameter values also yields estimates of the SEPIR model parameters.

As a second example, we fit the SL-MW model to the global annual atmospheric carbon dioxide level in parts per million (CO_2 ppm) from 1800 to 2022. We also fit the model to the world average Temperature (^{0}C). We then use the fits to predict the CO_2 and Temperature levels until 2100, showing what might happen under different scenarios.

2 SINGLE WAVES

Fitting the SL function to the observations of one wave of the process of interest forms the basis for fitting several SL functions to MW data. We therefore start by summarising the description of the SL function given in Dye et al (2020).

The SL function takes the form:

$$D(t,\boldsymbol{\theta}) = a\frac{e^{b(t-\tau)}}{[1+e^{\frac{1}{2}(b-d)(t-\tau)}]^2} = a\frac{e^{d(t-\tau)}}{[1+e^{-\frac{1}{2}(b-d)(t-\tau)}]^2} \tag{1}$$

where $D(t,\boldsymbol{\theta})$ is the prevalence or incidence of the quantity of interest, this being the number of active Covid-19 cases in Example 1 and CO_2 in Example 2. All four parameters are readily interpretable. The parameter a, is close to the maximum value of $D(t,\boldsymbol{\theta})$, whilst τ indicates the location of the maximum. The parameters b and d, respectively indicate the rates of rise and decline of $D(t,\boldsymbol{\theta})$.

As shown in Equation 1, b and d are mathematically identical.

We write $\boldsymbol{\theta} = (\sigma, a, b, d, \tau)$ where σ is the standard deviation of an individual observation. As dependence of $D(t,\boldsymbol{\theta})$ on the parameters is obvious, $\boldsymbol{\theta}$ is usually (but not invariably) omitted: so that we simply write: $D(t)$.

We assume that $b > 0$ and $d < 0$ so that the term $(b-d)/2$ in the denominator is always positive. $D(t) \cong e^{b(t-\tau)}$ when $t \to -\infty$, so that b gives the exponential rate of increase of $D(t)$ as t increases from $-\infty$, and $D(t) \cong e^{d(t-\tau)}$ when $t \to \infty$, so that $d < 0$ gives the exponential rate decrease. Either form of $D(t)$ in Equation 1 can thus be used when fitting to data, with the signs of b and d showing which role they have.

The explicit maximum value of $D(t)$ is:

$$D_{max} = ar^{(\frac{2r}{1+r})}(1+r)^{-2} \tag{2}$$

where $r = -b/d$. The maximum point is at

$$t_{max} = \tau + \frac{2}{b-d}\log(-\frac{b}{d}). \tag{3}$$

Equations (2) and (3) show that, when b and $-d$ are close in value, a will be close to the maximum of D, and τ wil be close to the maximum point.

As described in Dye et al (2020), the SL function can be fitted to data using the method of maximum likelihood (ML) using Nelder-Mead search. This latter needs initial parameter values to be provided. This is not always straightforward, especially with multimodal functions, particularly when the number of parameters is large. A very attractive feature with the SL function is that, its readily interpretable parameters enables the user to intervene and interactively select appropriate initial parameters which are close the best. We can simply visually examine the data to obtain an initial estimate of its position and size. The rates of increase and decrease can be roughly estimated to give starting values for b and d. An initial value for a can then be obtained using Equation (2), whilst the horizontal position of maximum and Equation (3) can be used to give the initial value of τ.This enables Nelder-Mead to reliably obtain an accurate optimum.

In applying ML optimization, we include an extra parameter σ, the standard deviation of the observational error. Estimation of σ is included in the ML method, so that the quality of the fit is also assessed.

We turn now to the MW situation.

3 MULTI-WAVE

3.1 Function Formula and Data Examples

We assume a very simple additive functional form for when there are N waves. Namely:

$$D(t, \boldsymbol{\theta}) = \sum_{i=1}^{i=N} a_i \, e^{b_i(t-\tau_i)}\{[1 + e^{\frac{1}{2}(b_i-d_i)(t-\tau_i)}]^{-2}\}, \tag{4}$$

where $\boldsymbol{\theta}$ comprises all the parameters σ, $\{a_i, b_i, d_i, \tau_i\}$ $(i = 1,2, \dots, N)$, and typically, but not invariably, the τ_i are in increasing order $\tau_1 < \tau_2 \dots < \tau_N$ indicating the (approximate) location of each of the maxima of the waves. We give two data samples where $D(t, \boldsymbol{\theta})$ might be fitted.

The first, Example 1, comprises the observed Daily New Cases of Covid-19 from 17 March 2022 to 3 November 2022. Figure 1 shows a chart of the sample.

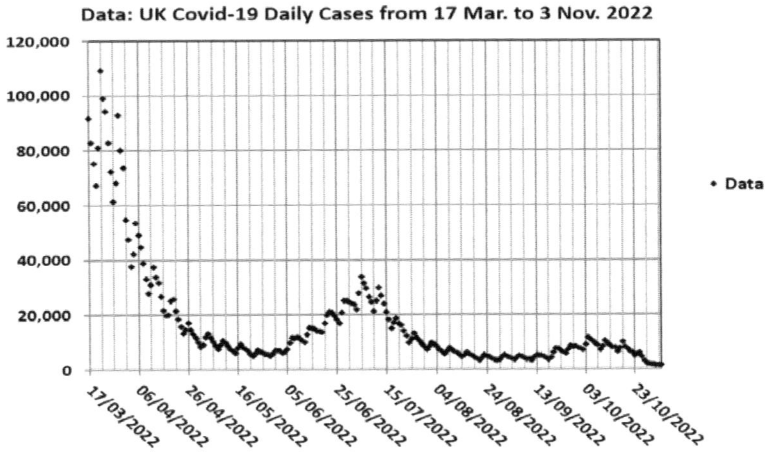

Figure 1. *Covid-19 observations from 17 March to 3 November 2022*

Example 2, comprises the observations of the global CO_2 level, in parts per million (CO_2 ppm), from 1800 to 2022. A plot of this sample is shown in Figure 2

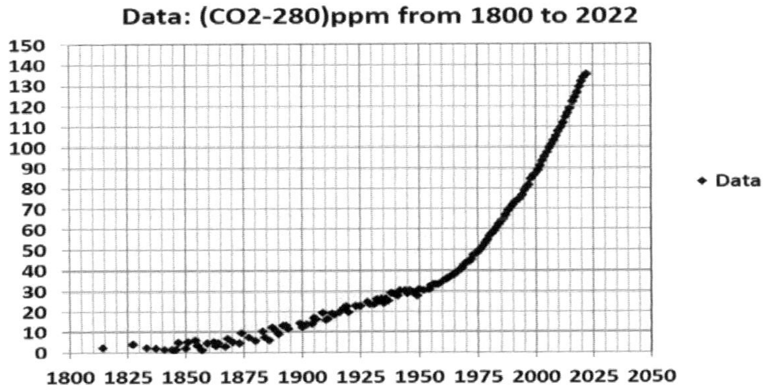

Figure 2. *CO_2 –282 ppm Level from 1800 to 2022*

There is a noticeable peak in the sample plot where a wave is particularly prominent, but this of course depends on the juxtaposition of the waves, so that where a wave peaks may not be obvious from the full sample plot. In Example 1, one might conjecture that there are three waves. In Example 2, the number of waves is not so clear if there are one or two waves.

We consider our proposed fitting method next.

3.2 The Basic Method

The proposed method of fitting $D(t)$ to data takes two stages.

In the first stage the data is grouped into separate sections, with each representing the position where the data is most dependent on a particular wave. The division process could be made automatic, but if speed is not important, and this is the case in our two examples, then it is simplest for the user to carry out this process by eye. As already mentioned, in Example 1 the number of waves looks likely to be three, whilst the choice is not so obvious in Example 2. In this latter case the initial slope variation suggests an earlier wave that becomes dominated by a second wave. Here more adjustment is needed to fix the position of the first wave and our final choice turns out to be two waves.

With a choice of the number of waves and their likely positions made, a single wave model is separately fitted to each section of the data. There are four parameters, a, b, d and τ for each section so that in Example 1, with three waves fitted, the total number of parameters is 13 as we assume the same standard deviation σ for all waves. Varying σ is possible, but there was no evidence this was needed in this case. Similarly for Example 2 the total number of parameters in the full fit is 9.

In the second stage the MW model of equation (4) is then fitted to the entire data set using, as the initial parameter estimates, the values obtained in the first stage for each of the single waves.

To find the maximum when using search methods like Nelder-Mead, a good choice of initial parameters is extremely important, particularly when the parameter count is high, when the search domain is high dimensional. The problem is particularly demanding when the function is multimodal. In our case, visual evidence of a good fit of a regression line to data, together with confidence level assessment, provides reassurance that our method is satisfactory.

In the next section we study the Examples in more detail.

4 FITTING METHOD

4.1 Covid-19 Example

The fitting problem is amenable to being tackled visually, and the general procedure is very flexible. There are no hard and fast rules and, depending on the data, adjustments are easy to make. Our simple overall approach was to divide the fitting into two steps. We consider Example 1.

As the first step we make a visual inspection of the sample. This suggests choosing 3 waves with waves 1,2 and 3 contributing respectively to data points in the ranges $S_1 = \{1, 86\}$, $S_2 = \{87,172\}$, $S_3 = (173,243\}$, where the sample size is $n = 243$. We then separately fit the single wave SL model of $D(t)$, as given in Equation (1), to each of the data sets S_j, $j = 1,2,3$. The fitting process is described in Cheng et al (2020), Section 2 and in Dye et al (2020), Supplementary Materials, and will not be repeated here. Figure 3 gives the results of this step.

The easy selection of likely parameter values, based on the shape and location of data sub-samples, allows charting tools to be developed for obtaining a good initial fit. Such initial values provide a starting point for the Nelder-Mead optimization. Charting tools enable the sensitivity of the fit to changes in parameter values to be examined at any stage of the fitting process; though this is not really necessary, as bootstrapping provides more reliable quantified evidence of fit.

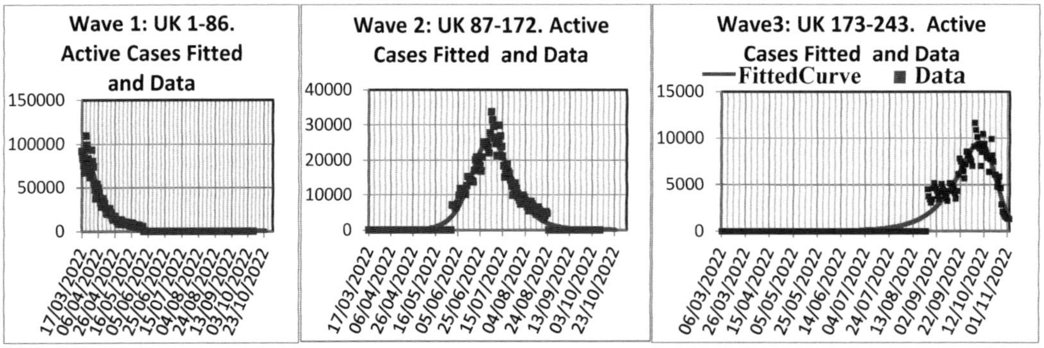

Figure 3. *The three fitted waves for the Covid-19 sample*

The fitted parameter values are given in Table 1 below.

Table 1: *Parameter estimates obtained by dividing full sample into 3 subsamples and fitting a single wave separately to each subsample.*

Parameters	1-86: Wave 1	87-172 Wave 2	173-243 Wave 3
σ	4490	1630	748
a	254000	101000	27500
b	0.215	0.103	0.0492
d	-0.0596	-0.0713	-0.174
τ	44629	44741	4485

Once the single waves have been fitted, we proceed to the second step which is to fit the N=3, Multi-wave $D(t)$ of Equation (4) taking $\theta = \{\sigma, a_1, b_1, d_1, \tau_1,\ a_2, b_2, d_2, \tau_2,\ a_3, b_3, d_3, \tau_3\}$ as the vector of initial parameter values. The subscripts correspond to the number of the fitted Wave.

The initial multi-wave $D(t)$ obtained, using the initial parameter values of Table1, is shown (in green) in the chart of Figure 4. Fitting a wave to each separate data sample gives good separate individual fits, but does not guarantee a good combined fit. In the present case the last segment of the overall curve, corresponding to Wave 3, is noticeably different from the fit to Wave 3 shown in the third plot of Figure 3. Nevertheless, the initial parameter estimates are a good starting point.

Figure 4 also shows that in our case the final fit (red line) obtained for the full sample, where it will be seen that the optimization greatly improves the overall fit.

The optimized parameter estimates are shown in Table 2.

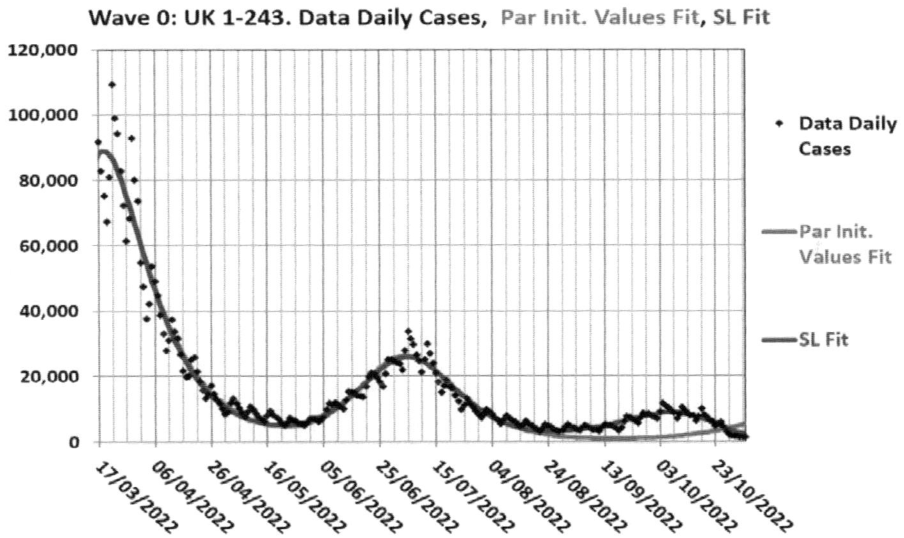

Figure 4. *The N = 3 Multi-wave fit to the full Covid-19 sample*

Table 2:
Parameter estimates obtained by an N=3 Multi-wave to the full Covid-19 sample

Parameters	$i = 1$	$i = 2$	$i = 3$
σ	4610		
a_i	27000	102000	28300
b_i	0.198	0.0817	0.0483
d_i	-0.0637	-0.0874	-0.135
τ_i	44630	44747	44852

4.2 CO_2 Example

Visual inspection of the data sample of Figure 2 suggests fitting two waves in Figure 5.

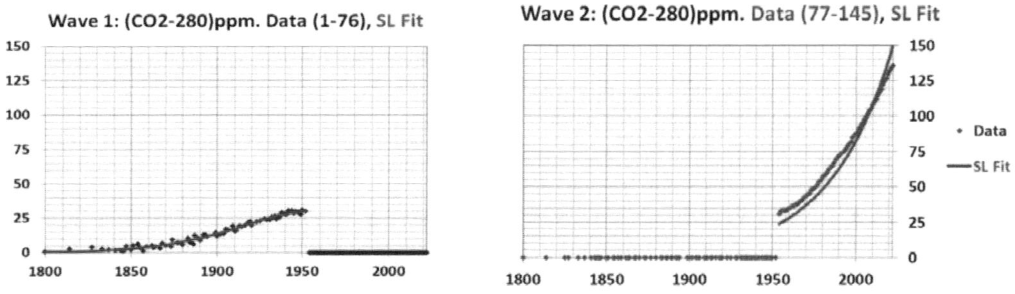

Figure 5. *The two Single Wave fits for the (CO_2 -280)ppm sub-samples: S_1=1-76, S_2=77-145.*

Because the SL wave rises from and then usually falls to zero, we have therefore subtracted 280ppm from the CO_2 observations, so that the lowest value is just above zero. This allows D(t) to achieve a good fit. This value could be treated as an unknown parameter, but given that this limit is little changed over the years little would be gained in estimating it.

Table 3: *Parameter estimates obtained by dividing full GlobalCO_2 sample into 2 subsamples and fitting a single wave separately to each subsample.*

Parameters	1-76: Wave 1	77-145 Wave 2
σ	1.46	8.66
a	111.0	675
b	0.0394	0.0269
d	-0.0196	-0.784
τ	1939.0	2078.0

Use of these single wave parameter estimates as initial values produces the *N=2* Multi-Wave fit shown in Figure 6 below.

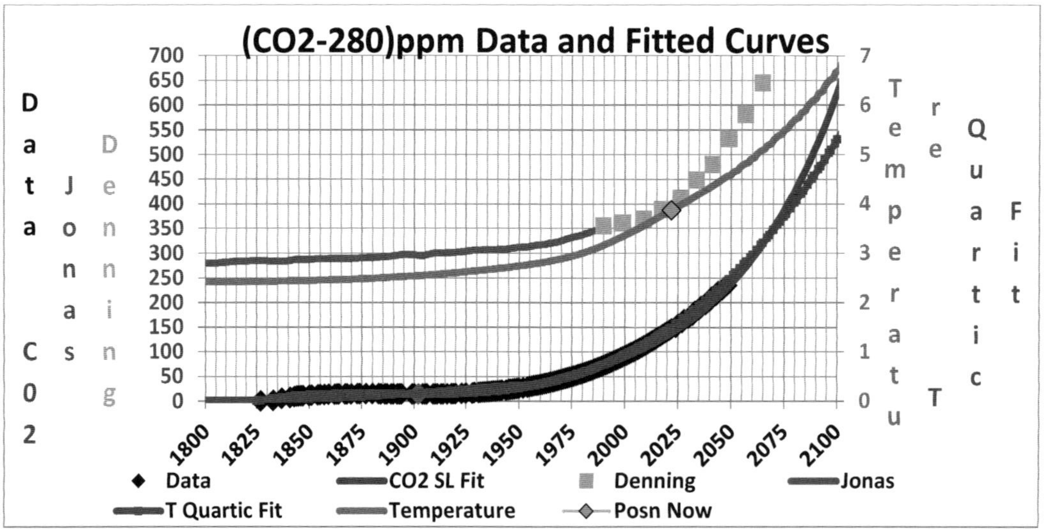

Figure 6. *The N=2 Multi-wave fit to the full Global CO_2 ppm sample and the corresponding estimate of the Temperature (light green), using a linear approximation with $D(t, \boldsymbol{\theta})$ the mantissa, estimated by the SL method, see Subsection 4.3. Also shown the fit using the T-Quartic method, the Jonas fit (the most optimistic) and the Denning estimate. The single point is the current Temperature.*

4.3 Temperature Calculation

In this subsection we explain the calculation of the Temperature curve from 1800 to 2022 plotted in Figure 6. We make the simple first order linear assumption that $T(t)$, the average global temperature in degrees Centigrade, is

$$T(t) = \alpha + \beta[C(t) - 280]. \tag{5}$$

From 1800 to 2022, $C(t)$, the global concentration level in ppm at time t increased by about 48% from 280 ppm to 420 ppm, an increase of (420-280) =140, whilst $T(t)$ increased by about 1.5 degrees to the present 13.9 degrees. We can set as the initial base value of $T(1800) = 0$, so that from Equation 5, we have

$$\alpha = 0 \text{ and hence } \beta = \frac{T(2022)-T(1800)}{C(2022)-C(1800)} = \frac{1.5}{140} = 0.0107 \approx 0.01.$$

The actual Temperature plot replaces $C(t)$ by $D(t, \widehat{\boldsymbol{\theta}})$.

Jonas derives a formula

$$RC_y = 5.35 * ln\left(\frac{C_y}{C_0}\right) - j\,[(T_0 + TC_y - 1)^4 - T_0{}^4]$$

showing the yearly change in CO_2 in terms of the corresponding change of year. Simply as a reflection of this quartic formula, but without using any of the considered reasoning given in (Jonas, 2015) we used a corresponding quartic polynomial giving the reverse calculation:

$$C(T) = a + b(T - f) + c(T - f)^2 + d(T - f)^3 + e(T - f)^4.$$

It will be seen in Figure 6 that the calculation gives a comparable result to the SL fit. The difference of the curve based on Equation (5) and the other fits arises from different base T values used as the initial year.

5 ERROR ESTIMATION

5.1 Bootstrap Analysis

We use bootstrap analysis to calculate confidence levels of the fitted quantities. The method is well-known (see for example, Cheng, 2017; Dye et al., 2020, Cheng et al., 2020), so full details are not repeated here.

In summary, to estimate confidence levels for $D(t)$, we generate a parametric bootstrap sample ,$E(t_i)$ i= 1,2,...,n, where

$$E(t_i) = D\left(t_i, \widehat{\boldsymbol{\theta}}\right) + \varepsilon_i,\ i = 1,2, \ldots, n$$

and $\widehat{\boldsymbol{\theta}}$ is the ML estimate of $\boldsymbol{\theta}$, and

$$\varepsilon_i \sim NID(0, \sigma^2),\ i = 1,2, \ldots, n,$$

is a random sample of mutually independent distributed, normal pseudorandom variables with variance σ^2.

This is carried out B times so that we have

$$\{\ E^{*(j)}(t_i) = D^{(j)}\left(t_i, \widehat{\boldsymbol{\theta}}\right) + \varepsilon_{(i)}^{(j)}, i = 1,2, \ldots, n\ \},\ j=1,2,\ldots,B$$

with the asterisk denoting that the observation is bootstrapped.

We can now estimate the parameters from each of the samples giving the bootstrap parameter estimates and bootstrap functions:

$$\widehat{\boldsymbol{\theta}}^{*(j)};\ \{ D^{*(j)}(t) = D(t,\widehat{\boldsymbol{\theta}}^{*(j)})\ \ i = 1,2,\ldots, n \},\ j=1,2,\ldots,B.$$

From these a confidence interval can be obtained for each parameter using the ranked values of that parameter.

Note however that this only gives the confidence interval level at each t_i separately. The confidence level is reduced if several different t_i are considered simultaneously. A more complicated method is needed using maximized likelihood regions. See Cheng (2017) for details. We could have used the simpler method in this paper, but instead we made a modification to model the dependence of an observation on previous observations. Specifically the observations are assumed to be a first order autoregressive process : Thus we set

$$E(t_i)=\alpha E(t_{i-1}) + \beta\varepsilon_{(i-1)},\ i = 1,2,\ldots, n.$$

The standard deviation of an observation is therefore not σ, but is s, which is approximately

$$s = \frac{\beta}{1-\alpha}\sigma$$

Thus values of $\alpha = 0.75, \beta = 0.25$, would make $s = \sigma$.

In the experiments reported here we actually varied the value of β slightly to remove possible bias in the estimate of σ. This is not of great practical consequence as our results are only for discussion and are not for use in practice. In any case our interest is centred on the SL parameters, and a well-known property is that the estimate of σ is asymptotically independent of the other parameters.

5.1 Covid-19 Bootstrap Results

Figure 7 illustrates the results of a representative set of twelve bootstrap scatterplots of pairs of parameters, obtained from $B = 50$ bootstraps. These indicate the typical scatter.

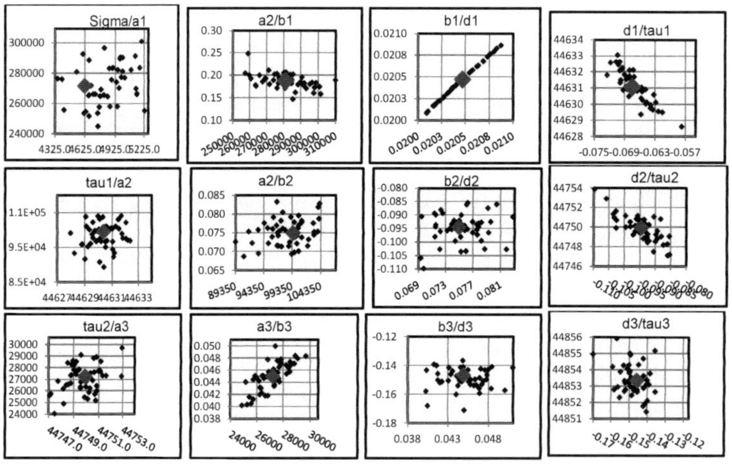

Figure 7. *Some Bootstrap Scatterplots*

Figure 8 gives shows the resulting $D(t)$ plots of the MLE and the Upper and Lower CI curves with confidence level 90%.

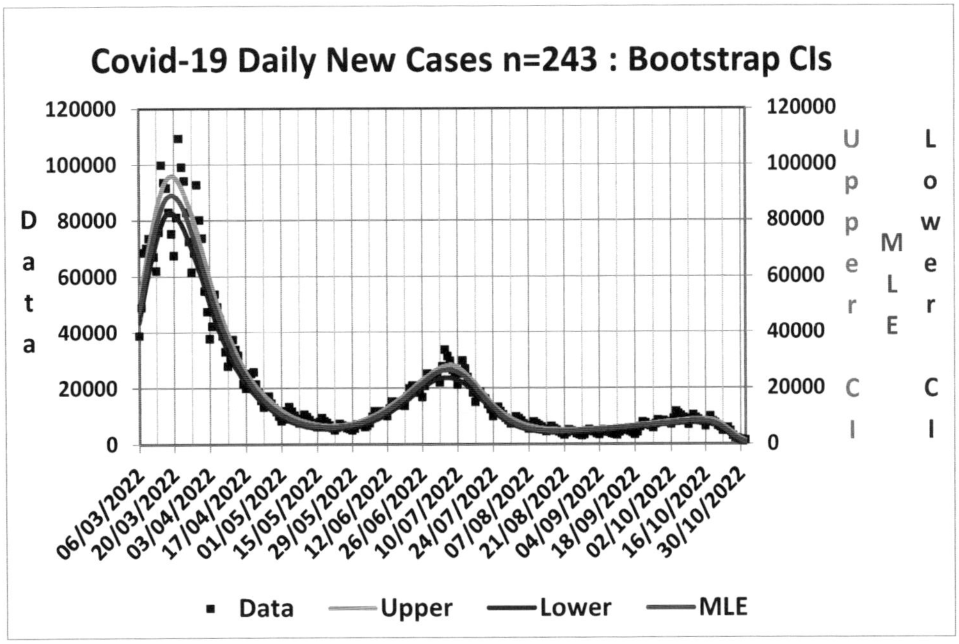

Figure 8. *The N=3 Multi-wave fit to the Covid-19 sample*

5.2 CO_2 Bootstrap Results

Figure 9 illustrates the results of a representative set of eight bootstrap scatterplot pairs of parameters obtained from $B = 50$ bootstraps. These indicate the typical scatter.

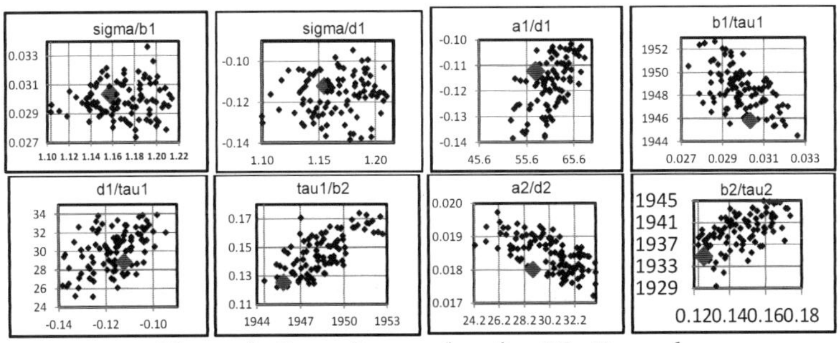

Figure 9. *Some Scatterplots for CO_2 Example*

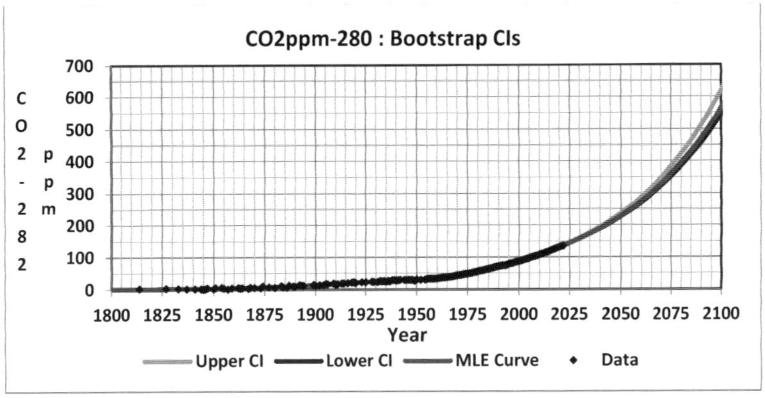

Figure 10. *The N=2 Multi-wave fit to the CO_2 sample*

6 IMPLICATONS OF CO_2 SCENARIO

To understand the impact that *Homo sapiens* has had on the earth's climate it is necessary to deal with very large numbers. The weight of the earth is 5.97×10^{24} kg of which 9×10^{19} kg is carbon in the earth's crust. The carbon in plants weighs 4.50×10^{14} kg very much less than in the earth's crust. The carbon in 1 ppm CO_2 in the atmosphere weighs 2.12×10^{12} kg. Before the industrial revolution, the concentration of CO_2 in the atmosphere fluctuated, between about 190 ppm and 270 ppm for 800k years, with a period of about 100k years, while the rises tended to be faster than the declines. The weight of carbon in the atmosphere was therefore 4.87×10^{14} kg, about the same as the weight of carbon in plants, fluctuating from about 4.03×10^{14} to 5.72×10^{14} kg.

Since the industrial revolution the concentration of CO_2 in the atmosphere is now about 415 ppm (Figure 6) as the result of burning fossil fuels, so that the weight of carbon has increased to 8.80×10^{14} kg or about twice weight of carbon in plants. The consequence of this is that the earth's atmosphere has already experience an increase in temperature of about 2°C (Figure 6) with the prospect that, if the concentration of CO_2 continues to increase, following the current trend, the temperature could increase by 6°C above preindustrial levels which would be catastrophic for our survival. We therefore need to sequester CO_2 on a massive scale.

In order to return to pre-industrial levels of atmospheric carbon and therefore temperatures, we need to remove about 4×10^{14} kg of carbon from the atmosphere or 1.5×10^{10} kg every day from now until the year 2100. The present world population is 8 billion. This means, on an individual basis, we each have to ensure that 2kg of carbon is returned to Earth each day.

The biomass of rain forests is 500 metric tons per hectare or 50 kilograms/m^2 so that one would have to expand the worlds rain forest, which currently occupies a 7 million square metres, by about 12k square meters every day until the end of the century. Efficient ways to sequester carbon from the atmosphere are desperately needed.

A recent discussion suggests that systematically scattering iron-rich dust onto target areas in oceans around the world could sequester perhaps 3×10^{13} kg carbon per year, or about 10^{11} kg per day, if the world's deep oceans were to be treated annually. Using a new method of carbon capture it seems that one could remove carbon at a cost of US\$0.5/kg of carbon which means that it would cost about 10^9 US\$ per day, every day from now until 2100 while the worlds GDP is 2.7×10^{10} per day. These studies suggest that it may be technically feasible to sequester carbon on the necessary scale but this would have to be done on an extraordinary scale and much more efficiently than is now possible. The long term survival of our species depends on it.

7 SUMMARY

The main purpose of this paper has been to illustrate how the SL function can be used to represent data with a multi-wave form; with the main geometric features of each wave easily related to the parameters of a single SL function. These single SL functions provide a good starting point to enable a multi-wave SL function to fitted to the full data sample.

Our Covid-19 example is such a case. Though not discussed here, the SL parameters can be reparametrized with transformed parameters representing infection levels and transition rates between he compartments of a compartment model, See Cheng et al (2020).

It would be of interest in further work to link the SL parameters in the CO2 example to parameters on which climate change depends.

We are investigating a problem in cognitive science where the multi-wave SL function is used to analyse data relating the speed of writing mathematical symbols with mathematical ability. We hope to report this elsewhere.

For further work, it would be of interest to examine the recent new area where the simulation process is the so-called Digital Twin of a real process (See for example Lichtenstern and Kerber (2022); for more examples go to https://informs-sim.org/wsc22papers/by_area.html). The multi-wave SL function could be used in the simulation twin that is used, to examine and optimize the performance of a real-time process so that it follows the best trajectory possible.

REFERENCES

Cheng R C H (2017). *Non-Standard Parametric Statistical Inference*. Oxford University Press, Oxford.Cheng R. C. H. and Williams B. G. (2022) Uses of the Skew-Logistic function for multi-wave functions. https://medrxiv.org/cgi/content/short/2022.12.19.22283694v1

Dye C, Cheng R C H, Dagpunar J and Williams B G (2020). The scale and dynamics of Covid- 19 epidemics across Europe. *R. Soc. Open Sci.* . **7**: 201726 https://doi.org/10.1098/rsos.201726

Jonas, M. (2015) The Mathematics of Carbon Dioxide, Part 1. https://archive.ph/FC7zE

Lichtenstern, I. and Kerber F. (2022). Data-based Digital Twin of an Automated Guided Vehicle System. https://informs-sim.org/wsc22papers/by_area.html#ptrack224

AUTHOR BIOGRAPHIES

RUSSELL CHENG retired from the University of Southampton in 2007 where he had been Head of the Operational Research Group, having held previous positions at Cardiff University and the University of Kent at Canterbury. https://www.southampton.ac.uk/maths/about/staff/rchc.page

BRIAN WILLIAMS is Senior Research Fellow at the South African Centre for Epidemiological Modelling and Analysis (SACEMA) having held the position of Epidemiologist at the World Health Organization from which he retired in 2008.

Proceedings of the Operational Research Society Simulation Workshop 2023 (SW23)
C. Currie and L. Rhodes-Leader, eds.
https://doi.org/10.36819/SW23.018

PLEDGES MODEL: AN INNOVATIVE TOOL TO MANAGE CARBON BUDGET DISTRIBUTION ACROSS THE EU27 MEMBER STATES.

Dr. Ilaria Perissi

Global Sustainability Institute
Anglia Ruskin University
183 East Road
CB1 1PT, Cambridge
ilaria.perissi1@aru.ac.uk

Prof. Aled Jones

Global Sustainability Institute
Anglia Ruskin University
Address183 East Road
CB1 1PT, Cambridge
aled.jones@aru.ac.uk

ABSTRACT

PLEDGES (Pledge Limits Evaluation for Decarbonization: Goals of the EU27 Strategy) is a new simulation-based tool that for the first time attempts to include in its scope the whole carbon budget distribution across all European Union member states (EU27). The model is built using System Dynamics and presents a radial structure which interconnects the EU27 carbon budget with each Member State (MS) budget. MSs are connected with each other, allowing for the assessment of emissions compensation strategies in terms of pledge quotas assigned to each country in the case of a deviation from an established "cap policy". In PLEDGES, the cap policy is implemented to comply with the Green Deal decarbonization objectives in 2030 (-55% emissions vs 2005) and 2050 (carbon neutrality). PLEDGES is easily customized, quick to respond and easy to use, which allows policymakers to review decarbonization policy and compensation strategies as frequently as necessary and to analyse their implication on MSs concerning the Green Deal objectives.

Keywords: carbon budget, system dynamics, green deal, decarbonization, modelling

1 INTRODUCTION

Each EU country has different barriers and opportunities for decarbonization. However, these are not exploited nor explored by the current EU Green Deal (EU climate neutrality by 2050) objectives or by the long-term mitigation targets established under the United Nations Conference of the Parties (COP) agreements. Current policies mainly focus on emission reduction targets and do not take into account cumulative emissions or the speed at which the transition to lower carbon can be achieved. Considering these factors are critical to the success of policies that can achieve a global temperature target, the PLEDGES model aims to fill this gap, providing EU policy-makers with a tool able to downscale EU and world climate mitigation policy to national levels. It sets a European carbon budget and allocates an allowance for carbon emissions to each EU region, according to a set of eligibility criteria based on an effort-sharing approach (Raupach *et al.*, 2014).

The model includes a number of novel features:

1) PLEDGES is uniquely able to explore the impact of mitigation actions of individual European countries, as the currently used tools, i.e. Convergence and Contraction (Meyer, 2007), C-ROADS (Climate Interactive, 2017) World Climate (Government of United Kingdom-Department for Business Energy & Industrial Strategy, 2016), MEDEAS (Solé *et al.*, 2020), use aggregate regions and do not give any information or data about individual country's carbon budget.

2) PLEDGES configures the European Union as a set of 27 GHG emissions stocks (the countries), whose flows must be regulated through the adjustment of socioeconomic variables (e.g. capability to spend GDP for mitigation, increasing investment in renewable energy, low carbon habits of citizen),

without exceeding a predetermined cap, in order to achieve the Green Deal objectives (-55% of emissions by 2030, carbon neutrality by 2050).

3) PLEDGES allows the reallocation carbon budgets between MSs in a dynamic way, within the context of Effort Sharing Regulation (ESR) and Emission Trading System (ETS) decarbonization strategies.

PLEDGES can still be modified and expanded, depending on the availability of new data or new information, but the current version provides a solid basis to serve as a framework for the European scale model.

As for point 3), PLEDGES incorporates carbon budget restrictions compatible with the 2 °C of Global Warming by 2050 commitment of the Paris Agreement. It is also set to accomplish the 55% package, assuring -55% of emission rate in ESR and -43% of emission rate in ETS by 2030 in comparison to 2005. These goals/limits are rarely considered together in the literature or in other modelling tools focused on European Union.

The results illustrate the potential of the model: the consideration of feedback and interrelations between submodules allows PLEDGES to respond to any deviation from the assumption of a yearly cut of emissions (for example, -55% emissions rate in 2005 by 2030) to include decarbonization scenarios with unexpected increases in emissions. For instance, in the case of an energy shortage due to the recent geopolitical situation in Ukraine (gas supply, fuel prices) resulting in short term changes to energy production sources including an increase in the use of coal.

Despite the innovative objective of simulating the carbon budget distribution and their temporal evolution in respect to the Green Deal objectives, the authors recognize that PLEDGES has limitations, especially regarding the lack of dynamics associated with the use of a carbon price. Nevertheless, in the present form, PLEDGES helps in planning compensation strategies at the national level, estimating the commitment that each MS should expect in the face of unexpected increases in emissions that might seriously compromise the overall EU decarbonization objectives.

2 METHODOLOGY: THE PLEDGES MODEL

PLEDGES has been designed by applying System Dynamics, which facilitates the integration of knowledge from different perspectives as well as the feedback from different subsystems. The PLEDGES model file is developed with Vensim Professional ® (2018) and released as ".mdl" file for proprietary software; a published and public version of the PLEDGES model for those users that do not have the proprietary version of Vensim will be available as ".vpm" which can be used with the Vensim Model Reader. Vensim Model Reader is an application that allows read-only access to models created with Vensim. Model Reader is free so that others can simulate and analyze the models without using proprietary Vensim itself.

PLEDGES simulations run from 2020 to 2050 and it has a radial tree structure constituted by the emissions budget assessed for EU as a whole and 27 sub-modules representing each member state. At the moment, PLEDGES has been structured with the main module (main window) connected to 27 sub-modules with a total of 28 windows, plus a display data window (Figure 1). Each submodule (national view) has been programmed with approximately 53 variables for each MS, using more than 1400 variables in total. The model consists of a modular and flexible structure, where each module (national view) can be expanded/simplified/replaced by another version or submodule and new modules can be easily added due to the radial structure. The model is informed by employing an excel file containing the input variables: Vensim invokes an excel file during the simulation set-up so that the model changes the variables' values or variables selection before the active simulation begins. This input file is called "histEU.xlxs".Detailed description of the input file will be reported after the following detailed descriptions of the model. The model script is reported in Appendix A.

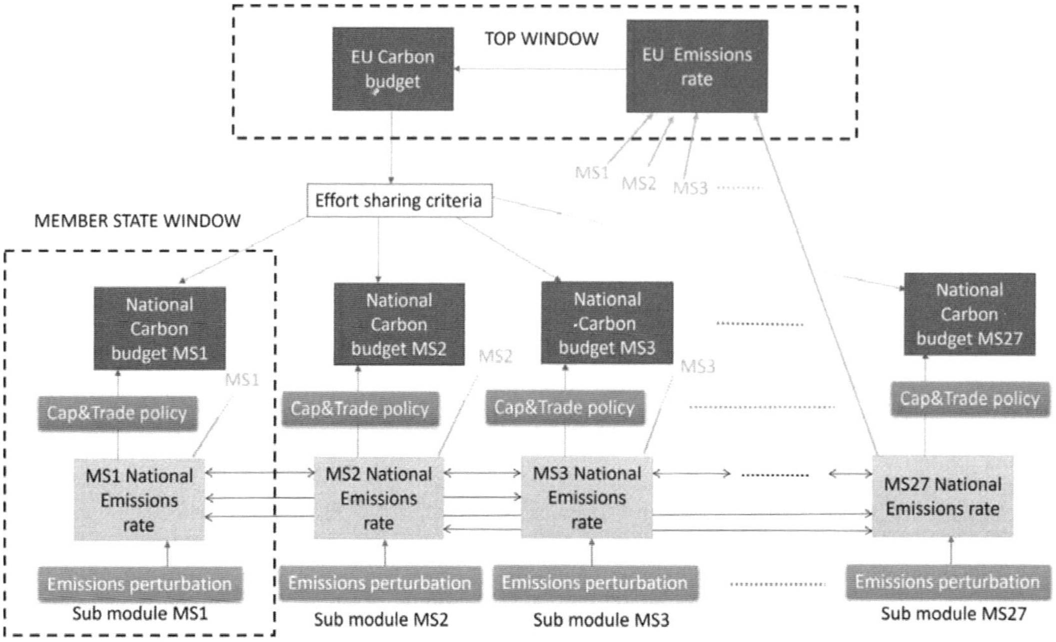

Figure 1. *Overview of the structure of the PLEDGES model.*

2.1 Top Window: EU27 as a whole

PLEDGES is structured as a top-down tool with a radial architecture, where the top level is represented by the Carbon Budget (EU27-CB) assigned to the European Union. The assessment of this carbon budget has been discussed in the recent work of Perissi and Jones (submitted to "Carbon Management"). In brief the EU budget was obtained by a linear projection of historical European Union emissions and constrained by 2030 and 2050 Green Deal objectives. This emission projection represents the optimal decarbonization pathway for EU27 and it is set as a "flow", and the ideal decay (decay 27) of the carbon budget level assigned to the stock called "EU27" in figure 2, which shows the upper window of the model.

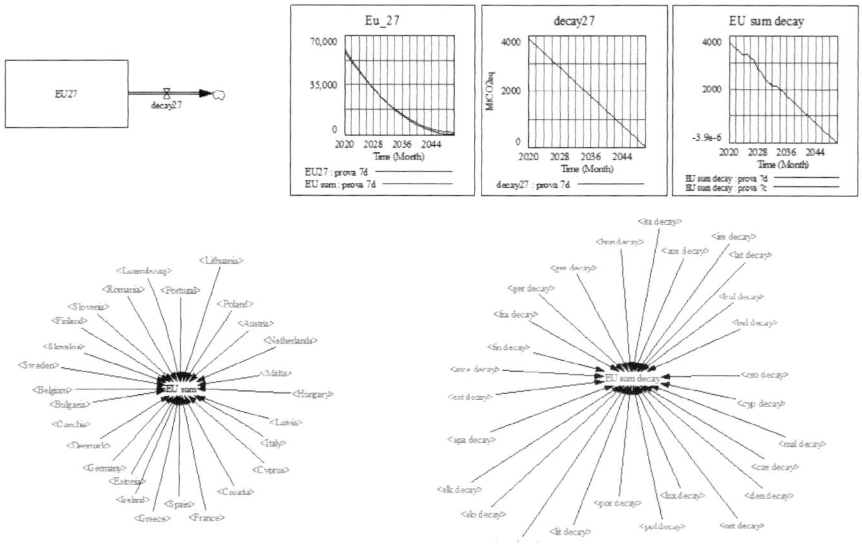

Figure 2. *Top window of PLEDGES concerning the carbon budget assignment to EU27 as a whole.*

The carbon budget declines linearly (EU_27 in Figure 2) and the sum of each MSs carbon budget is reported in each sub-module correspondent to each Member State. The top window in Figure 2 also shows the "EU sum decay" variable, which checks that the individual emissions flow from each country corresponds to the decay27 variable, except for any perturbation introduced by an unexpected event. Both EU sum and EU sum decay have no role in the dynamics, but are just references to check the model behaviour.

2.2 Member States windows

For each MS a dedicated window has been added as a submodule. At the present state of the art, within each MS it is possible to focus on 5 interconnected structures:

1) The EU27 budget distribution for that MS (stock)
2) The "emission rate" (flow) feedback
3) The emission rate policy choice
4) The perturbation tool in the policy
5) The distribution tool of perturbation among the MSs

The same structure is repeated for each MS. Figure 3 shows this structure for Belgium. A detailed description of each group of variables and their use is in the following paragraphs.

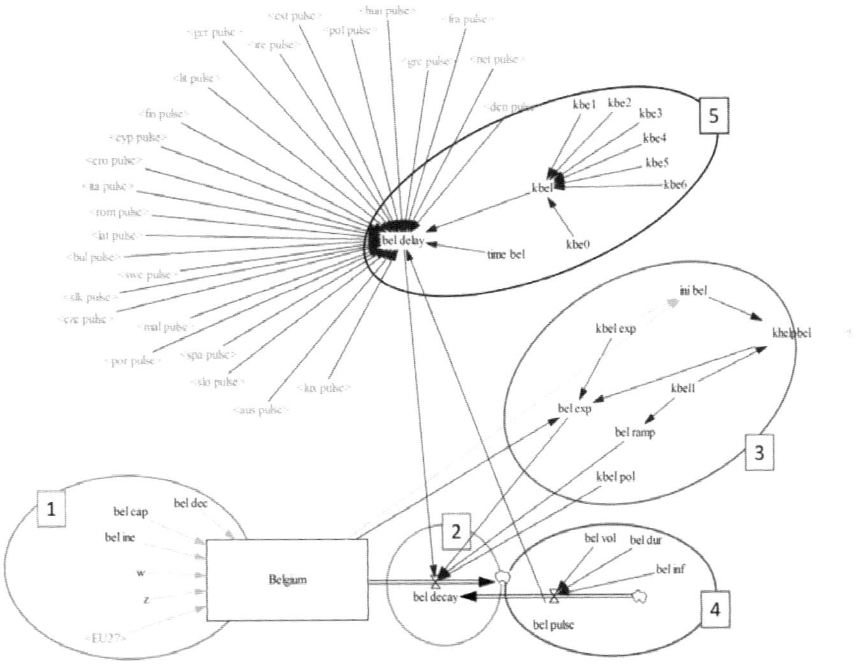

Figure 3. *Example of Member State's window in the PLEDGES-UL model (Belgium). The other member states have identical implementation.*

2.2.1 EU27 carbon budget distribution for a Member State (group 1)

Group1 variables assign a national carbon budget to the MSs, portioning the carbon budget set in the top window for EU27. Three shadow variables from the Top window represent the connection of MS's carbon budget with the EU27. The national quotas are obtained according to the effort-sharing approach, represented by the following equation:

$$(1-z)*(w*(bel\ dec*EU27)+(1-w)*(bel\ cap*EU27))+z*(bel\ ine*EU27) \quad (1)$$

Where z and w are parameters between 0-1; bel dec, bel cap and bel ini are the distribution terms respectively for decoupling, capability, and inertia (Perissi and Jones, 2022a). All those parameters are set in the histEU.xlxs file.

2.2.2 Member State emissions rate feedback (group 2)

The Top window of the model deals with the planned emission rate between 2020 and 2050. Here the national carbon budget (here for Belgium "bel decay") is set using feedback from the other MSs utilizing all the shadow variables called "bel delay". The delay variables aim to assign a recovery carbon quota due to an unexpected increase of emissions in one or more MSs, reshaping the MSs emission rate policy to achieve the EU27 carbon budget. The delay variables are discussed in group 5.

2.2.3 Member State emission rate policy choice (group 3)

Two ways are implemented in the model as a possible trend in national emission projections toward 2050 carbon neutrality: a linear decay (bel ramp) and exponential decay (bel exp). They can be selected through "kbel pol" in the excel file (kbel pol =1, exp; kbel pol =2, ramp). The variable "ini bel" sets the initial values for the exponential decay; kbel exp, kbel1 and khelpbel, are auxiliary variables to correctly set in "bel decay" the desired output curve.

2.2.4 Member State policy perturbation (group 4)

This part introduces a perturbation in the planned emission rate. This perturbation is intended to simulate an unexpected deviation from the planned decarbonization policy (the previously set ramp or exp decay). For instance, Germany recently reactivated coal-powered plants to produce electricity and preserve gas supplies due to the recent geopolitical situation in Ukraine. Or, oppositely, events like the pandemic brought a fast emission decrease, but a possible emission rebound must be considered and managed. This perturbation has been set as a "pulse", which Vensim builds as a square wave. The pulse magnitude and duration are set utilizing three parameters: bel dur, the pulse duration; bel inf, the year the pulse starts; and bel vol, the amplitude of the pulse.

2.2.5 Perturbation distribution tool among the Member States (group 5)

A perturbation happening in a certain country (for example, Germany in figure 4), is quite unlikely to be immediately compensated for by the other MSs. For this reason, a delay time is set in the model to simulate the time necessary to reduce MS emissions to compensate for the extra emissions. A portion of this delayed compensation is also given to the country where the perturbation takes place.

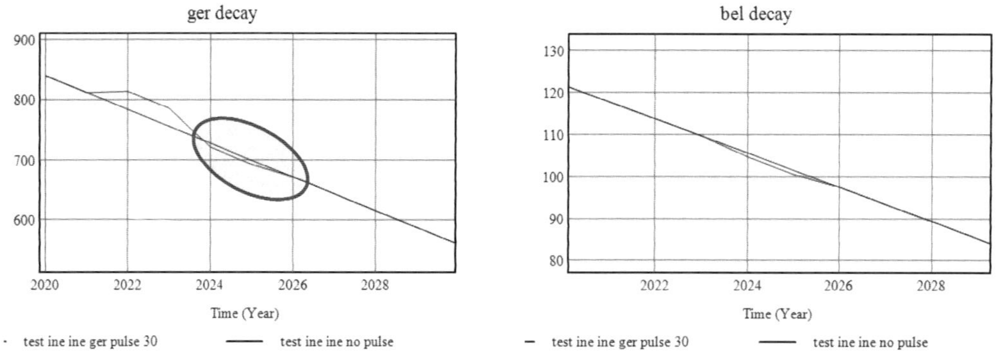

Figure 4. *Perturbation on Germany(30 Mtons/year 2022 and 2023) emission rate and example of emissions recovery from other countries (here shown Belgium) after a 2 years delay (2024 and 2025).*

Each country window incorporates the perturbation from another country: for instance, in the Belgium model, group 5 assesses how Belgium will recover of a portion of the perturbation in Germany (with a delay). The portion is assigned with the same criteria used for the carbon budget: inertia, capability, decoupling and a customized distribution, with the use of the parameters kbe1, kbe2, kbe3, kbe4 and kbe5 and kbe6. Particular attention is needed for the kb5 and kb6, which assign a capability to pay (kb5) or a decoupling (kb6) criteria nut not with the same rationale as the carbon budget. Indeed, it is important to underline that the capability index for Carbon Budget assigns less carbon budget to those countries that are richer or have a good economic decoupling. Here, the perturbation is redistributed with the capability to pay or to a decoupling index criterion according to the principle that the higher the GDP, the higher the ability to pay and therefore the higher the portion of perturbation to be assigned. Finally, Kbe0 selects which of the previous criteria (inertia, capability, decoupling) will be used. The delay is set by means of "bel delay", where it is possible to specify the year in which the recovery will start.

3 PLEDGES MODEL IN PRACTICE: EXAMPLE OF SIMULATION

3.1 Dealing with an unexpected increase in emissions

Despite the effort put into accelerating the energy transition to renewables and a low carbon economy, a sudden increase in emissions is more frequent than hoped. Quite recently, for instance, the decision to shut several large nuclear reactor sites in Germany (by 2022) after Fukushima disaster, has led to a resurgence in gas and coal use. Moreover, the recent geopolitical situation in Ukraine (Pereira *et al.*, 2022), slowed down gas consumption almost everywhere across the EU: again, Germany fired up coal plants as Russia turns down the gas (Germany is highly reliant on Russian gas).

On the other hand, those increases stimulated measures like the recent agreement signed by Member States on a voluntary reduction of natural gas demand of 15% for winter 2022-2023 (European Council, 2022). This reduction however, is across all MSs and does not guarantee emissions reduction as some countries could resort to relying on coal or other fossils.

All the previous examples represent some real cases that evidence the need for a flexible tool to manage emissions over all sectors and especially over all EU27 countries. An overview of all the MSs positions toward the decarbonization path (Perissi and Jones, 2022b) while dealing with an unexpected increase in emissions is vital to properly plan future decarbonization policies.

For example, the following simulation designs a short-term scenario where coal stations are turned on, with a specific objective, as for Germany, to reduce gas dependence. The Agora Energiewende recently (Meza, 2022) estimated 20 to 30 million tons of additional emissions over the whole year. Hypothesizing that Germany and the EU27 need 2 years to plan alternative provisioning, the impact on decarbonization policy will be 60 MtonCO2eq of extra emissions to be recovered. The following recovery scenarios have been explored under an effort-sharing approach (inertia, capability/decoupling, customized). In all these simulations:

- The EU27 carbon budget is split among the MSs according to the inertia term, which is the relevant criteria driving the MSs emissions footprints in EU
- The decarbonization caps in compliance with the Green Deal objectives are set as a linear decay from 2019 historical emissions and 0 emissions in 2050
- An increase of emission (perturbation) was simulated as a "pulse" with an amplitude of 30 MtCO2eq according to the Agora Energiewende assessment, starting in 2022 and finishing in 2024 (2 years), as the estimated time necessary to propose and implement new actions
- The delay to emissions recovery is set to start in 2024 for each MS, and the recovery measure was set to last 2 years.

3.2 Inertia-based criteria for distribution of emissions perturbation

In this first simulation, inertia criteria allocate the MSs carbon budgets and also redistribute the perturbation across all the member states. The coal stations turning on is supposed to be an extraordinary and transitory measure to lower the dependence on natural gas and is considered to last for 2 years while other measures are implemented to turn the coal stations off again in 2024. By hypothesis, the increase of emissions due to coal firing in Germany is recovered in the two years after the coal stations are turned off, i.e. 2024 and 2025. The results of these settings are reported in figure 5. The variable EU sum decay aims to work as a comparison variable between the desired cap trajectory (in black) and the results from the unexpected perturbation (in blue). While the increase of the emissions is only in Germany (red circle), the recovery part (red dashed circle) is the result of splitting the previous peak in smaller quotas redistributed as a decrease in the caps of all MSs between 2024 and 2025.

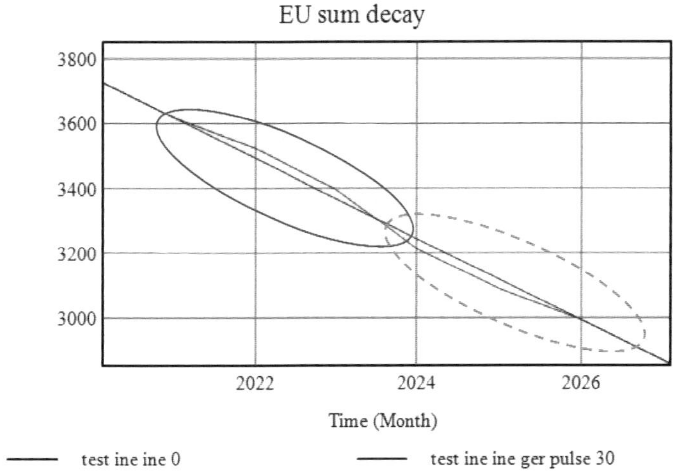

Figure 5. *In black: EU27 caps trajectory according to the Green deal. In blue: perturbed trajectory, first increase due to Germany coal station turning up in 2022 and 2023 and planned recovery of emission in 2024*

In this simulation, PLEDGES splits the recovery among MSs on an inertia basis, a criterium provided by the inertia index defined in Perissi and Jones (2022a). Inertia criteria reflect the emissions burden of the country: the higher the historical emissions the higher the quota assigned to be recovered. This accounts for proportionality between the burden and the capability to recover the burden itself: a country that has a low emissions impact cannot recover a quota as high as the one for a bigger emitter. In principle, this also spurs a faster emission reduction for those countries that are big emitters. The quotas obtained by perturbation redistribution are assessed in each country's delay variable, whose value is available by exporting the simulation from Vensim dataset.

This is just one of the possible scenarios, where all the MSs plan a compensation strategy at the same time between year 2024 and year 2026. Other scenarios are possible, some countries might delay their compensation strategy for a few years, or they might pay off their part immediately if they have saved allowances. However, even though Inertia based split represents a reliable first assessment criterium to distribute pledges quotas, it does not consider other important contributions that can be provided by the economic situation of each country, in terms of "ability to pay for mitigation".

3.3 Capability to pay and economic decoupling criteria for distribution of emissions perturbation

In this simulation, the economic parameters are considered to allocate perturbation quotas across all the MSs. Economic parameters are "capability to pay" and "economic decoupling". As a brief reminder, capability to pay is assessed using a capability index based on countries' GDP, which assigns less carbon budget to those countries that have a higher ability to pay. But here the logic of capability is the opposite: the higher the GDP, the higher the portion of perturbation to be assigned. This is done by defining the following index

$$CI_{pertMSj} = \frac{\frac{GDP_{capMSj}}{GDP_{capEU}}}{\sum_{1}^{j} \frac{GDP_{MSi}}{GDP_{capEU}}} \quad (2)$$

where GDP_{capMSj} is the GDP capita per MS, GDP_{capEU} is the GDP capita in EU27 (CI states for "Capability Index").

This index (evaluated with the last available data for GDP in 2019, applied by variable kxx5) suggests that bigger quotas of perturbation should be compensated by those countries that are richer (higher GDP). Results of this simulation show a different distribution (figure 6, blue columns) of the recovery quotas based on capability criteria in comparison to the one obtained with the inertia criteria (Figure 6, orange columns).

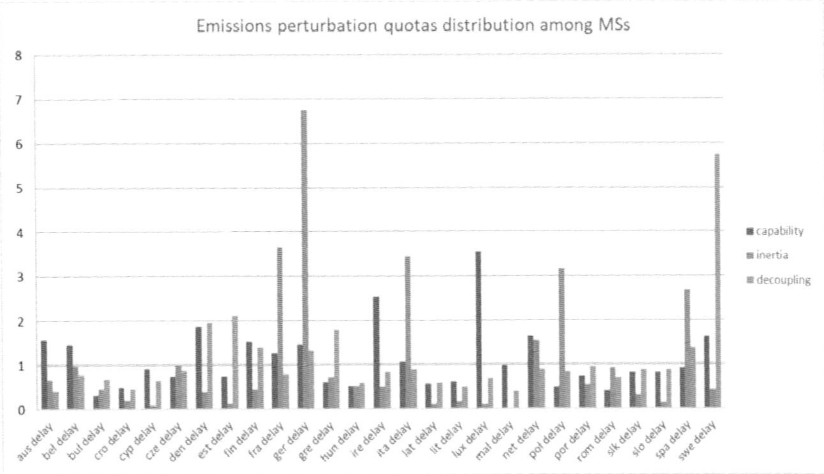

Figure 6. *Recovery quotas for MSs according to inertia and capability to pay criteria established with the PLEDGES model.*

In the same manner, the degree of decarbonization of the economy must be considered to establish the "level of carbonization" of the capability to pay. This assessment can be carried out with the decoupling index by Perissi and Jones (Perissi and Jones, 2022a). As for the capability, decoupling-based recovery quotas must be portioned considering that the higher the decoupling, the higher the pledge quota that can be assigned to the countries whose economy is less carbonized. Thus, a revised index for portioning decoupling-based pledge is defined as follow:

$$DI_{perMSjdec} = \frac{\frac{DI_{msj}}{DI_{EU27}}}{\sum_1^j \frac{DI_{msj}}{DI_{EU27}}} \quad (3)$$

where DI_{MSj} is the Decopuling Index per MS. This is applied using variable kxx6 (with data from 2019).

The final split of quotas among the MSs by inertia capability and decoupling criteria, is summarized in figure 6. The results show very different allocations based on the capability to pay and decoupling in comparison to the inertia distribution, showing also that small emitter can account for a good capability to pay for mitigation and a good decoupling in comparison to the bigger emitters whose main drive of decarbonization pledges must be proportioned to their historical emissions (inertia). Indeed, splitting the EU27 budget based only on the Economic term would dramatically lower carbon budgets for those countries, creating higher costs which may not be feasible. The analysis of MSs in respect to their inertia versus economic factors, results in the following groups:

- *inertia is higher than both capability and decoupling quotas*: Czechia, France, Germany, Italy, Poland, Romania, Spain

The pledge assigned by inertia is dominating, this is the group of the bigger emitters in EU27; except for Czechia which has a comparable economic pledge, the other countries should provide compensation quotas based mainly on inertia; Czechia could even pledge a bit more due to its good decoupling and capability to pay quotas, assessed as the same order of magnitude of the inertia pledge.

- *inertia is lower than capability higher than decoupling quotas* Austria, Belgium, Netherland; *and inertia is lower than decoupling but not of capability quotas*: Bulgaria, Greece, Hungary

These second groups includes countries where one of the two economic capability pledges (capability or decoupling) is higher than the inertia pledges. This suggests that even the inertia predominates as initial quotas, a further pledge can be evaluated and charged due to their good economic ability (5-10% more, for instance).

- *inertia is lower than both decoupling and capability quotas* Croatia, Cyprus, Denmark, Estonia, Finland, Ireland, Latvia, Lithuania, Luxembourg, Malta, Portugal, Slovakia, Slovenia and Sweden

These countries show a lower historical burden and higher potential pledges due to their economic factors (evaluated in 2019). Provided that the inertia burden is the starting pledge, all these countries should be pledged more than that established on an inertia basis (10-25% more, for instance).

Thus, at this stage of the simulation, we can assess that extra emissions experienced from Germany in 2022-2023 can be potentially recovered by other MSs (and partially by Germany too) between 2024-2025 based on pledges distributed mainly according to inertia criteria as a first assessment.

3.4 Customized criteria for distribution of emissions perturbation

The revised distribution can be simulated and verified with a "customized quota setting" in the input file "histEU27" at the variable kxx4. This is an empty column that can be filled by the user with customized indexes, constituted by a weighted mix of the previous indexes (inertia, capability, decoupling) or with index and criteria beyond those indexes (provided that the sum over all the portioning indexes will be equal to 1). Sticking with the previous example, a reviewed distribution of the initial inertia quotas can be assessed as follow:

- Group1: quotas are set equal to inertia index (kxx2).
- Group2 and Group3: quotas are evaluated using the inertia index increased by 5%
- Group 4: quotas are evaluated using the inertia index increased by 10%

The obtained recovery quotas are in Table 1 and shown in Figure 9.

Table 1. Recovery quotas for the MSs are based on effort sharing approach.

Recovery quotas	**capability**		**inertia**		**decoupling**		**customized**	
	2024	2025	2024	2025	2024	2025	2024	2025
aus delay	1.571651	1.571651	0.663488	0.663488	0.413948	0.413948	0.696663	0.696663
bel delay	1.460744	1.460744	0.976877	0.976877	0.788614	0.788614	1.025721	1.025721
bul delay	0.309557	0.309557	0.454407	0.454407	0.66556	0.66556	0.477127	0.477127
cro delay	0.479427	0.479427	0.194109	0.194109	0.463178	0.463178	0.21352	0.21352
cyp delay	0.915685	0.915685	0.079239	0.079239	0.642283	0.642283	0.087163	0.087163
cze delay	0.741954	0.741954	0.998551	0.998551	0.868162	0.868162	0.998551	0.998551
den delay	1.873135	1.873135	0.379704	0.379704	1.943943	1.943943	0.417675	0.417675
est delay	0.734584	0.734584	0.119525	0.119525	2.115486	2.115486	0.131477	0.131477
fin delay	1.524621	1.524621	0.446219	0.446219	1.405869	1.405869	0.490841	0.490841
fra delay	1.265253	1.265253	3.645894	3.645894	0.776994	0.776994	3.828189	3.828189
ger delay	1.467061	1.467061	6.730932	6.730932	1.33178	1.33178	6.730932	6.730932
gre delay	0.59981	0.59981	0.718638	0.718638	1.790744	1.790744	0.75457	0.75457
hun delay	0.524703	0.524703	0.523354	0.523354	0.57317	0.57317	0.549521	0.549521
ire delay	2.539631	2.539631	0.505993	0.505993	0.832061	0.832061	0.556592	0.556592
ita delay	1.055723	1.055723	3.453004	3.453004	0.891662	0.891662	3.453004	3.453004
lat delay	0.562257	0.562257	0.093236	0.093236	0.57317	0.57317	0.102559	0.102559
lit delay	0.613849	0.613849	0.166254	0.166254	0.50262	0.50262	0.182879	0.182879
lux delay	3.540952	3.540952	0.100652	0.100652	0.682946	0.682946	0.110717	0.110717
mal delay	0.976403	0.976403	0.021588	0.021588	0.393	0.393	0.023747	0.023747
net delay	1.645355	1.645355	1.544881	1.544881	0.886455	0.886455	1.622125	1.622125
pol delay	0.487851	0.487851	3.157928	3.157928	0.82172	0.82172	3.157928	3.157928

por delay	0.731425	0.731425	0.54533	0.54533	0.956187	0.956187	0.599863	0.599863
rom delay	0.404319	0.404319	0.916454	0.916454	0.705813	0.705813	0.916454	0.916454
slk delay	0.813201	0.813201	0.322078	0.322078	0.871321	0.871321	0.151158	0.151158
slo delay	0.813201	0.813201	0.137416	0.137416	0.871321	0.871321	0.151158	0.151158
spa delay	0.927267	0.927267	2.674605	2.674605	1.378723	1.378723	2.808335	2.808335
swe delay	1.628157	1.628157	0.429645	0.429645	5.752713	5.752713	0.472609	0.472609

No decrease in the revised inertia index distribution, simply means that with these new quotas, the compensation will be overachieved, leaving unused allowances to be saved for the future. If the plan will be accomplished by all the MSs, the unexpected increase of 60 MtCO2eq over two years in Germany will be then compensated with around 61.8 MtCO2eq, leaving around 2 MtCO2eq as reserves. Moreover, inertia criteria are not exhaustive in describing for instance the condition of MSs that are experimenting with stable emissions or even an increase in respect to the historical trend. This is the case of the recent post-pandemic condition: almost all EU27 countries have an emissions increase so the pledge must be redistributed and traded considering also MSs emissions projections. This is also possible more frequently than on a yearly plan.

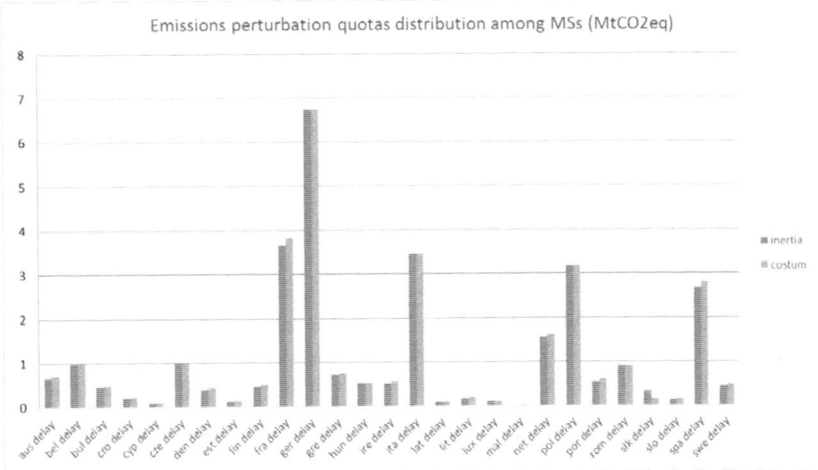

Figure 7. *Customized compensation quotas distribution using customized indexes based on inertia principle revisited under the light of the MSs economic features (ability to pay and decoupling)*

4 CONCLUSION

The PLEDGES model is a new simulation based tool that for the first time attempts to include in its scope the whole carbon budget distribution across all EU27 member states. The model is built with a radial structure which interconnects the EU27 carbon budget with each Member State. Member states are also connected to each other, allowing for the simulation of emissions compensation strategies in terms of pledge quotas to be assigned to each country in the case of a deviation from an established "cap policy". In PLEDGES, the cap decarbonization policy is implemented to comply with the Green Deal decarbonization objectives (-55% emission in 2030 from 2005, carbon neutrality by 2050).

In the vision to share emissions burden and plan emissions compensation strategy, the model adopts an "effort sharing" approach based on a mixture of inertia, capability to pay and the degree of MSs economy carbonization (decoupling). As an example, the impact of the recent turning on of coal stations in Germany has been simulated and studied with PLEDGES and a compensation strategy is released in terms of recovery quotas to be assigned and potentially traded to each MS (including Germany) for the next few years.

The simulation shows, at present, that the inertia principle plays a major role in distributing pledges, especially for those countries which have low historical emissions burden. The economic factors (capability and decoupling), that were evaluated as further effort-sharing principles cannot yet be equally implemented, they can mainly be used to adjust the inertia distribution. However, the inertia criterium alone shows limits, especially in the present context of emissions rebounding due to post Covid19 pandemic increase in economic activity.

Despite the complexities in managing emission scenarios and decarbonization policy, PLEDGES is easily customized, quick to respond and easy to use, which allows the review of possible decarbonization policy and compensation strategies as frequently as necessary to analyze their implication on MSs concerning the Green Deal objectives.

ACKNOWLEDGMENTS

This research was funded by PLEDGES project European Union's Horizon 2020 MSCA IF GA ID: 101023109.

A APPENDICES

PLEDGES code

REFERENCES

Climate Interactive (2017) *C-ROADS*, *Climate Interactive*. Available at: https://www.climateinteractive.org/tools/c-roads/.

European Council (2022) *COUNCIL REGULATION on coordinated demand reduction measures for gas 2022/0225 (NLE)*, *Consilium Europa*. Available at: https://www.consilium.europa.eu/en/press/press-releases/2022/07/26/member-states-commit-to-reducing-gas-demand-by-15-next-winter/.

Government of United Kingdom-Department for Business Energy & Industrial Strategy (2016) *UK carbon budgets*. Available at: https://www.gov.uk/guidance/carbon-budgets.

Meyer, A. (2007) 'Contraction and convergence', *Global Commons In* [Preprint].

Meza (2022) *Resurgent coal power could cause 30 million tonnes of extra emissions in Germany – think tank*, *Clean Energy Wire*. Available at: https://www.cleanenergywire.org/news/resurgent-coal-power-could-cause-30-million-tonnes-extra-emissions-germany-think-tank#:~:text=Estimates%20from%20Berlin%20think%20tank,first%20half%20of%20the%20year

Pereira, P. *et al.* (2022) 'Russian-Ukrainian war impacts the total environment', *Science of The Total Environment*, 837, p. 155865. Available at: https://doi.org/10.1016/j.scitotenv.2022.155865.

Perissi, I. and Jones, A. (2022a) 'Influence of Economic Decoupling in assessing carbon budget quotas for the European Union'. arXiv. Available at: https://doi.org/10.48550/ARXIV.2211.11322.

Perissi, I. and Jones, A. (2022b) 'Investigating European Union Decarbonization Strategies: Evaluating the Pathway to Carbon Neutrality by 2050', *Sustainability*, 14(8). Available at: https://doi.org/10.3390/su14084728.

Raupach, M.R. *et al.* (2014) 'Sharing a quota on cumulative carbon emissions', *Nature Climate Change*, 4(10), pp. 873–879. Available at: https://doi.org/10.1038/nclimate2384.

Solé, J. *et al.* (2020) 'Modelling the renewable transition: Scenarios and pathways for a decarbonized future using pymedeas, a new open-source energy systems model', *Renewable and Sustainable Energy Reviews*, 132, p. 110105. Available at: https://doi.org/10.1016/j.rser.2020.110105.

AUTHOR BIOGRAPHIES

ILARIA PERISSI received a BSc (Hons) Physical Chemistry from the University of Florence (IT) in 2001. She completed her PhD in Material Science at the same University in 2009. She is currently a Marie Curie International Fellow at the Global Sustainability Institute of Anglia Ruskin University Dr Perissi has an extensive background in this field for the past 6 years: https://aru.ac.uk/people/ilaria-perissi

ALED JONES is the inaugural Director of the Global Sustainability Institute at Anglia Ruskin University and Professor of Sustainability. His research focuses on risks and opportunities in finance and governments from global resource trends such as food, energy and water: https://aru.ac.uk/people/aled-jones

Proceedings of the Operational Research Society Simulation Workshop 2023 (SW23)
C. Currie and L. Rhodes-Leader, eds.
https://doi.org/10.36819/SW23.019

LOCATION OF CHIEF POLICE OFFICERS IN THE STATE OF PERNAMBUCO, BRAZIL, USING OCBA

M.Sc. Yifu Wei

University of Leeds and University of Manchester
Woodhouse, Leeds LS2 9JT, UK
pmywe@leeds.ac.uk

Dr. Carlos Lamas-Fernandez

University of Southampton
Southampton Business School, University Rd,
Highfield, Southampton SO17 1BJ, UK
c.lamas-fernandez@soton.ac.uk

Dr. Walton P. Coutinho

Federal University of Pernambuco
Department of Technology,
Avenida Marielle Franco, s/n,
Nova Caruaru, Caruaru 55014-900, Brazil
walton.coutinho@ufpe.br

Dr. Christine Currie

University of Southampton
School of Mathematics, University Rd,
Highfield, Southampton SO17 1BJ, UK
christine.currie@soton.ac.uk

ABSTRACT

Over the past few decades, the escalation of violence made public security take centre stage on the Brazilian political and social agenda. This paper focuses on the Brazilian state of Pernambuco, one of the most violent states in Brazil according to national indexes. We study a location problem originating from the interaction between different police institutions. We aim at determining the location of chief civil police officers responsible for crime registration and investigation. This work applies Discrete Event Simulation (DES) in order to model the reporting process of the police when criminal activity happens. The DES model is used as a basis for an Optimal Computing Budget Allocation (OCBA) algorithm in order to improve the overall simulation efficiency for finding the best locations. Our numerical experiments show that the proposed framework could potentially help increase the operational efficiency of police forces in Pernambuco's cities.

Keywords:

Public security, Location problem, Simulation-optimisation

1 INTRODUCTION

Public security is a major factor of concern for any government as it involves the protection of citizens, organisations, institutions and societies. Over the past few decades, the escalation of violence made public security take centre stage on the Brazilian political and social agenda. As the largest country and the major economy in South America, Brazil faces serious public security problems. In 2021, according to the Brazilian Forum on Public Security (BFPS), there were 22.3 Intentional Lethal Violent Crimes (ILVC) per $100,000$ Brazilian citizens, which include intentional homicides, robbery followed by murder, grievous bodily harm followed by death and deaths during police intervention (BFPS 2022).

This article focuses on the Brazilian state of Pernambuco. With an estimated population of 9.6 million people, this is the second most populated state of Brazil's north-east region. Pernambuco is made up of 185 municipalities (cities and towns) of which only around 20% have a population of 50 thousand people or higher. Figure 1a shows the demographic distribution per city/town in Pernambuco as of the most up-to-date 2010 population census. One can observe that only a small portion of cities in Pernambuco have a high population concentration.

In 2021, Pernambuco registered 33.85 ILVCs per 100,000 citizens, which is considerably higher than the national average (SDSPE 2022a). In contrast, the amount of governmental funding to combat violent crime is insufficient (BFPS 2022), with Pernambuco reportedly having only 16,422 police officers, roughly 39% less than the necessary minimum fixed in the state's Ordinary Law 12.544 (Brasil 2004).

Faced with a lack of investment and tight budgets, the state's police forces must seek operational efficiency in order to serve society in a satisfactory way. In this paper, we study a location problem originating from the interaction between different state police institutions. More specifically, we aim at determining the location of chief civil police officers, responsible for crime registration and investigation, in Pernambuco's territory such that the protection of its cities is optimised, subject to uncertainties in travel times, service times and the incidence of crimes. A more detailed definition of this problem will be provided in Section 2.

This work applies Discrete Event Simulation (DES) in order to model the reporting process of the police when criminal activity happens. The DES model is used as a basis for an optimization framework that decides the best location of chief police officers. Furthermore, we employ an Optimal Computing Budget Allocation (OCBA) algorithm in order improve the overall simulation efficiency for finding the best locations (Li et al. 2017). Computational experiments are performed using an open data set of ILVCs in the state of Pernambuco as a case-study. Our numerical experiments show that the proposed framework is capable of efficiently locating chief police officers and could potentially help increasing the operational efficiency of police forces in Pernambuco's cities.

The remainder of this article is organised as follows. In Section 2, we introduce a detailed definition of the problem, discussing its challenges. In Section 3, we review the literature related to our problem definition. Section 4 presents the proposed solution method and in Section 5 we discuss the results of our numerical experiments. Finally, Section 6 concludes this paper and presents some future research directions.

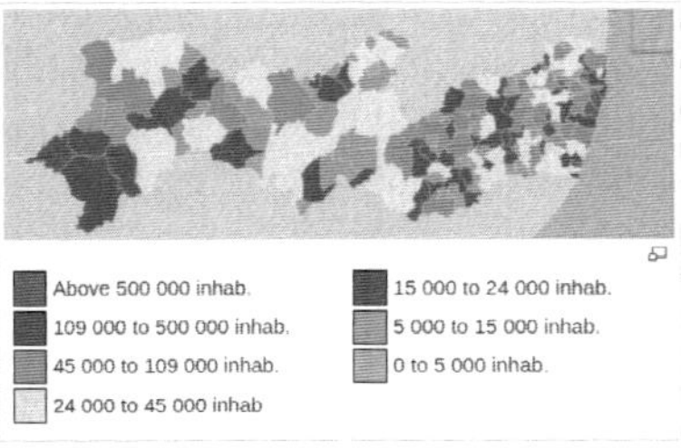

(a) Population in Pernambuco, adapted from (Drayton, Wellber 2010).

(b) Map of the region covered by the 23rd Battalion, comprised by twelve cities.

Figure 1: Maps of the state of Pernambuco and one Military Police Battalion.

2 PROBLEM DEFINITION

In this section, we discuss in more detail how public security is tackled in Brazil. Next, we use this information to introduce the location problem originating from the interaction of different police forces in Pernambuco.

Law enforcement in Brazil is the responsibility of eight main institutions, but we will focus on the interaction between the State Military Police and Fire Brigade (Military Police or MP, for short) and the Civil Police (CP) and its effects on the protection of Pernambuco's citizens. The MP is a police institution at the state level in charge of ostensive policing and preserving public order. Usual MP activities include street patrolling and apprehending suspects of criminal activities. Such suspects must be handed to the CP's custody, more specifically to the Commissioner of Police or Chief Police Officer (CPO). From the point of view of the MP, Pernambuco's territory is divided into a number of battalions and companies. In turn, each battalion or company is responsible for a number of cities. Currently, there are 38 battalions and companies in Pernambuco, which are in charge of protecting the state's 185 cities (Military Police of Pernambuco 2022). One should notice, however, that large urban areas such as Pernambuco's capital Recife (a big city with around 1.6 million citizens), are subdivided into more than one battalion (Recife is in fact divided into 3 battalions). This is not the only difference between large cities and smaller (maybe rural) towns. Usually, due to the concentration of violence in the urban centres, the number of police officers and patrolling vehicles with respect to the population size is much bigger in larger cities than in smaller ones. Some smaller towns in Pernambuco with 10-20 thousand citizens, for example, are served by only one or two police teams. For the sake of simplicity, in this paper we consider that each police team is composed of between three and four police officers in a patrolling vehicle. In addition, a town is denoted *unprotected* if all of its MP teams have been deployed either to attend to a call or to respond to any other duties not related to their ostensive policing tasks (i.e., patrolling).

The CP is organised in a different way. Each one of the 185 cities in the state of Pernambuco has at least one CP station under the command of a CPO. However, 65 out of the 217 CP stations in Pernambuco lack a dedicated CPO (Guerra 2021), meaning that a single CPO can be in charge of several towns and small cities at the same time. In addition, many of these towns lack the 24 hour support of a CP station. In fact, many CP stations not located in the big cities do not have the support of a CPO during weekends and out-of-hours shifts. For such cities this means that, if a crime happens and the attention of a CPO is needed, the MP police team that is deployed to handle the crime needs to travel to the nearest CPO location within its battalion in order to report this crime. Most towns and small cities in Pernambuco lack enough police teams for reporting and protecting their territory. Therefore, minimising the time MP teams spend away from their cities while their population is unprotected is necessary.

Roughly 80% of Pernambuco's municipalities can be considered as small towns, municipalities with up to 50 thousand people according to IPEA (2008). Figure 1a shows that such small towns cover a large portion of Pernambuco's territory, meaning that, in general, their few police teams are responsible for the protection of large areas. Let us consider, for instance, the MP's 23rd batallion depicted in Figure 1b. During weekends and out-of-hours shifts, if the CPO is located in *Afogados da Ingazeira* and a crime happens in *São José do Egito*, the MP team in charge of the crime in the latter must travel to the former in order to report the crime to the CPO. That means a 2h 8min (112.6km) return journey from *São José do Egito*. This time may vary if uncertainty is taken into account regarding travel and *service* times. By service time, we mean the time a CPO takes to register a crime. If there is only one team on patrolling duty in town during this time, that means the population of *São José do Egito* is considered *unprotected* while this MP team is away.

Following the discussion above, in this paper we are interested in defining the most suitable location for CPOs during weekends and out-of-hours shifts such that the unprotected time of the cities under the responsibility of its respective battalion is minimised. Among Pernambuco's MP 38 battalions and companies, our study focuses on those battalions covering three or more towns, where the amount of MP teams is insufficient. This reduces our sample to 23 battalions and companies. Uncertainty is taken into account since, in this situation, crimes, travel and service times variability can affect the absence of MP teams in their designated locations.

3 RELATED WORK

In this section, we review the relevant literature related to the subject of this paper. First we review the literature on probability distributions applied in criminology research. Second, we cover the most relevant applications of discrete event simulation and OCBA on public security.

3.1 Probability Distributions Applied in Criminology Research

The prediction of crime occurrences has a positive impact on public security and has been the subject of many scientific studies. Traditional methods involve the application of regression models on historical datasets in order to forecast the occurrence of crimes (Osgood 2000). According to Garnier et al. (2018), the overall crime rate of a specific geographic location is one of the most common measures that can be inferred from historical data. Among the most usual probability distributions to characterize crime, the Poisson distribution has been frequently applied for modelling crime rates (Osgood 2000). The Poisson distribution describes the probability of a given number of discrete events happening in a given time interval and geographical region (Banks et al. 2014).

Several studies have applied the Poisson distribution for crime forecasting under different contexts. For example, Diefenbach and West (2001) analysed the relationship between several parameters including age, gender, education, and criminal rates to predict the occurrence of crimes based on the Poisson distribution. Wang et al. (2016) applied a Poisson-Gamma mixture model, a.k.a. negative binomial model, together with a GPS technology for crime rate inference at the neighbourhood level in Chicago, US. Garnier et al. (2018) built a comprehensive modelling framework for criminal prediction using the Poisson distribution for predicting individual crime rates based on socio-economic aspects. Curiel et al. (2018) developed a new technique to measure crime rates in Mexico based on low frequency data and a high degree of geographic concentration by means of a Poisson distribution. Hu et al. (2018) showed that the number of crimes obeys a Poisson distribution even when the crime rate is relatively low.

Even though the majority of prediction methods that can be found in the literature apply the Poisson distribution, it is possible to employ alternative models. For instance, Britt et al. (2018) built a model based on the zero-inflated negative binomial distribution. The results of their analysis showed that the proposed method performs better than traditional models in criminal data analysis. Kim and Lee (2017) described a case study in New South Wales focusing on robust estimation of crime occurrence based on zero-inflated Poisson autoregressive models. The paper suggests that such models are appropriate for specified scenarios in terms of criminology. Finally, Kang and Kang (2017) proposed a crime forecasting method that considers environmental context information using multi-modal data fusion. The authors employed a deep neural network with feature-level data fusion.

3.2 Discrete Event Simulation and OCBA

DES models have been widely used in the literature for problems regarding criminology and public security. In this paper however, we do not intend to exhaustively review the literature on this topic. Hence, we will focus on papers that apply DES for police resource allocation. Next, we cover the literature concerning the application of simulation-optimisation, with a focus on OCBA, in different areas such as healthcare and public security, among others.

Brooks et al. (2011) developed a DES model for scheduling police officers in order to improve metrics such as response delay, cross-sector calls for service and officer utilisation. A US police station was used as a case-study allowing a thorough analysis of the system based on real-world data. Srinivasan et al. (2013) applied DES for a similar purpose. In this paper, a model was developed to analyse staffing levels and potential scheduling alternatives. The proposed model helped dimensioning the patrol workforce so that certain performance criteria concerning public security could be met. Haque et al. (2017) developed a DES model with the objective of exploring adaptive allocations for patrol police officers. The proposed DES approach was capable of determining how many patrol officers should be deployed in each station and shift so that the efficiency of crime response was improved under different scenarios. Meier and Vitor (2021) built a DES model combining mechanisms of crime generation and handling in order to overcome human traffic. The model was proposed with the purpose of assisting governmental and private organizations by guiding public security policies and reallocating financial funds to the police force.

Simulation-optimisation (SO), a.k.a. optimization via simulation, is one catalogue of optimization techniques based on the output of simulation (Olafsson and Kim 2002). Within the SO framework, there are a number of well-established algorithms including KN++, OCBA, Response Surface Models (RSM), gradient-based procedures, random search, sample-path optimisation, and metaheuristics (Fu et al. 2005). SO is a widely used methodology that supports decision-making in several areas (Tekin and Sabuncuoglu 2004).

Only a few papers have applied SO-based frameworks for public security. Furtado et al. (2009) applied an ant colony optimisation algorithm and a genetic algorithm in order to solve the problem of criminal-gateway allocation optimisation. In this work, gateway means the initial location of criminals during a simulation process. That is, the proposed method modelled the calibration of a simulation as an optimisation problem to optimise the matching between real-world and simulated data. Zhang and Brown (2014) used DES for patrol district design. More specifically, the paper applied a RSM for finding optimal or sub-optimal districting plans aimed at balancing workload among officers and minimising average response times. The authors show that the proposed methodology was capable of generating efficient plans. In addition, the method is also shown to work for agent-based simulations.

Algorithms for Ranking and Selection (R&S) aim at selecting the best option from a fixed and finite set of alternatives based on their probability of success (Fu, 2015, Lee et al., 2004). Several methods can be found in the literature which follow the R&S paradigm, for instance, Optimal Computing Budget Allocation (OCBA) and Expected Value of Information (EVI) (Hong et al. 2021) and KN++ (Hong and Nelson 2001). In this paper, we will focus on OCBA, which works by optimally allocating simulation budget to potential alternatives so that the best one can be identified with the highest probability (Chen and Lee 2010, Lee et al. 2010, Fu 2015).

3.3 Contributions of This Paper

This paper studies a real location problem on the context of public security. Uncertainties in travel times, service times and the incidence of crimes are taken into consideration, making the approach more realistic and applicable in real-world scenarios. This paper adds to the existing literature by providing an optimisation-simulation approach to address the problem. Our analysis focuses on the influence of the location of chief police officers on the usage and efficiency of military police forces. The results of the numerical experiments demonstrate the effectiveness of the proposed framework and has the potential to improve the operational efficiency of policed in the face of limited budgets and increasing crime rates. We believe this paper provides a relevant and important case-study that can be used to optimize the location of police officers in other regions and countries.

4 METHODOLOGY

In this section, the main methodological steps applied in this research are discussed in more detail. These include statistical tests performed on the source data, the DES model that evaluates the performance of the police system and the OCBA algorithm employed for selecting the best location of CPOs.

4.1 Problem Formulation and Assumptions

Our problem consists of finding the optimal locations of a battalion's CPOs such that the expected total unprotected time of cities and military police travel within the battalion's designated region is minimised. Let i denote an arbitrary battalion in Pernambuco and J_i the set of cities in battalion i. The binary decision variable x_j takes value 1 when the CPO is located at city $j \in J_i$ during weekends and out-of-hours shifts, and 0 otherwise. The function $U(j,J_i)$ computes the unprotected time of cities in battalion i if a CPO is located at city $j \in J_i$. We recall that by unprotected time, we mean the length of time in which all police teams of a given city are busy. One can notice that this definition considers that citizens are not in protection even if no crimes happen while the city's police teams are busy. Analogously, the function $T(j,J_i)$ represents the total time the military police of battalion i require to report crime when the CPO is located at city $j \in J_i$. For a given battalion i, the objective function can be written as

$$\min \quad \sum_{j \in J_i} \big(\alpha_1 U(j,J_i) + \alpha_2 T(j,J_i)\big) x_j$$

Since battalions are independent, one can solve each optimisation problem individually. To approximate the values of $U(j,J_i)$ and $T(j,J_i)$ we use a simulation model (Section 4.3), which makes the following main assumptions:

- a As soon as a crime happens in some city, any available MP team of that city is assigned to handle it immediately. The time for handling a crime includes the time spent dealing with the crime plus travel time to the crime location and to and from the nearest CPO location.
- b We assume that only one CPO is available per battalion.
- c The travel times follow a lognormal distribution;
- d The time a CPO takes to register the crime follows an uniform distribution;
- e The CPO registers crimes individually, and if more than one MP team is present to report a crime, they queue with a first in, first out (FIFO) rule;

f Criminal events are assumed to follow a Poisson distribution based on historical criminal data;
g A police team can only respond to crimes happening in its own city/town. It means that if a crime happens in a given city and its police teams are busy, this crime is "queued" until a police team is available for handling it. Queued crimes are organised following a FIFO rule;
h Battalions in Pernambuco are also independent. Therefore, MP teams can only report to CPOs located within their designated battalions.

4.2 Distribution Fitting

In order to predict the occurrence of crimes, historical crime data has been collected from Pernambuco's MP official sources (SDSPE 2022b). The collected data consists of a list of all crimes that happened in the state between January 2014 and August 2021. Out of the 38 battalions in Pernambuco, we focus on 23 located away from the big centres. From this list, only the date and location (town/city) of crimes are relevant to this study. Since we are only interested on ILVCs, crimes that need immediate attention from the police, any other type of crimes were removed from the list. However, one should notice that the available dataset do not include crime time stamps, which are necessary for our study. In the following, we discuss the procedure that was adopted in order to circumvent this issue.

A second private dataset of crimes was collected directly from Pernambuco's 23rd MP battalion. This dataset consists of all crimes that happened between January 2016 and December 2020 in the 12 towns that compose this battalion. In addition to the crimes' location and date, a time stamp of each crime is included in this dataset. By assuming that for each time interval (0h – 1h, 1h – 2h, ..., 23h – 0h) the crime occurrence rate follows a Poisson distribution, it is possible to estimate a Poisson daily crime occurrence rate by summing up each interval's rates for each town. Based on this information, which can be calculated for each city in the 23rd battalion, it is possible to estimate hourly crime occurrence rates for the remaining cities in Pernambuco by assuming they also follow the same time variations. The resulting data was then used to generate crimes during out-of-hours shifts and weekends within our DES model.

4.3 Discrete Event Simulation

In this paper, the proposed DES model is a critical step to implement the SO framework that decides the location of CPOs. For a given CPO location $j \in J_i$ within battalion i, the DES model computes the mean unprotected time based on the occurrence of random crimes and stochastic travel and service times. Figure 2 shows a visual representation of the DES model developed in this article.

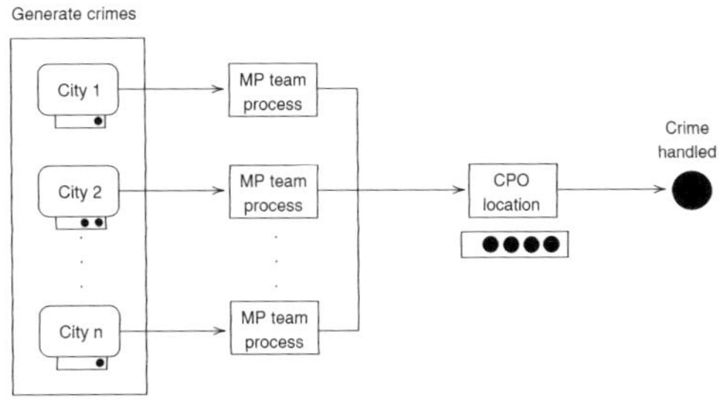

Figure 2: Schematic representation of the proposed DES model.

Our DES model works as follows. Since the battalions are independent, we narrow the scale of the DES model to a single battalion at a time instead of the entire state. At each simulation run, a different location is assigned to the CPO within the current battalion. Assuming that the occurrence of crimes is a Poisson process, at every time instant crimes are generated for each city. For every city where crimes happened, the necessary number of MP teams are dispatched to deal with the crimes. Crimes are queued in a given city if it does not have enough MP teams to deal with all generated crimes. Queued crimes are represented by black dots below each city in 2. Theoretically, the capacity of the queue is infinite.

Individual unprotected times for each city are incremented when all of their respective MP teams are busy. The number of MP teams for each city has been estimated from an incomplete dataset provided by Pernambuco's MP. Each MP team not located in the same city as the CPO needs to travel to the CPO location in order to register the crime. Uncertain travel times based on real data are taken into account in this case. In addition, CPOs can only register one crime at a time and the time taken to register the each crime is also uncertain. Therefore, a queue with infinite capacity of MP teams waiting to meet the CPO can be formed. This queue is also represented by black dots below the CPO block in Figure 2. After the registration process, a legal investigation is opened by the delegado and the crime is considered to be *handled* by the MP team. One should notice however that the return trip to an MP's original location (town/city) still counts as unprotected time if that city does not have any available (free) MP teams.

We have set the length of the simulation to 3600 minutes, which represents weekend shifts from 7pm on Friday to 7am on Monday. The resolution of simulation time has been set to 5 minutes, meaning that our simulation runs for 720 time units. As for a single running of the model, it outputs the unprotected time counted in the 720 units of time. Our DES model has been implemented using `Python` by means of its `Simpy` library.

4.4 OCBA Algorithm

In this section, we consider the problem of finding the best location for the CPO such that the mean unprotected time of a battalion is minimised. The main challenge in this problem is that the objective is measured using the output of the proposed DES, therefore it is subject to stochasticity. In summary, the proposed OCBA algorithm begins by allocating a fixed number of DES replications to each alternative CPO location, in order to obtain initial information about each alternative's mean and variance. OCBA then determines the most critical locations, represented by those which are both close to being the best in terms of average unprotected time and/or highly variable. Next, more DES replications are assigned to the critical locations, until the full computing budget is realised.

Monks (2021) provides a general `Python` implementation of OCBA which is used in our computational experiments. We refer the interested reader to Calverley et al. (2021) and Currie and Monks (2021) for more details about OCBA.

Let us define as k the number of possible CPO locations (cities j in J_i) within battalion i and a computational budget B, meaning that we can run at most a total of B replications during the experiment. Let us also define Δ as the number of DES replications allocated at each step of the algorithm. As we run n_0 replications for each of the k possible CPO locations during the initialisation phase, $B - kn_0$ must be a multiple of Δ. After these initial replications, the OCBA algorithm will then iteratively keep allocating the remaining budget to the different alternatives, with the aim of improving the simulation efficiency and maximising the probability of correct selection.

5 NUMERICAL EXPERIMENTS

In this section, we present the results of our computational experiments. The total budget used in our simulations, amounted to 10000 replications per alternative, that is $B = 10000k$. The geographical areas of the battalions differ, and hence the number of cities they cover (the value of k) is different. We have only optimised battalions that cover three or more cities. The values of Δ and n_0 have been set to $\Delta = n_0 = 100$. Travel times for each journey were randomly sampled from a lognormal distribution, where the mean is set to duration estimations obtained with OpenStreetMap data and the Open Source Routing Machine (OSRM) software, and the standard deviation was set to 2.5% of the mean. Service times were sampled from an uniform distribution $U[1,2]$ (in hours). Table 1 presents a summary of our results for the state of Pernambuco. In this table, column **Batt** shows the identification of each individually solved battalion in Pernambuco optimising for unprotected time only ($\alpha_1 = 1, \alpha_2 = 0$) and for travel and unprotected time ($\alpha_1 = \alpha_2 = 1$). Column **Alt** indicates the number of cities (alternatives solved) in that battalion. The column **Location** presents the best location found for the CPO for the corresponding objective, and **Reps** the number of replications OCBA allocated to that alternative. Columns $\overline{U}$ and $\overline{T}$ show the average total unprotected time and travel time of the best alternative, respectively.

According to the simulation results shown in Table 1, for most battalions in Pernambuco, the CPO locations that minimise the total average unprotected time are not in the largest cities/towns (as often happens in practice). Instead, the best location is usually in smaller towns. For example, the 23rd

Table 1: Best CPO locations for the battalions of Pernambuco that cover more than three cities, minimising unprotected time (columns 3-6) and unprotected and travel time (columns 7-10). Times are in hours.

		Minimise unprotected time				**Minimise unprotected and travel time**			
Batt	Alt	Location	Reps	$\overline{U}$	$\overline{T}$	Location	Reps	$\overline{U}$	$\overline{T}$
10bpm	12	Palmares	55439	5.57	6.86	Palmares	34862	5.57	6.86
10cipm	5	Tamandaré	15460	0.1	3.3	Barreiros	24685	0.13	3.23
11cipm	7	Jupi	29099	4.38	3.62	Lajedo	22232	4.48	1.93
14bpm	7	Triunfo	32607	0.59	2.54	Serra Talhada	20957	0.7	0.71
15bpm	5	Tacaimbó	14786	0.88	6.02	Belo Jardim	20429	1.19	4.86
1cipm	4	Itacuruba	18570	0.22	2.43	Belém do S. Fran.	19626	0.22	0.95
21bpm	7	Chã de alegria	30657	2.47	24.14	Vitória de S. Antão	26789	2.73	12.33
22bpm	9	St. Maria do Cam.	33729	4.18	6.72	Surubim	21888	4.39	3.4
23bpm	12	Brejinho	54980	1.44	2.08	S. José do Egito	32105	1.45	1.6
24bpm	6	Jataúba	23786	1.95	30.74	St. Cruz do Cap.	17932	2.04	9.83
26bpm	4	Itamaracá	18135	0.13	7.2	Igarassu	12412	0.17	2.02
2bpm	14	Nazaré da Mata	45847	8.85	15.7	Carpina	45078	9.15	13.07
3bpm	10	Venturosa	25882	0.36	6.47	Arcoverde	23040	0.61	3.21
4bpm	13	Bonito	58259	30.61	71.83	Caruaru	33362	31.86	18.99
4cipm	4	Jatobá	10820	0.24	0.59	Petrolândia	21052	0.37	0.4
5bpm	3	Dormentes	9962	0.12	52.51	Petrolina	4046	0.4	0.5
6cipm	5	Cumaru	15803	0.98	6.42	Limoeiro	21663	1.54	1.9
7bpm	7	Santa Cruz	22945	0.74	2.66	Ouricuri	29648	0.82	1
8bpm	6	Salgueiro	21418	0.92	0.44	Salgueiro	16381	0.91	0.43
8cipm	4	Alagoinha	14906	0.74	2.98	Pesqueira	15606	0.82	0.69
9bpm	15	Garanhuns	50757	3.49	3.87	Garanhuns	29082	3.49	3.88
9cipm	3	Trindade	9743	0.01	2.17	Araripina	9837	0.01	0.86
	Avg	-	*27890.5*	*3.14*	*11.88*	-	*22850.5*	*3.32*	*4.21*

battalion shown in Figure 1b contains 12 cities. *Afogados da Ingazeira* is the largest town in this battalion, but the optimal CPO location found by our algorithm model is *Brejinho*, a small town located in the north corner of the 23rd's territory that has a single police team. Nonetheless, this changes when adding the time travel time to the optimisation. In the 23rd battalion example, the best location switches to *São José do Egito*, a larger nearby city, as this reduces travel. A more extreme example happens in the 4bpm battalion, where ignoring travel time in the optimisation locates the CPO in the small city of *Bonito* (circa 38000 inhabitants, one police team) and creates a large travel burden for the teams in *Caruaru* (circa 369000 inhabitants).

6 CONCLUSIONS

This paper focused on a location optimization problem in the Brazilian state of Pernambuco. Our approach aimed at minimising the average total unprotected time of the cities of the battalion, as well as the travel time of police teams. First, the administrative structure and operational process of Brazilian police forces were introduced. Such discussion allowed us to introduce a location problem that arises from the interaction between the military and civil police institutions in Pernambuco.

A DES model was developed with the purpose of computing the average total unprotected time of a battalion for a given CPO location. In order to improve the probability of choosing the best alternative for each battalion, we employed an off-the-shelf OCBA algorithm (Monks 2021). Numerical experiments were performed in order to test our framework. Our results indicate that the location of the CPO can have a strong influence in the usage of police forces, and that minimising unprotected time alone can generate large travel times. Further research might address the optimisation as a multi-objective problem, studying the trade-off between these measures.

In addition, some of our assumptions do not allow a precise representation of the reality in Pernambuco. As a result, a more comprehensive and efficient DES model might be needed in future studies.

REFERENCES

Banks, J., J. S. Carson II, D. M. Nicol, and B. L. Nelson. 2014. *Discrete event system simulation*. Prentice-Hall.

BFPS Brazilian Forum on Public Security 2022. "*Anuário Brasileiro de Segurança Pública*". Available at https://forumseguranca.org.br/wp-content/uploads/2022/06/anuario-2022.pdf?v=5, Accessed 20 November 2022.

Brasil 2004. "*Lei Nº 12.544, de 30 de Março de 2004 - Fixa o efetivo da Polícia Militar de Pernambuco, e dá outras providências.*". *Diário Oficial do Estado de Pernambuco*.

Britt, C. L., M. Rocque, and G. M. Zimmerman. 2018. "The analysis of bounded count data in criminology". *Journal of Quantitative Criminology* 34 (2): 591–607.

Brooks, J. P., D. J. Edwards, T. P. Sorrell, S. Srinivasan, and R. L. Diehl. 2011. "Simulating calls for service for an urban police department". In *Proceedings of the 2011 Winter Simulation Conference (WSC)*, 1770–1777. IEEE.

Calverley, J., C. Currie, B. S. Onggo, T. Monks, and M. Higgins. 2021. "Simulation optimisation for improving the efficiency of a production line". In *Proceedings of the Operational Research Society Simulation Workshop*, 137–144.

Curiel, R. P., S. C. Delmar, and S. R. Bishop. 2018. "Measuring the distribution of crime and its concentration". *Journal of quantitative criminology* 34 (3): 775–803.

Currie, C., and T. Monks. 2021. "Tutorial on optimisation via simulation: How to choose the best set up for a system". In *Proceedings of the Operational Research Society Simulation Workshop 2021 (SW21)*, edited by Fakhimi, Robertson, and Boness, 35–41. Birmingham, UK: The Operational Research Society.

Diefenbach, D. L., and M. D. West. 2001. "Violent crime and Poisson regression: A measure and a method for cultivation analysis". *Journal of Broadcasting & Electronic Media* 45 (3): 432–445.

Drayton, Wellber 2010. "File:População pernambuco". Available at https://commons.wikimedia.org/wiki/File:Popula%C3%A7%C3%A3o_pernambuco.png, Acessed 01 December 2022. This work is licensed under the Creative Commons Attribution 4.0 International License. To view a copy of this license, visit http://creativecommons.org/licenses/by/4.0/.

Fu, M. C., F. W. Glover, and J. April. 2005. "Simulation optimization: a review, new developments, and applications". In *Proceedings of the Winter Simulation Conference, 2005.*, 13–pp. IEEE.

Furtado, V., A. Melo, A. L. Coelho, R. Menezes, and R. Perrone. 2009. "A bio-inspired crime simulation model". *Decision Support Systems* 48 (1): 282–292.

Garnier, S., J. M. Caplan, and L. W. Kennedy. 2018. "Predicting dynamical crime distribution from environmental and social influences". *Frontiers in Applied Mathematics and Statistics* 4:13.

Guerra, Raphael 2021. "*Descaso: 30% das delegacias em Pernambuco não têm delegados titulares*". Available at https://jc.ne10.uol.com.br/colunas/ronda-jc/2021/11/14354971-descaso-30-das-delegacias-em-pernambuco-nao-tem-delegados-titulares.html, Accessed 01 November 2022.

Haque, K. M., V. C. Chen, and B. L. Huff. 2017. "A discrete-event simulation model for adaptive allocation of police patrol". In *IIE Annual Conference. Proceedings*, 1169–1174. Institute of Industrial and Systems Engineers (IISE).

Hong, L. J., W. Fan, and J. Luo. 2021. "Review on ranking and selection: A new perspective". *Frontiers of Engineering Management* 8 (3): 321–343.

Hu, T., X. Zhu, L. Duan, and W. Guo. 2018. "Urban crime prediction based on spatio-temporal Bayesian model". *PloS one* 13 (10): e0206215.

IPEA Brazilian Institute of Applied Economic Research 2008. "*População das cidades médias cresce mais que no resto do Brasil*". Available at https://web.archive.org/web/20090819081149/http://www.ipea.gov.br/003/00301009.jsp?ttCD_CHAVE=5499, Acessed 01 December 2022, Archived from original 19 August 2009.

Kang, H.-W., and H.-B. Kang. 2017. "Prediction of crime occurrence from multi-modal data using deep learning". *PloS one* 12 (4).

Kim, B., and S. Lee. 2017. "Robust estimation for zero-inflated Poisson autoregressive models based on density power divergence". *Journal of Statistical Computation and Simulation* 87 (15): 2981–2996.

Li, J., W. Liu, G. Pedrielli, L. H. Lee, and E. P. Chew. 2017. "Optimal computing budget allocation to select the nondominated systems – A large deviations perspective". *IEEE Transactions on Automatic Control* 63 (9): 2913–2927.

Meier, S., and F. Vitor. 2021. "Developing a Discrete Event Simulation Model to Overcome Human Trafficking". In *IIE Annual Conference. Proceedings*, 7–12. Institute of Industrial and Systems Engineers (IISE).
Military Police of Pernambuco 2022. "PMPE Unidades Operacionais".
Monks, Thomas 2021. "sim-tools: fundamental tools to support the simulation process in python". http://doi.org/10.5281/zenodo.4553642. Zenodo.
Olafsson, S., and J. Kim. 2002. "Simulation optimization". In *Proceedings of the winter simulation conference*, Volume 1, 79–84. IEEE.
Osgood, D. W. 2000. "Poisson-based regression analysis of aggregate crime rates". *Journal of quantitative criminology* 16 (1): 21–43.
SDSPE Secretariat for Social Defence of Pernambuco 2022a. "*Evolução Anual dos Números de Vítimas de CVLI* em Pernambuco por Município.". Available at https://www.sds.pe.gov.br/images/indicadores/CVLI/ANUAL_POR_MUNIC%C3%8DPIO_CVLI.pdf, Accessed 31 October 2022.
SDSPE Secretariat for Social Defence of Pernambuco 2022b. "*Indicadores criminais em Pernambuco*". Available at http://www.sds.pe.gov.br/estatisticas, Accessed 07 September 2022.
Srinivasan, S., T. P. Sorrell, J. P. Brooks, D. J. Edwards, and R. D. McDougle. 2013. "Workforce assessment method for an urban police department: Using analytics to estimate patrol staffing". *Policing: An International Journal of Police Strategies & Management*.
Tekin, E., and I. Sabuncuoglu. 2004. "Simulation optimization: A comprehensive review on theory and applications". *IIE transactions* 36 (11): 1067–1081.
Wang, H., D. Kifer, C. Graif, and Z. Li. 2016. "Crime rate inference with big data". In *Proceedings of the 22nd ACM SIGKDD international conference on knowledge discovery and data mining*, 635–644.
Zhang, Y., and D. Brown. 2014. "Simulation optimization of police patrol districting plans using response surfaces". *Simulation* 90 (6): 687–705.

AUTHOR BIOGRAPHIES

YIFU WEI is a Ph.D. student at the University of Leeds and the University of Manchester in the Growing skills for Reliable Economic Energy from Nuclear (GREEN) programme. He finished his masters degree in MSc. Supply Chain Management and Logistics at University of Southampton in April 2022. He has graduated in Logistics Management by the Chongqing University of Post and Telecommunications in China. His mains research interests lie within cutting and packing, heuristic and metaheuristic algorithms and applications in nuclear fission and fusion reactors.

CARLOS LAMAS-FERNANDEZ is a Lecturer in Business Analytics/Management Science in Southampton Business School. He completed his PhD in Operational Research in the Southampton Business School in 2018. Prior to that, he had worked as a research fellow in NIHR ARC Wessex and as a scientific developer at ETH Zurich. In his research, Carlos is interested in using optimisation techniques in areas such as sports, health care, transportation and logistics. He has developed both heuristic and exact methods for cutting and packing, vehicle routing and facility location problems.

WALTON P. COUTINHO is an Assistant Professor at the Department of Technology of the Federal University of Pernambuco, Brazil. He received his B.Sc. degree in Mechanical Production Engineering from the Universidade Federal da Paraíba and his M.Sc. degree in Production Engineering (Operational Research) from the same institution. He obtained his Ph.D. degree in Operational Research on September 2018 at the School of Mathematics of the University of Southampton. His main research interest is in the development of optimisation algorithms for drone routing and trajectory optimisation.

CHRISTINE CURRIE is a Professor of Operational Research in Mathematical Sciences at the University of Southampton, UK, where she also obtained her Ph.D. She is Editor-in-Chief for the Journal of Simulation. Christine was co-chair of the Simulation Special Interest Group in the UK Operational Research Society until September 2013. Her research interests include simulation optimisation, mathematical modelling of epidemics, optimal pricing and applications of simulation in health care.

Proceedings of the Operational Research Society Simulation Workshop 2023 (SW23)
C. Currie and L. Rhodes-Leader, eds.
https://doi.org/10.36819/SW23.020

SIMHEURISTIC AND LEARNHEURISTIC FOR SOLVING STOCHASTIC AND/OR DYNAMIC PORTFOLIO OPTIMIZATION PROBLEMS

Mr. Yuda Li
Dept. of Computer Science, Universitat Oberta de Catalunya
Barcelona, Spain
yli1@uoc.edu

Dr. Onur Polat
Dept. of Applied Statistics and Operations Research, Universitat Politcnica de Valncia
Alcoy, Spain
Dept. of Public Finance, Bilecik Şeyh Edebali University
Bilecik, Turkey
opolat@upvnet.upv.es

Dr. Angel A. Juan
Dept. of Applied Statistics and Operations Research, Universitat Politcnica de Valncia
Alcoy, Spain
juanp@upv.es

Dr. Laura Calvet

Dept. of Telecomm. and Systems Engineering, Autonomous University of Barcelona
Sabadell, Spain
Laura.Calvet.Linan@uab.cat

Dr. Renatas Kizys

Dept. of Banking and Finance, Southampton Business School, University of Southampton
Southampton, United Kingdom
r.kizys@soton.ac.uk

ABSTRACT

Constructing portfolio by proper asset selection to maximize return and minimize risk has been considered an essential task for investment activities. Rich portfolio optimizations with realistic constraints are NP-hard problems and are commonly solved using metaheuristics. However, financial markets are characterized by their high volatility and uncertainty, and metaheuristics do not fully account for these random and/or dynamic components, which renders them unrealistic in the presence of heightened uncertainty and dynamism in financial markets. Therefore, this paper proposes a simulationoptimization approach specifically, a simheuristic algorithm to deal with the stochastic version of the problem and a learnheuristic algorithm for solving the dynamic version of the problem. Computational experiments are performed on a benchmark instance to illustrate the advantages of the proposed methodologies and analyze how the solutions change in response to a different degree of stochasticity, dynamism, and minimum required return.

1 Introduction

Financial decisions have the uppermost significance in the creation of wealth, enhancement of welfare standards, and sustainable economic growth. They play an essential role in providing funds to firms, transforming ideas and resources into profitable projects, and eventually serving social benefits for societies. Such improvements are characterized mainly by the formulation of optimization problems in financial economics.

Developed by Markowitz (1952), the portfolio optimization problem (POP) consists of the investment decision as a strategy of (i) determining financial assets; (ii) computing the appropriate weights allocated to those financial assets in the desired portfolio return, and allocating a minimum level of risk. The POP is carried out through a quadratic objective function that aggregates the weighted covariances of the associated asset returns, which is then minimized subject to the desired portfolio return. Predominantly, the exact methods have been employed to solve this basic version of the POP. However, exact methods become inefficient when dealing with realistic and large-scale combinatorial optimization problems (COPs) due to their NP-hard nature. In this context, metaheuristics have been employed as alternative solution techniques to overcome these drawbacks. Metaheuristics are general solving procedures, that are able to solve near-optimal solutions to COPs in reasonable computing times. Particularly, metaheuristics have been extensively implemented to deal with complex portfolio optimization problems, where pre-assignment, quantity, and cardinality constraints are considered (Armananzas and Lozano 2005, Doering et al. 2019).

Pre-assignment constraints entail the pre-selection of some assets, regardless of their risk-return features. Quantity constraints prescribe the fractions attained to an asset in the portfolio within a floor and ceiling constraint. The quantity constraints render the administration of a portfolio manageable by minimizing lot sizes and lowering transaction costs (Cesarone et al. 2013). Finally, cardinality constraints dictate a lower and upper bound for the number of assets covered in the portfolio. Cardinality constraints not only limit the number of assets in the portfolio to cope with but provide a certain threshold of diversification (Gaspero et al. 2011).

In contrast to the real-world trading constraints, a plethora of studies assumes constant covariances. This work presents two more realistic versions of the POP: the stochastic POP (SPOP) and the dynamic POP (DPOP). In the SPOP the covariances are modeled as random variables, which may be estimated by historical data using certain statistical measures. Therefore, the predictions encompass a certain threshold of uncertainty (noise). In the DPOP, the covariances matrix is deterministic but non-static (i.e., while we know the estimated value for the covariance of each pair of assets, the specific value that this covariance takes is influenced by synergy effects of assets in the portfolio). We deal with the SPOP by implementing the simheuristic algorithm proposed by Kizys et al. (2022). Furthermore, we propose a novel learnheuristic algorithm, which is an integration of a machine learning algorithm with a well-known metaheuristic, the Variable Neighborhood Search (VNS) (Hansen et al. 2019), to address the DPOP. Both methodologies carry considerable benefits compared with the presumption of deterministic/static risk. A set of computational experiments is carried out to illustrate and validate the solving approaches.

We proceed with the study as follows: Section 2 reviews related studies in the field. Section 3 presents the mathematical model and the problem definition for the POP. Section 4 describes the SPOP and the simheuristic solving approach, provides computational experiments, and analyses the results. Section 5 provides details for the learnheuristic algorithm and analyzes the results of computational experiments. Finally, Section 7 draws the main findings of the work and discusses future research lines.

2 Literature Review

Seven decades ago, Markowitz (1952) proposed the first mathematical formalization of the idea of diversification of investments. This work postulated that an investor should maximize expected portfolio return while minimizing portfolio variance of return. Since then, multiple authors have contributed to the field of portfolio optimization by considering different risk measures, realistic constraints, and datasets, and solving methodologies (ranging from exact methods for small problem instances to approximate methods for bigger and more realistic problem instances) (Kolm et al. 2014). Portfolio optimization continues to be an interesting and challenging task that attracts the attention of researchers. The reader interested in recent reviews is referred to Milhomem and Dantas (2020), Kalayci et al. (2019), and Doering et al. (2019).

For instance, recently, Kizys et al. (2022) present a simheuristic algorithm (Juan et al. 2015) that integrates a variable neighborhood search metaheuristic with Monte Carlo simulation to address the portfolio optimization problem with pre-selection, quantity and cardinality constraints, as well as stochastic returns and noisy covariances modeled as random variables. Zhai et al. (2020) build a multiconstraint portfolio optimization model based on the second-order stochastic dominance rule, the higher moments of return series, and the Shannon entropy, in addition to other investment constraints. The authors employ the whale optimization algorithm to solve the problem, which is compared against other swarm intelligence optimization algorithms. The numerical results used data from the FTSE100 index stocks during 2018. Sehgal and Mehra (2020) proposed a model for robust portfolio optimization with second order stochastic dominance constraints, in which the input returns of each asset at every scenario is varied in a bounded and symmetric interval. The cutting plane algorithm is employed to deal with the optimization problem. The performance of the model is assessed by experimenting with datasets drawn from S&P 500, S&P BSE 500, Nikkei 225, S&P Global 100, FTSE 100, and BOVESPA index.

3 Problem Definition

This section provides a mathematical formulation of the constrained POP. It includes a description of the notation employed and the mathematical formulations, which are based on Kizys et al. (2022).

3.1 Formal Description

In the classical POP, there is a set $A = \{a_1, a_2, \ldots, a_n\}$ of n assets. Each asset a_i ($\forall i \in \{1,2,\ldots,n\}$) is characterized by an expected return r_i. The covariance between two assets a_i and a_j ($\forall i, j \in \{1,2,\ldots,n\}$) is represented by σ_{ij}.

A solution for the POP can be encoded as a vector $X = (x_1, x_2, \ldots, x_n)$, where each element $x_i (0 \leq x_i \leq 1)$ reveals the weight or fraction of the investment allocated to the asset a_i.

The main goal is to minimize the portfolio risk, while meeting the following constraint: the expected return has to be greater than an investor-given threshold R. The constrained version of the POP describe a more realistic scenario by including pre-selection, quantity, and cardinality constraints. Pre-selection constraints reveal whether an asset a_i has been pre-selected by the investor and, therefore, it has to be included in the solution ($x_i > 0$). They are described by means of the parameter p_i: $p_i = 1$ if a_i is pre-selected, and $p_i = 0$ otherwise. Quantity constraints set a lower and an upper bounds for the weights x_i, ε_i and δ_i ($0 \leq \varepsilon_i \leq \delta_i \leq 1$), respectively. Finally, the cardinality constraint specifies the minimum and maximum number of assets included in the solution, k_{min} and k_{max} ($1 \leq k_{min} \leq k_{max} \leq n$).

3.2 Mathematical Model

The constrained POP can mathematically be defined as follows:

$$\min f(x) = \sum_{i=1}^{n} \sum_{j=1}^{n} \sigma_{ij} x_i x_j \tag{1}$$

subject to:

$$\sum_{i=1}^{n} r_i x_i \geq R \tag{2}$$

$$\sum_{i=1}^{n} x_i = 1 \tag{3}$$

$$\varepsilon_i z_i \leq x_i \leq \delta_i z_i, \forall i \in \{1,2,\ldots,n\} \tag{4}$$

$$0 \leq \varepsilon_i \leq \delta_i \leq 1, \forall i \in \{1,2,\ldots,n\} \tag{5}$$

$$z_i \leq M x_i, \forall i \in \{1,2,\ldots,n\} \tag{6}$$

$$p_i \leq z_i, \forall i \in \{1,2,\ldots,n\} \tag{7}$$

$$k_{min} \leq \sum_{i=1}^{n} z_i \leq k_{max} \tag{8}$$

$$z_i \in \{0,1\}, \forall i \in \{1,2,\ldots,n\} \tag{9}$$

Equation (1) represents the objective function, which represents the riskiness of the portfolio and is to be minimized. Equation (2) forces that the expected return of the investment is equal or greater than the threshold R. Equation (3) guarantees that the portfolio investment equals existing and pre-defined resources. For each asset a_i, Equation (4) sets lower and upper bounds (ε_i and δ_i, respectively) for x_i in case the asset is selected. Whether the asset a_i is included in the solution or not is represented by means of an auxiliary variable ($z_i = 1$ if positive; $z_i = 0$ otherwise). Equation (5) bounds the lower and upper bounds by zero and one inclusive. Equation (6) relates variables x_i with z_i, ensuring that if x_i is greater than 0, then z_i is 1 (M is a large positive value such that $Mx_i \geq 1$ for all i if $x_i > 0$). The pre-assignment constraint is defined by Equation (7). It relates variable z_i with parameter p_i; if the asset a_i is pre-selected (i.e., $p_i = 1$), it must be included in the portfolio (i.e., $z_i = 1$). Equation (8) sets the minimum and maximum number of assets to be included in the portfolio. Finally, Equation (9) defines z_i as a binary variable.

4 Stochastic Portfolio Optimization Problem

In this paper, the difference between the POP and the SPOP resides the modeling of asset covariances. In the POP, they are represented by the expected values. However, the SPOP proposes a realistic stochastic uncertainty and hence suggests these as random variables. Therefore, in the second case, covariances (C_{ij}) in the objective function are considered random variables that follow a given probability distribution:

$$f(x) = \Gamma\left[\sum_{i=1}^{n}\sum_{j=1}^{n} C_{ij}x_i x_j\right] \tag{10}$$

here Γ represents a specific statistical measure (such as the mean, or the variance).

4.1 Simheuristics

Simheuristics are combinations of metaheuristics with simulation and have been employed extensively by scholars to deal with NP-hard stochastic optimization problems. In this vein, studies on vehicle routing problems with stochastic times (Taş et al. 2013, Wang and Lin 2013, Guimarans et al. 2018), flow-shop scheduling problems with random processing times (Juan et al. 2014, Wu et al. 2018, Lee and Kim 2022), and inventory routing problems with stochastic demands (Gruler et al. 2018, Raba et al. 2020) can be mentioned.

In this work, we employ the simheuristic algorithm proposed by Kizys et al. (2022), which integrates a VNS with Monte Carlo simulation (MCS). The number of neighbors K is set to 3, and a movement in each neighbor constitutes changing 25%, 35%, and 45% of the assets, respectively. The algorithm is described as follows:

1. First, an initial solution (*initSol*) is constructed by selecting the k_{min} assets that provide the highest rate of return, including the s assets pre-selected by the investor, then, the optimal weights allocated to each asset is determined by a quadratic programming solver.
2. A list *bestSols* is developed for storing the l best-found solutions regarding the expected risk. Thereafter, *initSol* is copied into *currentSol* and k is set to one. In the next step, the expected risk of *currentSol* is derived in terms of MCS, and the solution is stored in the *bestSols* list.
3. A new solution (*newSol*) is constructed by 'shaking' the *currentSol*. This procedure randomly erases a number of non-pre-selected assets in the solution and randomly introduces new assets until reaches k_{max}. Furthermore, a local search is employed for the solution. *newSol* is compared against *currentSol*. If the former is better concerning the risk corresponding to the deterministic version of the problem, the expected risk for the stochastic version is computed with sim_{short} runs. If the expected risk of *newSol* is lower than that of *currentSol*, then *newSol* replaces

currentSol, k is set to one, and *bestSols* is updated. If not, the solution is removed. If *newSol* is not better, k is increased in one unit if $k < K$ or set to one otherwise. This step is repeated until the maximum time constraint T_{loop} is reached.

4. Finally, for each solution in *bestSols*, a larger number of runs (sim_{large}) is simulated to obtain a more accurate estimation of expected risk.

5 Dynamic Portfolio Optimization Problem

In this section, we consider a dynamic version of the POP in which the covariances matrix is deterministic but non-static, i.e.: while we know the estimated value for the covariance of each pair of assets, the specific value that this covariance takes is influenced by the synergy effects of assets s_i and s_j ($\forall i, j \in \{1, 2, \ldots, n\}$) in the portfolio, i.e., some combinations of assets might increase or decrease the initial covariance estimates. Therefore the original deterministic objective function is transformed into:

$$\min f(x) = \sum_{i=1}^{n} \sum_{j=1}^{n} \sigma_{ij} x_i x_j s_i s_j \tag{11}$$

5.1 Learnheuristic

To deal with this dynamic environment, the paper employs a learnheuristic algorithm, which is a hybrid algorithm that combines metaheuristic algorithms with machine learning to deal with optimization problems with dynamic inputs (Calvet et al. 2017). In these optimization problems, the inputs (either located in the objective function or in the set of constraints) are not fixed in advance and they vary in a predictable way according to the current status of the solution. Therefore, learnheuristic relies on machine learning techniques to learn the relationships between inputs and solution characteristics from historical data and metaheuristic algorithms to find high-quality solutions using the predicted inputs. In this sense, these optimization problems represent an extension of the classical static optimization problems, in which all inputs are given in advance and are immutable. Learnheuristics have been employed in different application areas. Calvet et al. (2016) used learnheuristic to solve a multi-depot vehicle routing problem with the assumption that customers show a different willingness to consume depending on how well the assigned depot fits their preferences. To solve this problem, they employed machine learning models trained with historical data to estimate the demand of the customers depending on the assigned depot and included this prediction as part of an enriched objective function as a way to better guide the stochastic local search inside the metaheuristic framework. Arnau et al. (2018) employed learnheuristic for solving a vehicle routing problem with dynamic traveling times which depend on the structure of the routing plan. Bayliss et al. (2020) proposed a learnheuristic algorithm to solve an aerial-drone team orienteering problem with the travel times between targets depending on drones flight path between previous targets.

Pseudocode 1 illustrates a simple description of the basic learnheuristic framework for solving our DPOP. First, a white box learner is generated using a machine learning model trained with historical data, and an initial dummy solution is saved as the current best solution. Then, at each iteration, the VNS algorithm described in Subsection 4.1 is used to select the assets and construct the portfolio. Once the portfolio is constructed, the white box learner is employed to predict the synergy effect of assets in the portfolio, and the optimal weight of assets is allocated according to the objective function (Equation 11) using the predicted synergy coefficients. This iteration is repeated until the maximum time constraint T_{loop} is reached. Finally, the best solution obtained is modeled in the black box simulator, and the actual risk is obtained to compare with the white box predicted risk.

6 Computational Experiments

6.1 DPOP

The proposed algorithm was executed as a Python application. We used a standard personal computer, Intel Core i7 CPU at 2.5 GHz and 12GB RAM with Windows 10 to perform all tests. The same dataset and metaheuristic described in the previous section are used to conduct the experiments. In order to evaluate our learnheuristic approach, the set of asset combinations that can trigger synergies are created, the ID of these assets are: (4, 8), (21, 22, 30), (14, 15), (19, 20, 21), and (27, 29), with their respective

Algorithm 1 Basic structure of the learnheuristic.

```
learnheuristic(instance, T)
t ← 0
WhiteBoxPredictor ← getMLPredictor(instance)
bestSol ← dummySolution(instance)
while {t < T} do
    portfolio ← constructPortfolio(instance)
    newSol ← allocateWeight(portfolio, WhiteBoxPredictor)
    if cost(newSol) ≤ cost(bestSol)} then
        bestSol ← newSol
    end if
    t ← elapsedTime
end while
blackBox(bestSol)
return bestSol
```

synergies coefficients 0.9, 1.2, 0.8, 0.85, and 1.15. The dataset is tested under 3 different levels of dynamism (i.e. the original coefficient is multiplied by 0.8, 1.0, and 1.2, respectively). The maximum time constraint T_{loop} is set to 15. The white box learner of learnheuristic can be any machine learning model, in this paper we employed a decision tree regressor model. Table 1 summarizes the results of our experiment. The first column displays the required return, showing only the first and the last 10 values. Columns 2, 3, and 4 are the final risk of our learnheuristic approach under the low, medium, and high levels of dynamism. Columns 5, 6, and 7 detail the gaps between the final risk of learnheuristic approach and the final risk of the static approach under the 3 different levels of dynamism. The final risk of the dynamic solutions is subtracted from that of the static solutions, therefore, a negative gap represents an improvement in the final risk of the solution. These results are also presented as the right sub-plot of Figure 1.

Table 1: Hang Seng (Hong Kong) Stock Market with dynamic covariances.

Required Return	Risk			Gaps (%)		
	Low	Medium	High	Low	Medium	High
0.002861137	0.0006111	0.0005727	0.0005463	-7.20%	-14.20%	-19.55%
0.002941981	0.0006204	0.0005774	0.0005714	-7.29%	-31.16%	-16.43%
0.003022826	0.000613	0.0005847	0.0005445	-10.77%	-29.66%	-22.94%
0.003103671	0.0006031	0.0005802	0.0005459	-19.60%	-28.87%	-36.87%
0.003184516	0.000621	0.0005766	0.0005514	-19.87%	-26.90%	-24.65%
0.003265361	0.0006217	0.0005909	0.000564	-15.22%	-12.27%	-33.24%
0.003346206	0.0006093	0.0005786	0.0005557	-18.84%	-28.43%	-23.39%
0.003427051	0.0006004	0.0005917	0.0005655	-24.65%	-23.34%	-35.47%
0.003507896	0.0006208	0.0005971	0.0005559	-9.26%	-2.45%	-38.05%
0.010056641	0.0029361	0.0028098	0.0026863	0.00%	0.00%	0.00%
0.010137479	0.0030279	0.0028977	0.0027703	0.00%	0.00%	0.00%
0.010218315	0.0031239	0.0029895	0.0028581	0.00%	0.00%	0.00%
0.01029915	0.003224	0.0030854	0.0029498	0.00%	0.00%	0.00%
0.010379986	0.0033283	0.0031852	0.0030452	-20.94%	-26.38%	-32.19%
0.010460822	0.0034368	0.003289	0.0031445	0.00%	0.00%	0.00%
0.010541657	0.0035495	0.0033969	0.0032476	0.00%	0.00%	0.00%
0.010622493	0.0036664	0.0035087	0.0033545	0.00%	0.00%	0.00%
0.010703329	0.0037874	0.0036245	0.0034652	0.00%	0.00%	0.00%
0.010784164	0.0039126	0.0037443	0.0035798	0.00%	0.00%	0.00%
0.010865	0.0047755	0.0047755	0.0047755	0.00%	0.00%	0.00%
			Average:	-7.68%	-11.18%	-14.14%

As we can observe, the final risk of the dynamic solutions is always lower than the final risk of static solutions, the average gap of dynamic solutions is -7.68%, -11.18% -14.14% for low, medium, and high levels of dynamism, respectively. The performance of the static solutions has shown a clearly decreasing trend as the synergy effect of assets increases. Furthermore, the gap of solutions is higher when the required return is relatively lower, this is not surprising as a lower return would allow more choice of assets and a more diversified portfolio, leading to a higher probability of creating synergy effects. Thus, the learnheuristic approach carries considerable benefits compared with the assumption of static risk.

6.2 SPOP

The proposed algorithm was executed as a JAVA application. We run our algorithm for the experimental stock market data from the repository ORlib (http://people.brunel.ac.uk/~mastjjb/jeb/orlib/portinfo.html). These datasets were gathered from Datastream and cover global financial and macroeconomic data. In this work, we used the Hang Seng (Hong Kong) stock market index at a weekly frequency between March 1992 and September 1997. Missing stock data were discarded. The data constitutes the mean and the standard deviation of returns from the stock market, and the correlation coefficients for all possible pairs of assets. We follow Gaspero et al. (2011) and divide the portfolio frontier into 100 equidistant points on the axis indicating the portfolio expected return.

This benchmark dataset is deterministic. In order to perform our simheuristic algorithm, we considered the following modifications and assumptions.

1. Standard deviation, S_i follows a $LN(\mu_S, \sigma_S)$, where LN denotes a log-normal distribution, and μ_S and σ_S are the mean and the standard deviation, respectively, of the natural logarithm of S_i. We assume that they take the values σ_i and $c\sigma_i$, respectively, where σ_i represents the standard deviation of the variable, and c is an input.
2. Correlation C_{ij} follows a truncated normal distribution $TN(\mu_C, \sigma_C, lb, ub)$, where μ_C is the mean, σ_C is the standard deviation, and lb and ub are the lower and upper bounds, respectively. μ_C is the correlation between the returns on assets a_i and a_j, ρ_{ij}, and σ_C is an input. We set lb and ub to -1 and 1, respectively, to ensure that the correlation varies between -1 and 1.

$c(0.01, 0.025, 0.08)$ and $\sigma_M(\sqrt{0.00002}, \sqrt{0.0002}, \sqrt{0.002})$ were used to determine three different levels of stochasticity, from lowest to highest. Following Kizys et al. (2022), β is randomly selected from a uniform distribution with parameters 0.05 and 0.25. Furthermore, sim_{short}, sim_{large}, and T_{loop} are set to 2500, 12500, and 15, respectively.

We assume stochastic covariances and plot the relationship between the required return and the expected risk for three different levels of stochasticity (low, medium, and high) for Hang Seng stock market data as the left sub-plot of Figure 1. In this figure, we only consider the first and the last 10 values of required returns and compare the best-found solutions obtained for the deterministic/static and stochastic/dynamic environments.

Unsurprisingly, the higher the returns, the higher the risks (level of stochasticity). Moreover, negative gaps between stochastic and deterministic solutions reveal that the simheuristic approach provides considerable benefits compared with the assumption of constant expected risk. Furthermore, the performance of the BSS ameliorates when covariances are more uncertain.

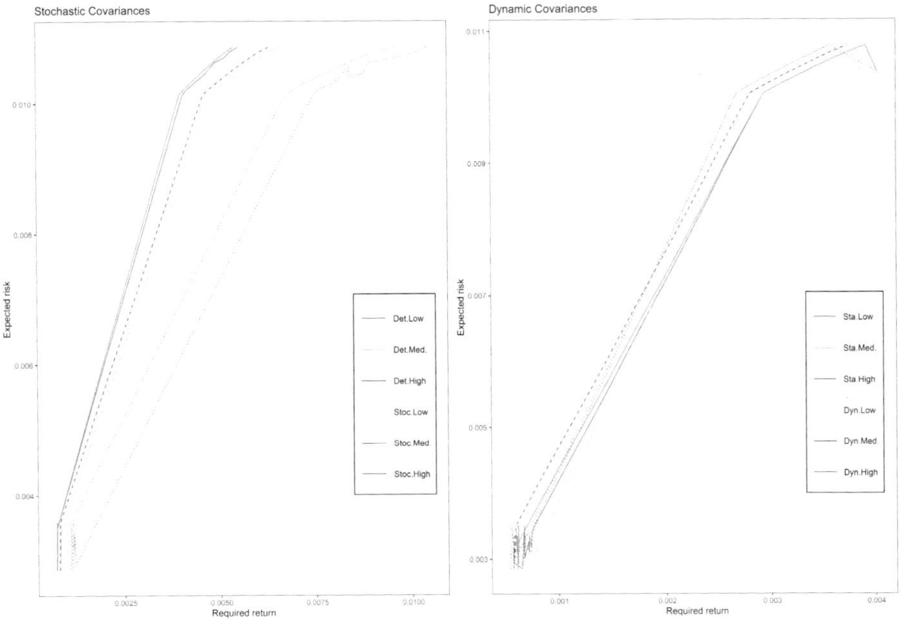

Figure 1: Risk curves for Hang Seng stock market with stochastic, dynamic covariances.

7 Conclusions and Future Research

While the literature on rich POP is extensive, most of them only considers deterministic versions of POP. However, real-life POP faced by investors is becoming increasingly complex. This can be due, among other factors, to the increasingly complex financial markets full of uncertainty and dynamism. On the other hand, advances in computing power over the last few decades have enabled the implementation of more powerful and faster algorithms, as well as the analysis of massive amounts of various types of data. As a consequence, hybrid approaches for addressing hard combinatorial and/or continuous optimization problems are becoming more popular. Therefore, this paper proposed two hybrid approaches for solving SPOP and DPOP, respectively. The first approach proposed is a simheuristic algorithm, a combination of metaheuristics with simulation for solving SPOP with stochastic covariances. Computational experiments showed considerable benefits compared to the traditional deterministic approach. The second approach is a learnheuristic algorithm, a combination of metaheuristics and machine learning for solving DPOP with dynamic covariances. Computational experiments showed that learnheuristic approach can provide significantly better results than the static metaheuristic approach.

In this paper, SPOP and DPOP are solved separately. However, in real-life scenarios, stochastic and dynamic components may coexist in the environment. Therefore, it is necessary to solve the problem with stochastic and dynamic components together. In future work, we plan to combine simheuristic and learnheuristic, i.e., hybridizing machine learning-simulation-optimization for solving optimization problems with both stochastic and dynamic components.

Acknowledgements

This work has been supported by the TÜBTAK 2219-A program, (1059B192200770).

REFERENCES

Armananzas, R., and J. A. Lozano. 2005. "A multiobjective approach to the portfolio optimization problem". In *IEEE Congress on Evolutionary Computation*, Volume 2, 1388–1395. IEEE.

Arnau, Q., A. A. Juan, and I. Serra. 2018. "On the use of learnheuristics in vehicle routing optimization problems with dynamic inputs". *Algorithms* 11 (12): 208.

Bayliss, C., A. A. Juan, C. S. Currie, and J. Panadero. 2020. "A learnheuristic approach for the team orienteering problem with aerial drone motion constraints". *Applied Soft Computing* 92:106280.

Calvet, L., J. de Armas, D. Masip, and A. A. Juan. 2017. "Learnheuristics: hybridizing metaheuristics with machine learning for optimization with dynamic inputs". *Open Mathematics* 15 (1): 261–280.

Calvet, L., A. Ferrer, M. I. Gomes, A. A. Juan, and D. Masip. 2016. "Combining statistical learning with metaheuristics for the multi-depot vehicle routing problem with market segmentation". *Computers & Industrial Engineering* 94:93–104.

Cesarone, F., A. Scozzari, and F. Tardella. 2013. "A new method for mean-variance portfolio optimization with cardinality constraints". *Annals of Operations Research* 205 (1): 213–234.

Doering, J., R. Kizys, A. Juan, A. Fit, and O. Polat. 2019. "Metaheuristics for rich portfolio optimisation and risk management: Current state and future trends". *Operations Research Perspectives* 6:100121.

Gaspero, L. D., G. D. Tollo, A. Roli, and A. Schaerf. 2011. "Hybrid metaheuristics for constrained portfolio selection problems". *Quantitative Finance* 11 (10): 1473–1487.

Gruler, A., J. Panadero, J. de Armas, J. A. M. Pérez, and A. A. Juan. 2018. "Combining variable neighborhood search with simulation for the inventory routing problem with stochastic demands and stock-outs". *Computers & Industrial Engineering* 123:278–288.

Guimarans, D., O. Dominguez, J. Panadero, and A. A. Juan. 2018. "A simheuristic approach for the two-dimensional vehicle routing problem with stochastic travel times". *Simulation Modelling Practice and Theory* 89:1–14.

Hansen, P., N. Mladenović, J. Brimberg, and J. A. M. Pérez. 2019. "Variable neighborhood search". In *Handbook of metaheuristics*, 57–97. Springer.

Juan, A. A., B. B. Barrios, E. Vallada, D. Riera, and J. Jorba. 2014. "A simheuristic algorithm for solving the permutation flow shop problem with stochastic processing times". *Simulation Modelling Practice and Theory* 46:101–117.

Juan, A. A., J. Faulin, S. E. Grasman, M. Rabe, and G. Figueira. 2015. "A review of simheuristics: Extending metaheuristics to deal with stochastic combinatorial optimization problems". *Operations Research Perspectives* 2:62–72.

Kalayci, C. B., O. Ertenlice, and M. A. Akbay. 2019. "A comprehensive review of deterministic models and applications for mean-variance portfolio optimization". *Expert Systems with Applications*.

Kizys, R., J. Doering, A. A. Juan, O. Polat, L. Calvet, and J. Panadero. 2022. "A simheuristic algorithm for the portfolio optimization problem with random returns and noisy covariances". *Computers & Operations Research* 139:105631.

Kolm, P. N., R. Tütüncü, and F. J. Fabozzi. 2014. "60 Years of portfolio optimization: Practical challenges and current trends". *European Journal of Operational Research* 234 (2): 356–371.

Lee, J.-H., and H.-J. Kim. 2022. "Reinforcement learning for robotic flow shop scheduling with processing time variations". *International Journal of Production Research* 60 (7): 2346–2368.

Markowitz, H. 1952. "Portfolio selection". *Journal of Finance* 7 (1): 77–91.

Milhomem, D. A., and M. J. P. Dantas. 2020. "Analysis of new approaches used in portfolio optimization: A systematic literature review". *Production* 30.

Raba, D., A. Estrada-Moreno, J. Panadero, and A. A. Juan. 2020. "A reactive simheuristic using online data for a real-life inventory routing problem with stochastic demands". *International Transactions in Operational Research* 27 (6): 2785–2816.

Sehgal, R., and A. Mehra. 2020. "Robust portfolio optimization with second order stochastic dominance constraints". *Computers & Industrial Engineering* 144:106396.

Taş, D., N. Dellaert, T. Van Woensel, and T. De Kok. 2013. "Vehicle routing problem with stochastic travel times including soft time windows and service costs". *Computers & Operations Research* 40 (1): 214–224.

Wang, Z., and L. Lin. 2013. "A simulation-based algorithm for the capacitated vehicle routing problem with stochastic travel times". *Journal of Applied Mathematics* 2013.

Wu, X., X. Shen, and Q. Cui. 2018. "Multi-objective flexible flow shop scheduling problem considering variable processing time due to renewable energy". *Sustainability* 10 (3): 841.

Zhai, Q., T. Ye, M. Huang, S. Feng, and H. Li. 2020. "Whale optimization algorithm for multiconstraint second-order stochastic dominance portfolio optimization". *Computational Intelligence and Neuroscience* 2020.

AUTHOR BIOGRAPHIES

YUDA LI is a predoctoral researcher at the ICSO research group at Universitat Oberta de Catalunya (Spain). He holds a BSc in Aeronautical Management from the Universitat Autonoma de Barcelona and a Msc in Computational Engineering and Mathematics from the Universitat Rovira i Virgili. His email address is yli1@uoc.edu.

ONUR POLAT is an Associate Professor in Public Finance at Bilecik University. He is also a postdoctoral research fellow at the Dept. of Applied Statistics & Operations Research at the Universitat Politcnica de Valncia. He received an MSc in Financial Mathematics from the Middle East Technical University in 2009. He completed his PhD in Economics at Hacettepe University in 2017. He held a visiting scholar position at Illinois State University. His website is https://sites.google.com/view/onurpolat/opolat

LAURA CALVET received a BSc on Applied Statistics from Universitat Autnoma de Barcelona (UAB) in 2012. She completed her PhD at the Open University of Catalonia in 2017. She is currently a lecturer of Supply Chain Management in the Department of Telecommunications and Systems Engineering at the UAB. Her website is: https://sites.google.com/view/lauracalvetlinan.

RENATAS KIZYS is an Associate Professor in Finance within Southampton Business School at the University of Southampton. He received BA and MA degrees in Economics from the University of Vilnius, Lithuania, and MSc and PhD in Economics from the University of Alicante, Spain. He also graduated from the Advanced Studies Program in International Economic Policy Research at the Kiel Institute for the World Economy, Germany. Prior to joining the University of Southampton, Renatas Kizys spent 8 years working at the University of Portsmouth, where he held various positions, ranging from Lecturer to Reader. He spent more than three years working as an Assistant Professor at the Technological Institute of Monterrey, Mexico. He held a P.K. Woolley Research Fellowship within the Department of Economics and Related Studies at University of York, in the UK. He has also held Visiting Academic positions in China, Germany, Lithuania, Indonesia and Spain. His institutional website address is https://www.southampton.ac.uk/people/5xmnbw/doctor-renatas-kizys and his email address is r.kizys@soton.ac.uk

ANGEL A. JUAN is a Full Professor in the Dept. of Applied Statistics & Operations Research at the Universitat Politcnica de Valncia (Spain). Dr. Juan holds a Ph.D. in Industrial Engineering and an M.Sc. in Mathematics. He completed a predoctoral internship at Harvard University and postdoctoral internships at the Massachusetts Institute of Technology and the Georgia Institute of Technology. His main research interests include applications of simheuristics and learnheuristics in computational logistics, transportation, and finance. He has published more than 130 articles in JCR-indexed journals and over 300 papers indexed in Scopus. His website address is http://ajuanp.wordpress.com and his email address is ajuanp@upv.es

Proceedings of the Operational Research Society Simulation Workshop 2023 (SW23)
C. Currie and L. Rhodes-Leader, eds.
https://doi.org/10.36819/SW23.021

HEALTH SYSTEMS SIMULATION MODELLING TO SUPPORT DECISION-MAKING IN COVID-19 PREVENTION AND CONTROL IN CARE HOMES

Dr. Le Khanh Ngan Nguyen

Department of Management Science
University of Strathclyde
Nguyen-le-khanh-ngan@strath.ac.uk

Prof. Susan Howick

Department of Management Science
University of Strathclyde
Susan.howick@strath.ac.uk

Dr. Itamar Megiddo

Department of Management Science
University of Strathclyde
Itamar.megiddo@strath.ac.uk

1 BACKGROUND

The COVID-19 pandemic has highlighted care homes' vulnerability to infectious disease outbreaks and the lack of context-specific best practice infection prevention and control (IPC) guidance for this setting. Care homes, where the majority of residents are elderly and have complex medical and care needs, have suffered devastating outcomes. Suspension of visiting due to COVID-19 outbreaks has also caused substantial unintended harms to the health and wellbeing of residents.

Since April 2020, we have collaborated with decision-makers from Health and Social Care Lanarkshire (HSCL) in Scotland, the Scottish Government, and the UK Government Department of Health and Social Care to provide evidence to support them in mitigating the impact of COVID-19 in care homes. We developed novel systems simulation models, which included agent-based models (ABM) and hybrid System Dynamics (SD) – ABM models, to help address the gaps in IPC knowledge and practice in this setting.

2 APPROACH

The initial objective of the ABM was to simulate the transmission dynamics of COVID-19 via contacts between individuals in a care home setting to evaluate the effectiveness of a range of intervention strategies relating to testing of staff and residents, using PPE, visiting policy, and cohorting (Nguyen et al. 2020 and Nguyen et al. 2021). When the vaccine rollout was launched in winter 2020, we adapted the model to explore the impact of different vaccination strategies and the impact of lifting routine testing of staff when vaccine coverages among staff and residents reach a certain threshold.

The objective of the integrated hybrid SD-ABM model was to understand the impact that temporary bank/agency staff, who work across multiple care homes, have on the spread of COVID-19 in care homes and to evaluate the effectiveness of a range of intervention strategies (Nguyen, Megiddo, and Howick, 2022).

We collected data from various sources. We interviewed care home stakeholders, including managers, staff in different roles, and had regular discussions with representatives from the Health and Social Care Partnerships, Public Health in Lanarkshire, and Scottish Government. We utilised these interviews and discussion to scope the problem, build the models, and design the intervention strategies. We also conducted literature reviews to obtain the values for parameters characterising the transmission of COVID-19 and the disease progression. Other parameters are based on national data (Scotland and UK) and regional data for North Lanarkshire where available. We gained confidence in the modules and the overall hybrid model using several approaches adapted from both SD and ABM practice.

3 IMPACT

At the time of our ABM study, no other published models considered elements specific to care homes, and interventions proposed by wider population models (e.g., closure and social distancing in schools) were not suitable for this setting—care homes act as a residence and staff interaction with residents is often unavoidable. We worked with partners in care homes and public health to model behaviour in these settings. To the best of our knowledge, our hybrid model is the first study that evaluated the effects of different interventions targeting bank/agency staff working across multiple care homes. Our work provided our partners with evidence that was more timely and easier to justify than guidance that was being provided, at the time, from central government.

Our models have improved understanding of what interventions work well in the environment of care homes, which has contributed to their implementation. Our early models, which we later published (Nguyen et al. 2020 and Nguyen et al. 2021), contributed to decisions on several interventions, including who to test in care homes and at what interval, creation of smaller cohorts of residents and staff, and development of visitation policy. HSCL stated that the decision to test staff weekly "potentially averted an estimated 9,250 COVID-19 cases among the 37,000 care homes residents in Scotland over a period of 3 months". The decision not to test residents but staff "has saved approximately £8.4 million in Scotland". Our work contributed to understanding the circumstances under which care homes could permit visiting, and the Scottish Government changing its visitation policy based on this evidence helped promote the mental health and well-being of residents and their families (SDG3). The modelling provided evidence underpinning decisions about effective interventions in care homes which, as stated by HSCL, "safeguarded the physical and mental health of residents, preventing unnecessary distress and reducing morbidity and mortality". The findings from our hybrid SD-ABM model have policy implications for care homes, which are heavily reliant on bank/agency staff due to staff shortages. The work is informing current policy and IPC guidance in these settings. Our work has received media attention which helps engage public audience and relevant stakeholders in meaningful conversations about the problem raised in the research.

REFERENCES

Nguyen, L. K. N., Howick, S., McLafferty, D., Anderson, G. H., Pravinkumar, J., Van Der Meer, R., & Megiddo, I. (2020). Evaluating intervention strategies in controlling COVID-19 spread in care homes: an agent-based model. *Infection Control and Hospital Epidemiology*. https://doi.org/10.1017/ice.2020.1369

Nguyen, L. K. N., Howick, S., McLafferty, D., Anderson, G. H., Pravinkumar, S. J., Van Der Meer, R., & Megiddo, I. (2021). Impact of visitation and cohorting policies to shield residents from covid-19 spread in care homes: an agent-based model. *American Journal of Infection Control*, *49*(9), 1105-1112. https://doi.org/10.1016/j.ajic.2021.07.001

Nguyen L.K.N., Megiddo I., Howick S. (2022) Hybrid simulation modelling of networks of heterogeneous care homes and the inter-facility spread of Covid-19 by sharing staff. *PLoS Computational Biology* 18(1): e1009780. https://doi.org/10.1371/journal.pcbi.1009780

Proceedings of the Operational Research Society Simulation Workshop 2023 (SW23)
C. Currie and L. Rhodes-Leader, eds.
https://doi.org/10.36819/SW23.022

EVALUATING THE ETHICAL AND SOCIETAL IMPACTS OF MODELLING PANDEMIC CRISES: EXPERIENCES FROM A WORKSHOP

Dr. Katrina Petersen

Trilateral Research
1 Knightsbridge Green, London,SW1X 7QA
Katrina.Petersen@trilateralresearch.com

Ms Susannah Copson

Trilateral Research
1 Knightsbridge Green, London, SW1X 7QA
Susannah.Copson@trilateralresearch.com

Dr. Anastasia Anagnostou

Modelling & Simulation Group,
Department of Computer Science, Brunel University London
Kingston lane, Uxbridge, UB8 3PH
Anastasia.Anagnostou@brunel.ac.uk

ABSTRACT

This paper describes a hybrid (virtual and online) workshop held as part of the EU STAMINA project that aimed to engage project partners to explore ethics and simulation modelling in the context of pandemic preparedness and response. The purpose of the workshop was to consider how the model's design and use in specific pandemic decision-making contexts could have broader implications for issues like transparency, explainability, representativeness, bias, trust, equality, and social injustices. Its outputs will be used as evidence to produce a series of measures that could help mitigate ethical harms and support the greater possible benefit from the use of the models. These include recommendations for policy, data-gathering, training, potential protocols to support end-user engagement, as well as guidelines for designing and using simulation models for pandemic decision-making. This paper presents the methodological approaches taken when designing the workshop, practical concerns raised, initial insights gained, and considers future steps.

Keywords: Agent-Based Simulation, Discrete-Event Simulation, Workshop, Ethics, Societal Impacts, Pandemics, COVID-19, Forecasting, Collaboration

1 INTRODUCTION

Models are a common tool for forecasting and exploring possible implications of different situations or decisions. This is especially the case in pandemic preparedness and response, where models become a valuable source of insights into these uncertain and infrequent crises. As predictive modelling is further folded into decision-making processes, it is important to consider the ethical and societal implications of using models within pandemic risk management decision-making processes. This paper describes a hybrid virtual and in-person *ethical and societal impact assessment workshop* developed as part of the European Commission H2020 project, STAMINA, to bring together pandemic practitioners and modellers to engage a series of ethical and societal issues in a semi-structured format.

Our aim with such a workshop was to set out a series of activities to collaboratively and cross-disciplinarily identify baseline understandings, assumptions, and challenges around a series of ethical issues that related to the specific models being developed in the project for pandemic risk management. To conduct the workshop, we developed methods that encouraged pandemic practitioners and model developers to engage in dynamic discussions and hands-on model play in order to identify, consider, and unpack the complexities of challenging issues such as inclusiveness, fairness, bias, representativeness, stigma, accountability, and uncertainty (Leonelli, 2016; Middlestadt et al 2016). From the results, we hope to produce a series of policy and data-gathering recommendations, suggestions for what kind of training would be most beneficial to deploy along with the models, potential protocols that would benefit end-users as they engage with how the models conceptualise pandemic risks, as well as future guidelines for designing and using models for pandemic decision-making. We found that approaching these issues in ways that allowed for interactive play with the models and their outputs elicited surprise in both the pandemic practitioners and modellers present as they discovered unexpected assumptions as well as new ways of asking questions about the quality of their models, data, and starting questions.

2 CONSIDERING ETHICAL AND SOCIETAL IMPACTS

Considering ethics and societal impacts acknowledges societal concerns, embedded values, and the potential impact of data and technology (Middlestadt et al, 2016). However, ensuring the societal and ethical issues are key considerations throughout the life cycle of research projects can be a challenge. They are abstract, hard to pinpoint (e.g. multiple definitions of fairness exist, all equally relevant yet clash with each other), and very contextual. For example, a 70/30 gender split could be representative of a group in one situation, but might point to biased results in another. Moreover, much of how ethical and social issues arise depends upon the use of a model by specific people in specific contexts (Nissenbaum, 2004). This is especially true when developing data-driven tools – like models – for pandemics (Boersma and Buscher, 2022). Exploring these implications requires hands-on, interactive, context- and situation-based approaches. One way to do this is through multi-disciplinary collaborative workshops that bring together those that use models with those that make them (Petersen et al, 2016; Petersen et al, 2015).

Interactive workshops, focused on ethical and societal issues, can enable intervention and change. The objective of such a workshop is to increase the reflexivity of the process and of decision-making, not to prescript or pre-define the results of an assessment process. The aim is to make designers and end-users aware of morally opaque features of design and use (Wright, 2011; Brey, 2000). It is proactive and considers future consequences and impacts of proposed actions. It encourages both users and developers to engage in critical scrutiny and to think about ethical and social considerations. Just as much, in the process, they each learn more about the problem domain, their assumptions, and their decision-making needs, opening debates about alternative trajectories. By the end, they should be able to reflect upon the potential impacts of different design moves, diverse user practices, and the possible kinds of knowledge generated and responsibilities gained considering the ambiguities of real-world experience (Leonelli, 2016). Such processes also contribute towards informed decision-making, protection of societal concerns, and overall effective strategy for increasing the benefits and sustainability for end-users (Reijers et al, 2018).

2.1 The Project and the Need to Consider these Issues

The STAMINA project researches pandemic preparedness and response technologies to better support first responders and national planners. It develops a wide range of tools, including predictive models among other (Bakalos et al, 2022). Three simulation models were used in the workshop that model different aspects of the COVID-19 pandemic. These include the dynamiC Hospital wArd Management (CHARM) Discrete-Event Simulation (DES) for hospital bed capacity management (Anagnostou et al, 2022), the Flu and Coronavirus Simulation (FACS) Agent-Based Simulation (ABS) that models the geospatial spread of the disease and the impact of preventive measures (Anagnostou et al, 2022), and the CoronAvirus Lifelong Modelling and Simulation (CALMS) ABS that considers long-term policy intervention effectiveness (Mintram et al, 2022).

The workshop was designed for STAMINA end-users using CHARM and/or FACS in their pilot trials and therefore were familiar with the models' structure and functionalities and not using CALMS so they were asked to comment on a model that they were not familiar with.

Four pandemic practitioner teams from across Europe and neighbouring countries participated in the workshop along with the model developers. These include practitioners from Lithuania and

Romania who worked with CHARM and FACS, from Tunisia who used only CHARM and from the UK who used only FACS in their pilot trials.

Before designing the workshop, we had already identified a series of ethical and societal issues drawing on literature reviews and early workshops with partners in which we discussed more abstract benefits and values. They showed some general trends of the need for better understanding of: 1) what data is to be collected and how this data is used in the models, 2) how to present and interpret the results, and 3) what role uncertainty plays in the models' outputs. It was also identified that training and hand-on experience was needed in order to gain trust on the models use and interpretation.

As we designed the workshop, we were presented with a challenge that is familiar to working during a pandemic: the need to work with a blend of participants attending in-person and online. All participants worked with models that were familiar with as well with models that they had not used before. The model developers, who all participated in-person, were supporting both in-person and online teams to use the models and run example scenarios.

In the next sections we discuss how we worked in this setting and the workshop findings.

3 WORKING THROUGH A HYBRID WORKSHOP

Our objectives for the workshop included understanding how to best use models to ensure inclusive benefits, collaboratively considering ethical impacts in and around pandemic predictive modelling and exploring how this relates to decision-making and model interpretation. The workshop adopted a hands-on and interactive approach, based upon a series of activities in which end-users from different regions and model developers interact with one another and work collaboratively. When designing the sessions, the aim was to elicit insights and recommendations for actions within and beyond the project to enrich the project's assessment of ethical and societal impact and further inform future model development.

What are the outcomes...
...for us:

- Validate, rewrite, and expand our recommendations for engaging ethics and societal issues raised by the design and use of the models in pandemic planning.
- Further enrich our understanding of the socio-ethical issues at play.
- Develop a list of potential future design, policy, protocol, and training needs.

...for the participants:

- Better understanding of the model and confidence in models.
- A list of their needs – data, training, definitions, additional knowledge – when using the models.
- Model results and interfaces that are more understandable and justifiable within their decision-making practices.
- Better communication practices, especially how their communication with models includes/excludes.
- Innovative ideas for how to improve decision-making or models for pandemic preparedness.

3.1 Methodology

In order to explore the relationship between predictive modelling and ethics, the workshop adopted overarching themes that did not explicitly use ethics terms in order to encourage participants to enter the activities with open minds. These included usability and explainability, communication and decision-making, and uncertainty. Each activity required a different approach. This was, in part, related to content: because of the challenges of eliciting ethical and societal impacts related to the themes. It was also partly related to the format: we needed to ensure those participating online remained active and had a voice, which is harder when sitting muted behind a screen.

Practically speaking, planning the activities had to consider a few key elements. First, we wanted to avoid situations where in-person discussions turned into a webinar for those online. This meant making sure all participants in each activity had an active role. Second, we were aware that being together in a room can build energy while being behind a screen can reduce it. Activities were broken into parts, no more than thirty minutes long, each new segment eliciting new kinds of thinking and input from all participants. Third, being online, a flat image on a big screen makes sitting around a table difficult. This meant large group activities were not going to be effective. Instead, we opted for small groups where the online participants were each tied to a different laptop, walking around the room with us, sitting face-to-face with those they were interacting with in more intimate interactions. Finally, such workshops ideally involve participants doing a lot of collaborative thinking and note-taking, often working with sticky notes, whiteboards, large sheets of paper, or drawing on printouts with markers. But if we had taken notes this way, online participants could not have interacted equally. To ensure they were included, we worked with online whiteboards and shared documents. Although seeing everyone group around laptops looked a bit unconventional in the physical room, this arrangement meant all involved could hear, speak, see, and write in the same hybrid space.

One additional workshop design consideration emerged around the practicalities of doing such a workshop. While sitting in a room it is easy to call on a quiet person, observe if a quiet person is taking their time to think, or assess if they are being hesitant to speak without an opening directed at them. It is also possible to readily observe body language to see, for example, if someone understands you or is getting bored, and adjust activities accordingly. The same is not as readily possible when participants are online, especially if cameras are turned off. In particular, it is difficult to know if silence heard is because a participant is listening or if they are having technical problems. To help alleviate this, we had a colleague who was not present in-person also participate online, jumping between groups and helping monitor that side of the virtual room to support when needed.

Thematically, we selected a series of ethical and societal issues previously identified as relevant for the models in the contexts of the pandemic practitioners' pilot scenarios. These included concerns around what makes good evidence, how models (and the data they use) are representative of the communities and risks they are representing, how the models support transparency and justifications, concerns about confirmation bias, fairness and bias in how the models outputs relate to the decisions they are evidence for, how the models represented vulnerability, and how working with models can impact stigmatisation of people with specific characteristics. These issues are very difficult to discuss, even more so with a mix of technical and non-technical partners. To help, we designed activities that addressed them indirectly, slowly approaching the issues, in some cases without even naming them, but by raising situations where insights related to them could emerge.

Each activity was structured around both interactive activities and playing with the models themselves. We divided participants into small groups, which were reshuffled regularly throughout and across the activities to capture different combinations of thought processes and expertise, and to keep the discussion fresh. We prepared questions to prompt ideas where needed, but encouraged free-flowing discussion between participants. We viewed the activities as learning opportunities for everyone, as each person had valuable knowledge that enriched the discussions and output of the workshop. Below are descriptions of some of the key activities.

3.2 Usability and Explainability

The first session aimed to engage participants in an interactive activity and discussion to identify how to engage aspects of transparency (and thus accountability, justification, understanding, and communication) the models in practice. We wanted to map assumptions the participants had about models, and therefore what model users and designers of them need to know in order to engage them in the most ethically and societally sound ways, and explore how their various assumptions impact user engagement and decisions with models. After an initial mapping, we asked the model-users present to work in a less familiar context: we had them look at the results from another participant's use of the same model, with that other participants' (different) data and starting questions, to see if what they had assumed was needed for understandability and explainability was sufficient or if additional elements emerged. As a whole, we aimed for a better understanding of 1) how models should/could be transparent about the processes and actions that underlie their design and outputs; and 2) how the models' decisions are being recorded so that decision-making processes can be explained either internally or externally.

We presented the group with an interactive whiteboard with a blank grid that displayed types of explainability and the predictive models that the workshop explored (see Figure 1). The figure aims at showing the richness of ideas under each theme. They were prompted with some starting questions:

- What do you need to know to understand the model? Why these features?
- What do you need from a model to be able to justify its output to others?

As they proceeded in the activity, follow questions included:

- How to avoid evidence from models being used to confirm pre-existing ideas (ensuring a model is read for what it means, not what a user might expect)?
- How can users be supported in seeing when difference matters in a model (from previous situations, across different regions, for different types of decisions)?

Participants then filled out ideas on virtual post-it notes as a quickfire mind-mapping exercise.

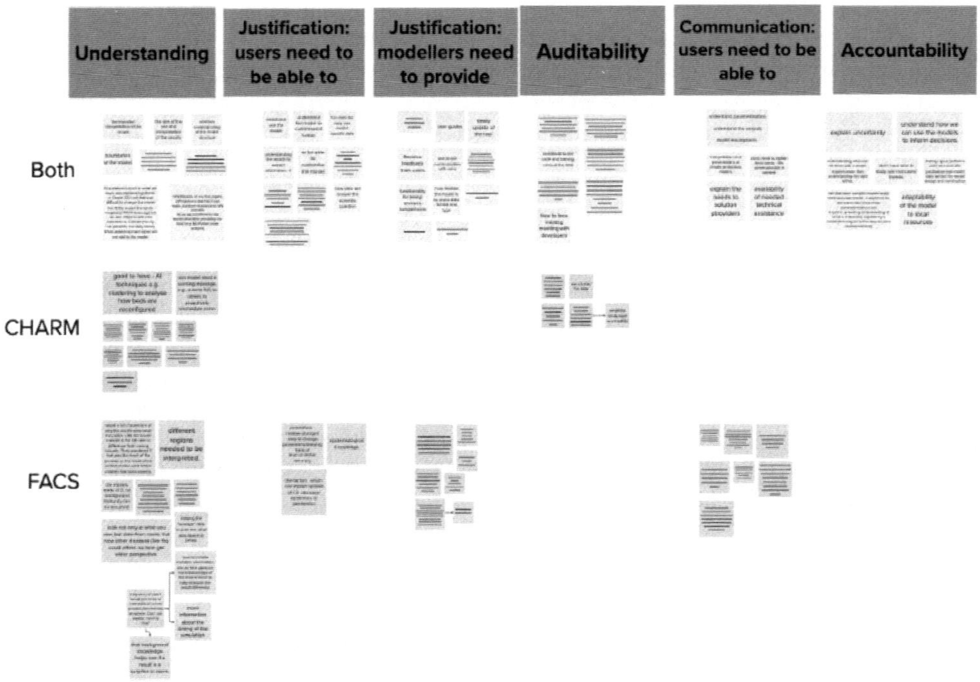

Figure 1 *Screenshot of the interactive transparency-themed whiteboard, populated with input from the session aiming at showing the richness of ideas elicited by the ethical themes.*

3.3 Communication and Decision-Making

The second session used the world café method. This is an activity in which participants circulate between 'stops' to discuss different issues. Each stop is short and timed, often around 5 minutes, aiming for just enough time to get some (incomplete) ideas down on paper. As people circulate between stops, they engage with different participants and build upon the ideas shared by previous groups. In our case, each 'stop' was a key term: quality, stigma, representative, vulnerability, forecasting. The purpose of this exercise was to co-define terminology commonly related to modelling that carries ethical implications. We wanted to understand where differences in definitions matter (both across practitioner regions, as well as between practitioner and modeller) and explore how these definitions could potentially affect the ethical outputs of decision-making processes. We aimed to:

- Learn how models can support or hinder shared baseline understandings.
- Identify what can be a co-defined common terminology or where differences matter.
- Build a list of societal and ethical implications from such misunderstandings.

Participants were split into groups of three and asked to fill in their definitions of keywords on a virtual whiteboard. Figure 2 shows the variety of ideas for each keyword. As in Figure 1, our aim is to show the richness of ideas rather than dive into the details of the activities. A timer was set for seven minutes for participants to discuss and make notes, and then each person was moved to a new word with a new group. There were five rounds to ensure different combinations of people and expertise, draw out different assumptions and ideas and cover all definitions. As a practical note, to make this work, we had to pre-assign breakout groups for each round to ensure that all participants were able to reach each station and work with different participants as they circulated. This made for much more time-intensive planning than a fully in-person world-café were participants self-select their movements.

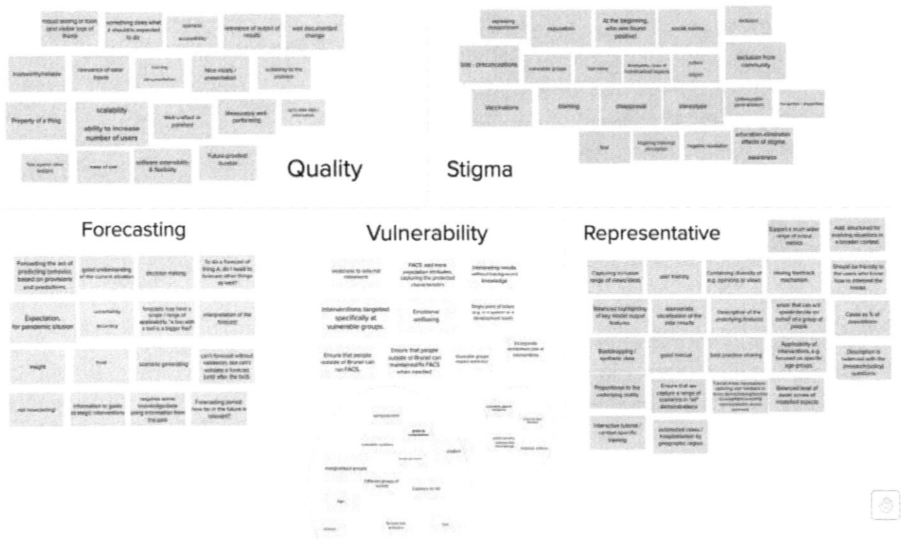

Figure 2 *Screenshot of the interactive world-café whiteboard, populated with input from the session aiming at showing the variety of the ideas of what defines its term.*

After this was completed, we asked the groups to focus on how they could fold the elements just identified as part of the terms 'vulnerability' and 'representative' into the model. Divided into small groups of three or four, participants not only interacted with each other, they had the opportunity to play with the models by asking modellers to change outputs, or, if the models could not be manipulated, they were encouraged to draw ideas on paper to imagining what a future model could be. In these sessions, participants were asked:

- What are the key differences or commonalities in the definitions?
- How do you know the model is legitimate towards these definitions?

Finally, they explored how the models could be enhanced, modified, or complemented in order to best address the definitions just created through the world café.

3.4 Uncertainty and What Could Go Wrong

This exercise began with a full-group brainstorm of 'worst case scenarios'. Participants were asked to consider what could go wrong with the models' use and what would be a negative outcome or failure of the model. Notes were taken on a virtual whiteboard. Going around the room and giving each person a chance to share, we collected answers to these questions:

- What could go wrong with how a model is used?
- What would be a negative outcome/failure with the model?
- What current problems remain?
- What new problems emerged?

Using the answers as a kind of scenario, we then moved on to how these worst cases point to sources of uncertainty and to consider the impacts that could arise from that uncertainty. These could be in model, data, or practice. Uncertainty, here, was intended as a proxy for talking through ethical and societal implications surrounding trust, equality, and social injustices. For example, we wanted to think about the implications for society when acting on the always incomplete data that make up pandemic information. We also wanted to consider what it might mean to have policy built upon a model that negatively affects a group or decreases public trust.

Participants were split into two groups, one familiar with FACS and the other familiar with CHARM. Both groups were paired with a modeller and facilitated by a moderator. The moderator led the group through an online worksheet to explore how these worst cases lead to sources of uncertainty in the context of their experience and respective model of focus, potential impacts that could arise from uncertainty and how to communicate about uncertainty and its risks to decision-makers and the public.

- What are some of the sources of uncertainty?
- What are the impacts of uncertainty?
- What risks could arise from that uncertainty?
  - *Are they equally spread across society?*
  - *Do they impact some groups/people more than others?*
  - *Does it change over peace time or wartime?*
- Can uncertainty lead to...
  - *Exclusion?*
  - *Bias?*
  - *Unfavourable generalisations?*
- What is needed to communicate about uncertainty and its risks?
  - To decision-makers?
  - To the public?
- How do you justify your use of the model with these uncertainties?

To encourage discussion, the moderator asked a series of questions designed specifically for each model to target ethical considerations. The group then reconvened to share key insights from the discussions and offer thoughts on what should be done next.

4 KEY FINDINGS

Through that engagement and co-design methodology, we were able to collaboratively produce evidence to build recommendations for how to address ethical and societal implications of pandemic models from both a technical partner and end-user perspective. These recommendations ended up being future steps for the project and models. We also were able to understand the value and challenges of such a format for conducting this work.

4.1 Ethical and Societal Insights

As part of the post-workshop analysis, findings were split into categories relating to the ethical impact assessment. They are as follows:

1. *Considering transparency, justification and accountability*

Over the course of the workshop, the case for training was made multiple times for varying purposes:. helping users understand model boundaries, data requirements and their limitations/biases, and understanding the specifics of the model they are using. It would also help develop user knowledge on parametrisation, model outputs and model assumptions.

Another key theme was communicating models and uncertainty. While we approached uncertainty as a way to consider trust, it also was key to transparency and explainability. The importance of communicating that models are predictive, not reality, was repeatedly emphasised. Ensuring this understanding may aid trust in the decisions made based upon model output. Promoting public understanding of imperfect but beneficial strategies was also found to be key. It was identified that the language of communications (e.g. 'uncertainty' and 'confidence') isn't right for public communication, as it means different things to trained modellers than it does to the public. This urged the importance of developing a better lexicon of non-scientific communication.

2. *Engaging challenges to representativeness*

Adding new parameters into models takes a lot of work for one small addition. Doing so can sometimes increase accuracy or nuanced analysis but it can also have the opposite effect, where moving away from abstraction can increase uncertainty. It can change nothing, make minimal impact, or even decrease the quality and accuracy of the results. Despite this, we often need additional parameters to get more developed insights for different communities. To aid this, it was repeatedly emphasised that models should be interpreted by experts to bridge the gap between the modeller, decision-maker and the public.

Another challenge is that some of the models, though intended to directly affect decisions about people, do not include people. For example, mapping hospital bed availability is about physical resources. But just having a bed does not mean there is medical staff to man the bed or that the bed goes to the most vulnerable. Different models are based on different assumptions, different priorities, and different data, all injected in slightly different ways. When they are combined discrepancies can emerge, and assumptions can be unpicked. Representativeness, thus, might have to come from building a framework of contextualisation around a model that gives it the decisions it is intended to support the qualities they need. As a whole, a better understanding of the processes that encompass the models and their data is needed to engage representativeness as a process.

3. *Considering stigma and differential vulnerability*

When discussing 'what could go wrong', participants raised an interesting tension when disaggregating data. While the importance of looking at demographic splits of data to understand differential impact was recognised, it was countered that doing so can draw accusations of stigmatising specific groups and carry the responsibilities and liabilities of doing so. It was then discussed whether such treatment of data and communities is beneficial or discriminatory; what could go wrong if we tried to better understand populations, but what could go wrong if we do not.

4. *Building trust*

Trust is built upon considered communication. The impact on trust of wrong or bad decisions as well as models that did not match with the data, was discussed extensively. This was particularly salient when discussing the impact on trust between decision-makers and models, decision-makers and the public, and public compliance with intervention measures. Building trust requires that modellers communicate clearly how models have been built, how they can use data in ways that improves quality, and – almost most importantly – the modellers expectations of how their model will be used. It also involves a pandemic practitioner to appropriately communicate how a model influenced their decisions, and what, other than the model, was used to make such a decision.

4.2 Methodological Insights

Running the workshop posed several practical challenges that should be considered for future hybrid workshops. Some of the issues are noted by other research in their experiences in running virtual workshops (Zimmermann et al., 2021; Tako and Kotiadis, 2021). First, sound was a problem. While we had a good quality all-room speaker and microphone – which was an absolutely necessity – when doing the smaller group activities we had to rely on laptop microphones and speakers which were often insufficient to capture only the sounds of the group or to be loud enough to be heard over the noise in the room. This was hampered by the occasional technical problems, such as one online participant suddenly not being able to unmute themselves, internet connectivity dropping for another, background noise from in-person surroundings, and the online whiteboard being inaccessible without additional driver downloads for one participant. In addition, we struggled to find online platforms that were free, accessible to all participants with their various local restrictions and connectivity limitations, and that were quick and easy to learn to use.

One unforeseen challenge was the fluidity of online participants. Online, people can move in and out of participation in different ways. For example, they can take the workshop on the go (travelling on trains with them) or they can switch who joins (swapping between two colleagues participating in the project). This posed particular challenges for activities where we had pre-assigned groups. In these cases, we would have benefits from an additional supporting leader, who could focus on addressing these live-time challenges without taking away time from notetaking or moderating.

Hands on working with the models also had its challenges. In-person and online participants were interacting with models that were set up on local machines. Further, all developers that supported the running of the experiments were participating in-person. Although, there were facilities to screensharing and remote controlling the models, there was a disparity at the level of model-user interaction between the in-person and online participants. In hindsight, we think that enabling online live use of the models would have benefit equally both cohorts. Nonetheless, this would have required more developing efforts.

5 CONCLUSION AND FUTURE WORK

In this paper, we have presented activities and lessons from a hybrid virtual and in-person workshop for exploring ethical and societal implications of models for pandemic planning and eliciting recommendations for how to act on them. While designed in ways that were particularly relevant to the research problem context and participant attendance situation, the value of such activities can be generalized and re-used/re-purposed in other situations. As a whole, we think that such activities can provide valuable insights if used iteratively in a project – as potential model users are learning about the models, developing their approaches to using the models and experimenting with how they fold into their decision making processes, as well as after they have had time to engage with models after they have been populated with participant-specific data for a participant defined scenario. Real challenges exist around keeping participants all connected in the same hybrid space and there are some limitations to how far models can be adapted in real-time to explore different ideas or recommendations as they emerge. Future work is still needed to understand how actionable some of the recommendations are, to assess the effectiveness of recommended trainings and protocols to mitigate ethics concerns, and to translate key insights into practical policy recommendations.

Key benefits to such approaches appeared for all participants. They included helping model users explore their own assumptions as they experienced how other participants used the same models; supporting modellers consider the implications of different design and data choices they could make as they further enhance their models; and providing the grounded evidence necessary for social science and human rights researchers to develop more concrete recommendations to push all involved to use models in ways that create the greatest ethical and societal benefits for all. It should be noted that this should not be treated as a detailed blueprint of how to run such a workshop as it is too specific to the project's needs. However, it can be built upon as the start of a framework for others to consider adopting as they further explore questions around ethical and societal impacts of models and forecasting.

ACKNOWLEDGMENTS

The research presented here is part of the STAMINA project, funded by Horizon 2020 grant n° 883441.

REFERENCES

Anagnostou A, Groen D, Taylor SJE, Suleimenova D, Abubakar N, Saha A, Mintram K, Ghorbani M, Daroge H, Islam T, Xue Y, Okine E and Anokye N (2022). FACS-CHARM: A Hybrid Agent-Based and Discrete-Event Simulation Approach for COVID-19 Management at Regional Level. In: B. Feng, G. Pedrielli, Y. Peng, S. Shashaani, E. Song, C.G. Corlu, L.H. Lee, E.P. Chew, T. Roeder, and P. Lendermann, (eds). *Proceedings of the 2022 Winter Simulation Conference (WSC)*. IEEE.

Bakalos N, Kaselimi M, Dulamis N, et al (2022). STAMINA: Bioinformatics Platform for Monitoring and Mitigating Pandemic Outbreaks. *Technologies* 10(3):63. https://doi.org/10.3390/technologies10030063.

Boersma K, Büscher M, and Fonio C (2022). Crisis Management, Surveillance, and Digital Ethics in the COVID-19 Era. *Journal of Contingencies and Crisis Management* **30(1)**:2- 9. https://doi.org/10.1111/jccm.12398.

Brey P (2000). Disclosive Computer Ethics. *Computers and Society* **30(4)**:10-16. https://doi.org/10.1145/572260.572264.

Leonelli S (2016). Locating Ethics in Data Science: Responsibility and Accountability in Global and Distributed Knowledge Production Systems. *Philosophical Transactions of the Royal Society A: Mathematical, Physical and Engineering Sciences* **374(2083)**:20160122. http://dx.doi.org/10.1098/rsta.2016.0122.

Mittelstadt B D, Allo P, Taddeo M, Wachter S, and Floridi L (2016). The Ethics of Algorithms: Mapping the Debate. *Big Data & Society* **3(2)**:2053951716679679. https://doi.org/10.1177/2053951716679679.

Mintram K, Anagnostou A, Anokye N, Okine E, Groen D, Saha A, Abubakar N, Islam T, Daroge H, Ghorbani M, Xue Y, and Taylor SJE (2022). CALMS: Modelling the Long-Term Health and Economic Impact of Covid-19 using Agent-Based Simulation. *PloS one* 17(8):e0272664. https://doi.org/10.1371/journal.pone.0272664.

Nissenbaum H (2004). Privacy as Contextual Integrity. *Washington Law Review* **79(1)**:119. https://digitalcommons.law.uw.edu/wlr/vol79/iss1/10.

Reijers W, Wright D, Brey P, Weber K, Rodrigues R, O'Sullivan D, and Gordijn B (2018). Methods for Practising Ethics in Research and Innovation: A Literature Review, Critical Analysis and Recommendations. *Science and Engineering Ethics* **24(5)**:1437-1481. https://doi.org/10.1007/s11948-017-9961-8.

Petersen K, Büscher M, Kuhnert M, Schneider S, and Pottebaum J (2015). Designing with Users: Co-Design for Innovation in Emergency Technologies. *In Proceedings of the ISCRAM 2015 Conference* Kristiansand, Norway. http://iscram2015.uia.no/wp-content/uploads/2015/05/4-7.pdf.

Petersen K, Oliphant R, and Büscher M (2016). Experimenting with the Ethical Impact Assessment. *In Proceedings of the ISCRAM 2016 Conference* Rio de Janeiro, Brazil. http://idl.iscram.org/files/katrinapetersen/2016/1364_KatrinaPetersen_etal2016.pdf.

Tako AA and Kotiadis K (2021). A Tutorial on Participative Discrete Event Simulation in the Virtual Workshop Environment. In: S. Kim, B. Feng, K. Smith, S. Masoud, Z. Zheng, C. Szabo and M. Loper, (eds). *Proceedings of the 2021 Winter Simulation Conference (WSC)*. IEEE.

Wright D (2011). A framework for the Ethical Impact Assessment of Information Technology. *Ethics and Information Technology* **13(3)**:199-226. https://doi.org/10.1007/s10676-010-9242-6.

Zimmermann N, Pluchinotta I, Salvia G, Touchie M, Stopps H, Hamilton I, Kesik T, Dianati K, Chen T (2021). Moving Online: Reflections from Conducting System Dynamics Workshops in Virtual Settings. *System Dynamics Review* 37(1):59. https://doi.org/doi: 10.1002/sdr.1667.

AUTHOR BIOGRAPHIES

KATRINA PETERSEN holds a PhD in Communication and Science Studies from the University of California, San Diego in 2014. She researches topics relating to social and ethical impact of new and emerging disaster technologies. She is a currently a Research Manager at Trilateral Research.

SUSANNAH COPSON holds a LLM International Human Rights Law from the University of Essex, 2020. Her research topics include the interplay between counterterrorism, new and emerging technologies and privacy rights. She is currently a Research Analyst with Trilateral Research.

ANASTASIA ANAGNOSTOU is a Senior Lecturer in the Department of Computer Science at Brunel University London and co-leads the Modelling & Simulation Group. She holds a PhD in Distributed Simulation from Brunel University London, 2014. She is the PI on the STAMINA Project.

Proceedings of the Operational Research Society Simulation Workshop 2023 (SW23)
C. Currie and L. Rhodes-Leader, eds.
https://doi.org/10.36819/SW23.023

NHS WORKFORCE PROJECTIONS 2022: THE ROLE OF THE NURSE SUPPLY MODEL

Dr. Siôn Cave

Decision Analysis Services
Grove House, Lutyens Close, Chineham Court,
Basingstoke, Hampshire, RG24 8AG, UK
SionCave@DAS-Ltd.co.uk

Dr. Nihar Shembavnekar

Health Foundation
8 Salisbury Square, London EC4Y 8AP, UK

nihar.shembavnekar@health.org.uk

Emma Woodham

Decision Analysis Services
Office 845. Spaces 8th Floor. The Programme
Building. 1 All Saints Street. Bristol
EmmaWoodham@DAS-Ltd.co.uk

Sandra Lewis

Decision Analysis Services
Grove House, Lutyens Close, Chineham Court,
Basingstoke, Hampshire, RG24 8AG, UK
SandraLewis@DAS-Ltd.co.uk

ABSTRACT

Nursing is the NHS's single largest staff group with over 300,000 full-time equivalent (FTE) registered nurses in the hospital and community sector. Nursing vacancies accounted for over a third of all FTE vacancies in NHS trusts in the quarter to June 2022. A lack of long-term planning and a coordinated workforce strategy has been acknowledged as a major factor for the shortfall. The Health Foundation commissioned Decision Analysis Services to develop a system dynamics model to represent the future supply of nurses across England. The model was designed to take a system-wide view of nurse supply and to consider second order effects The nurse supply model was used to generate projections of future nurse supply in England under three scenarios, with the results published in July 2022. The projections suggested that while the government appears to be on track to meet its 50,000 nurses target by 2023/24, this would still leave the NHS short of around 38,000 FTE nurses relative to projected demand in 2023/24.

Keywords: Workforce planning, System dynamics, Nurse, Nursing, National Health Service (NHS)

1 INTRODUCTION

Even before the COVID-19 pandemic, workforce issues were identified as the single biggest challenge for health and social care in England. Nursing, the NHS's single largest staff group with over 300,000 full-time equivalent (FTE) registered nurses in the hospital and community sector, is the key area of NHS workforce shortages. Nursing shortages, measured in terms of the vacancy rate in registered FTE nurse numbers, have been a recurrent theme for the NHS and have grown more prominent in recent years. In the quarter to June 2022, FTE registered nurse vacancies exceeded pre-pandemic highs and accounted for over a third of all FTE vacancies in NHS trusts.

High nursing vacancy rates are problematic as NHS trusts are forced to resort to a combination of bank and agency staffing to address shortfalls, which leads to increased costs for the service. One of the main reasons for the shortages in nursing, and in the NHS as a whole, is a lack of long-term planning around staffing levels and a 'boom and bust' approach linked to funding. The lack of high quality, robust and transparent projections of workforce supply and demand is a major factor underlying the lack of long-term planning and a coordinated workforce strategy. Existing modelling tends to take a relatively narrow view of nurse supply, not accounting for system-wide or second order effects.

To address this research gap, the Health Foundation's REAL (Research and Economic Analysis for the Long term) Centre commissioned Decision Analysis Services Ltd (DAS) to develop a nurse supply model (NSM) representing the whole of the nurse supply system (Cave et al, 2021). The nurse supply system is a series of interrelated markets with their own demand and supply schedules affected by a 'price' and other factors such as entry requirements to nurse education, nurse workload and immigration rules into UK and for other countries who recognise UK qualified nurses. The NSM will inform policy by enabling projections of future nurse supply in alternative scenarios exploring the impacts of changes in the factors that affect nurse recruitment and retention.

The NSM provides a set of tools that support the appraisal of alternative policies. In particular, it includes a conceptual framework which offers a bird's eye view of the system. This assists people to think about the effect of changes to the nurse supply system on the supply of nurses. It also includes a quantitative model that can be used to assess the impact of changes in policy based on a set of predetermined assumptions. The quantitative model is composed of two key elements: a system dynamics-based simulation engine which uses mathematical modelling to generate nurse supply projections with a 5 to 20-year time horizon and an R-Shiny-based data visualisation tool which enables projections to be viewed through interactive charts and tables and compared for further analysis.

In July 2022, The Health Foundation published their first set of projections using the NSM. The projections suggest that while the government appears to be on track to meet its 50,000 nurses target by 2023/24, the NHS would still be short of around 38,000 FTE Hospital and Community Health Service (HCHS) nurses and general practice nurses relative to projected demand (see Section 3 for further detail). In the longer term, in the current policy scenario, the NHS is projected to have a persisting shortfall of around 36,700 FTE nurses in 2030/31. The research found that in an 'optimistic' scenario, concerted policy action aimed at improving nurse retention and domestic training numbers can bridge the nurse supply-demand gap in the NHS HCHS sector by 2030/31. However, across all scenarios, the projections pointed to significant nurse supply-demand gaps in general practice and adult social care through this decade, highlighting the importance of comprehensive long term workforce planning.

This Paper provides a description of the NSM Simulation Model and how it was developed, an overview of the initial results that contributed to the NHS Workforce Publications and the next steps in terms development and use of the NSM.

2 DESCRIPTION OF THE NURSE SUPPLY MODEL AND THE MODEL DEVELOPMENT PROCESS

2.1 Description of the Nurse Supply Model

The NSM is a quantitative simulation model set within a conceptual framework that represents the nurse supply system.

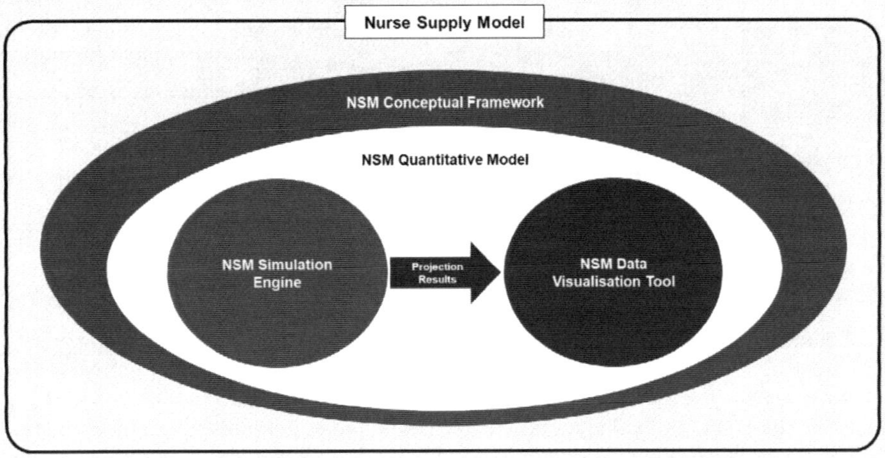

Figure 1 *Overview of the components of the Nurse Supply Model*

DAS adopted a collaborative approach to developing the NSM that generates long-term nurse supply projections and enables a wide range of forward-looking scenarios and policy levers to be investigated. For the purposes of this project, the 'long term' is a time period up to 20 years. The model has been co-designed with the REAL Centre and stakeholders so that it is widely acknowledged to be valid and fit for purpose. DAS has used the best and most efficient analytical tools based on the available data to ensure robust results and enable a handover of capability to the model's end users (principally the REAL Centre).

The conceptual framework takes a broad view of the nurse supply system and includes variables that are not quantified within the simulation model. For the purposes of the NSM the nurse supply system is considered to be composed of the nurse education market, the nurse labour market, qualified nurses not working as nurses, the international nurse labour market and the factors affecting these markets. The conceptual framework captures the key cause-and-effect relationships which impact nurse supply and integrates economics and stock and flow perspectives on the nurse supply system. This can be used as a tool to support scenario development and to set research agendas for quantifying relationships in the simulation engine.

The quantitative simulation model is composed of two key components, a simulation engine and a data visualisation tool. The purpose of the NSM simulation engine is to produce projections of nurse supply for England with a time horizon of up to 5 to 20 years. The purpose of the data visualisation tool is to enable the analysis and visualisation of the supply projections produced using the NSM simulation engine.

The simulation engine adopts the System Dynamics (SD) approach, implemented using Vensim[1]. SD enables complex systems to be better understood and their behaviour over time to be projected using computer simulation. SD has been used many times to support of strategic workforce planning (Cave and Willis, 2020). The NSM system dynamics model considers the main flows into and out of the nurse labour market, namely through degree education, international inflows and outflows and nurses leaving and re-joining the workforce, as illustrated in Figure 2.

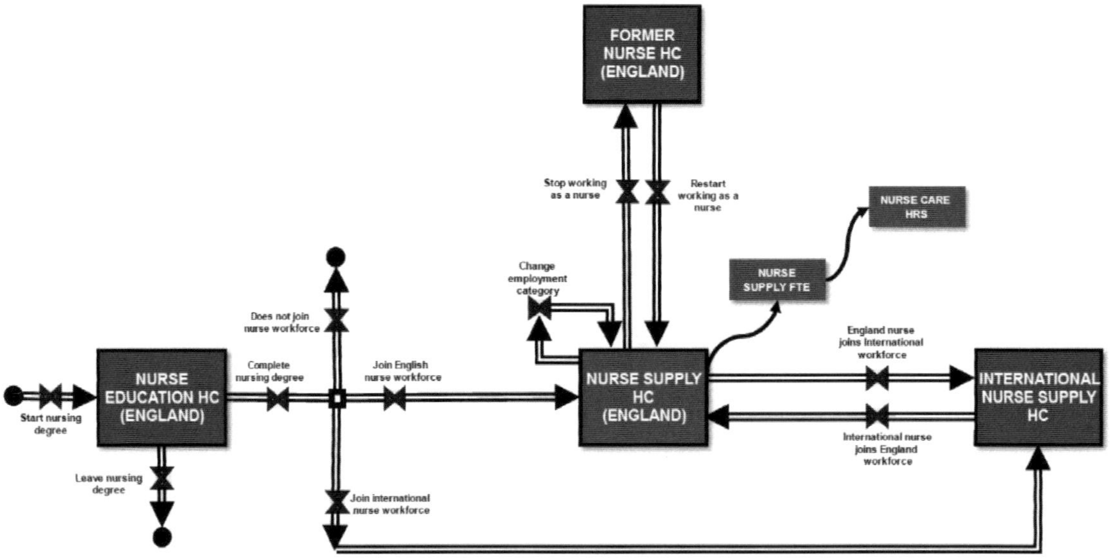

Figure 2 *Key stocks and flows representing nurse supply in the Nurse Supply Model*

The SD model is heavily segmented, with all the core stocks and flows representing age, gender, nationality and region. 17 different types of nurse are considered, for example Adult, Children and Learning Disability nurses. Midwives were not included in the scope of the model. The model contains

[1] https://vensim.com/

a series of 'scenario input' variables which are used to configure alternative scenarios. For example, a potential policy could be increasing the nursing student intake by 50% against current values for 3 years, followed by a 25% increase in the student intake thereafter combined with reducing student attrition by 1% a year for the next 5 years. It is in this way that the impact of alternative regulatory policies and instruments can be assessed. The model can also be run in Monte Carlo simulation mode, where input variables are represented as a distribution of values rather than a single value. It would also be possible to set the SD model up to undertake optimisation runs, where model inputs could be "flexed" to determine the best way to meet a desired supply line.

The SD model also includes numerous 'checks and balances' to maintain model robustness, for example system-wide mass balances and warnings if values exceed expected limits. All stocks and flows are initialised based on historic data. In addition, a wide range of stakeholders from across the nurse supply system were engaged through workshops and in-depth discussion to gain confidence in the model development process, model conceptualisation and the results from the projection model.

The NSM DVT enables the visualisation, analysis and comparison of supply projections produced using the NSM simulation engine. The DVT enables libraries of projection scenarios and historic data to be analysed, comparatively or using deep dive drill throughs. This is realised through a variety of different data visualisations, such as time series plots, bar charts and tables. The DVT enables results to be exported for use by other applications/models. The DVT has been developed using R-Shiny, an open-source statistical programming language.

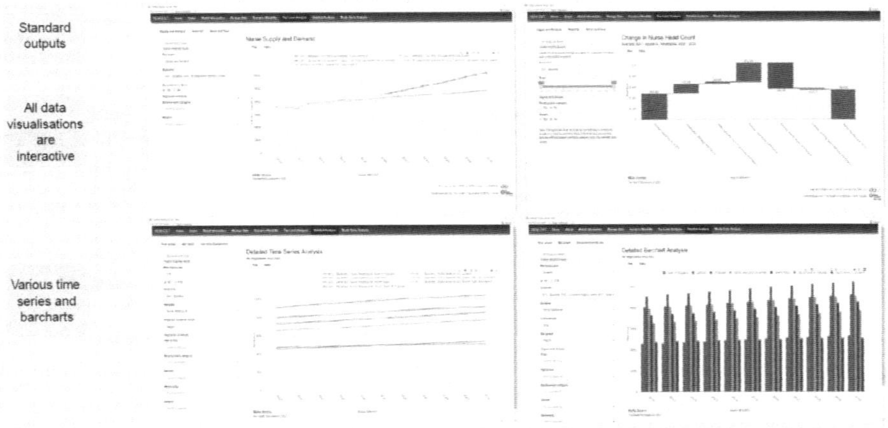

Figure 3 *Example screenshots from the Nurse Supply Model Data Visualisation Tool*

2.2 Nurse Supply Model Development Process

A collaborative and participatory approach was adopted to developing the NSM, with representatives from the various organisations in the nurse supply system involved. Over 25 organisations from across the nurse supply system were involved in the project.

Initially a series of rapid research tasks were undertaken, including a historic analysis of nurse supply in England, a review of the economics of nurse supply, a review of nurse supply modelling, a review of the nurse supply system and a review of the data landscape. These tasks were undertaken through reviews of the literature and stakeholder workshops.

The nurse supply model review (unpublished) considered 40 workforce projection models from the UK and overseas. This included models used by Health Education England (HEE), the Department of Health and Social Care (DHSC), NHS Wales and the Scottish Government. The review considered how workforce models are applied to strategic planning, identification of common approaches to workforce projection modelling, the pros and cons of available tools and methodology, identification of common issues or constraints and a review of data typically required for a model projecting nurse supply. A wide variety of simulation approaches were used, such as SD, Agent Based Modelling along with models implemented using MS Excel. The review concluded that existing modelling tends to take a relatively narrow view of nurse supply, not accounting for system-wide or second order effects, for example taking an explicit representation of the training pipeline and the movement of nurses between sectors.

This research was followed by the development of the NSM conceptual framework and the quantitative model. DAS developed a delivery plan to design, develop and test the model. An academic advisory panel from the University of Southampton provided oversight and scrutiny throughout the process. The panel was composed of experts in health care modelling and simulation from the University of Southampton: Dr Steffen Bayer, Professor Stephan Onggo and Professor Martin Kunc.

The conceptual framework was used to support two separate research tasks, namely a review of the economic factors that affect nurse supply (Derbyshire et al, 2021) and the impact of the first wave of COVID-19 on nurse supply (Woodham et al, 2021). The quantitative model was tested by the Health Foundation prior to final delivery and its subsequent use.

3 NHS WORKFORCE PROJECTIONS AND THE ROLE OF THE NURSE SUPPLY MODEL

As discussed in Section 1, in July 2022 the Health Foundation published their first set of projections using the NSM. This section provides further detail on these projections.

3.1 How the Nurse Supply Model was Used

The nurse supply model was used to generate projections of future nurse supply in England under three scenarios, described below. The research discussed how the supply of and demand for registered nurses might evolve in the years to 2023/24 (the end of the current parliament) and longer term to 2030/31.

1. 'Current policy' is a baseline scenario in which overall nurse supply between 2021/22 and 2030/31 largely grows in line with the trend up to 2019/20. It assumes no policy intervention beyond 2021/22 other than some continued growth in international recruitment to increase FTE nurse numbers.

2. In the 'optimistic' scenario, sustained policy action is undertaken to achieve increases in nurse supply through the three major supply channels: increases in the number of nurses in training, increased international nurse recruitment, and improved retention of existing nurse staff.

3. The 'pessimistic' scenario highlights key risks to future patterns of nurse supply, some of which are likely to have been exacerbated by the COVID-19 pandemic.

Section 2.3.4 of the Health Foundation's report illustrates how the nurse supply model was used to better understand the potential impact of the pandemic on student nurse attrition. It is relatively straightforward to identify the different areas of the pandemic's potential impact on the nurse workforce, based on current evidence. However, it is more difficult to describe the dynamics between different areas of impact and how they might affect the workforce in the future – this is where the nurse supply model, with its embedded system dynamics, added substantial value.

3.2 Nurse Supply Model projection results

The projections suggest that nurse supply in England's NHS HCHS can only close the gap with demand by 2030/31 in the optimistic scenario. Critically, this hinges on increases in future nurse supply through each of the major supply routes:

- Up to 2023/24, international recruitment will be the key policy lever for increases in nurse supply. For the longer term, however, it is critical policymakers aim to reduce or taper reliance on international supply, focusing instead on more sustainable and robust solutions based on increased domestic supply.

- Beyond 2023/24, sustaining recent increases in the number of student nurses in training will be key to supporting the optimistic scenario supply trajectory. This requires policy support and sustained investment in university capacity and in clinical training placements, with increased clarity over annual planning and budgets.
- Through the decade, retaining existing nurses will continue to be important to achieve overall increases in FTE nurse numbers. This calls for an improved retention policy, including promoting better work-life balance, career progression, leadership and training. It should also involve a renewed focus on responsive and fully funded NHS nurses' pay determination and the total reward offer, including incentives for childcare and adult care provision.

Figure 4 presents the longer term projections of FTE nurse supply and demand in the NHS HCHS in England up to 2030/31 under the three scenarios modelled: a 'current policy' scenario, an optimistic scenario and a pessimistic scenario. In the current policy scenario, the nurse supply-demand shortfall declines, falling from around 50,600 FTE in 2023/24 to around 30,300 FTE by 2030/31. However, in the optimistic scenario, nurse supply steadily increases more rapidly than projected growth in demand, potentially 'catching up' with projected demand in 2028/29 and yielding a surplus by the end of the decade. Even in this scenario, the 'catching up' needs time to build momentum and deliver sustained increases in supply. This highlights the importance of long-term workforce policy and planning if the NHS nursing shortfall is to be overcome.

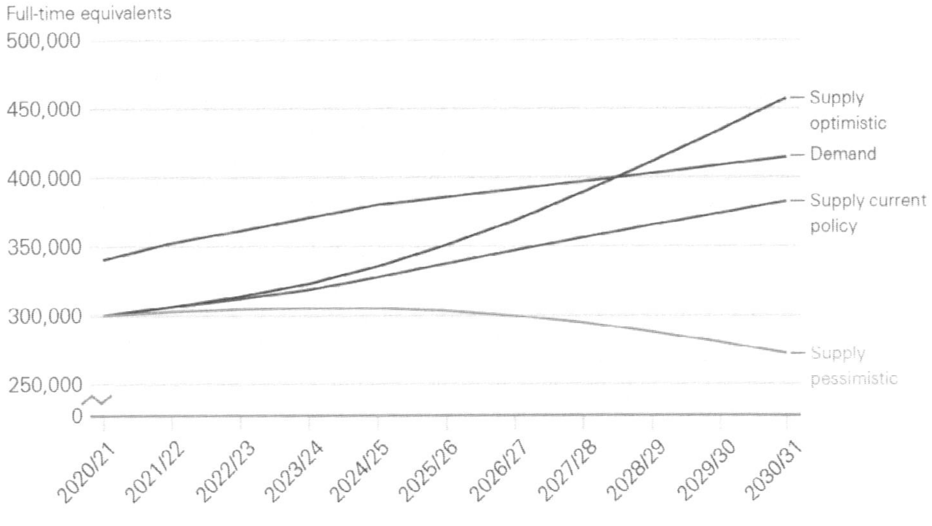

Figure 4 *FTE nurse supply and demand projections for the NHS HCHS in England, 2020/21–2030/31, under the three scenarios*

*Source: REAL Centre analysis using the nurse supply model based on a range of data sources. Note: At the time of writing, the model has updated data up to 2020/21. Numbers are rounded and estimates for 2021/22 onwards are projections.

The differences between the relative impact of the different supply channels are more prominent in the longer term than in the shorter term. Figure 5 presents a waterfall chart highlighting these channels for the 'current policy' scenario up to 2030/31. The model also facilitates comparisons of the relative significance of these channels across scenarios. For example, the assumptions on international nurse recruitment result in a projection of around 37,500 FTE additional nurses from outside the UK joining the nursing workforce in England between 2020/21 and 2030/31 in the optimistic scenario relative to the current policy scenario.

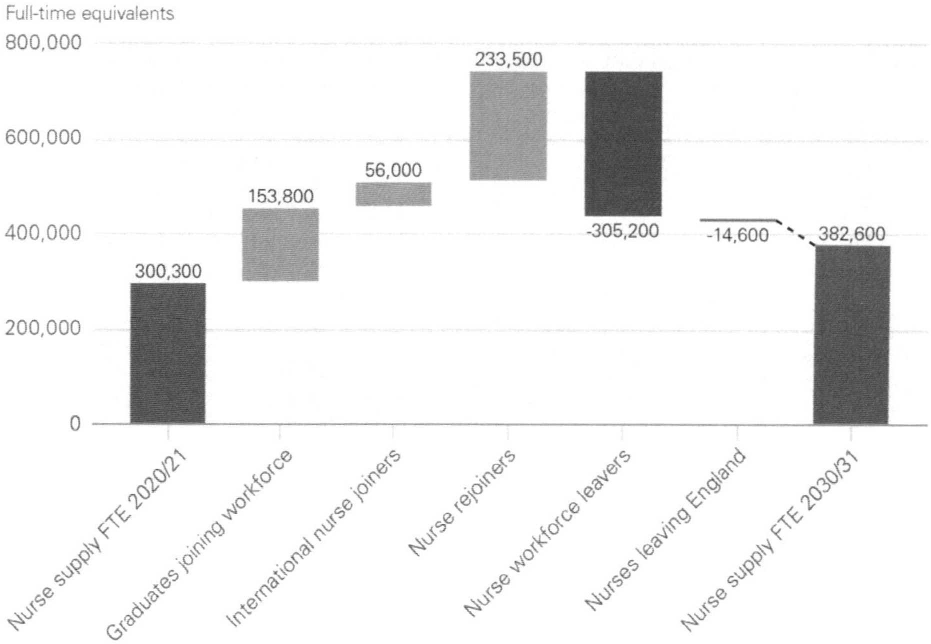

Figure 5 *FTE nurse supply projections in the NHS HCHS in England in the 'current policy' scenario, 2020/21 – 2030/31, waterfall chart**

*Source: REAL Centre analysis using the nurse supply model based on a range of data sources. Note: At the time of writing, the model has updated data up to 2020/21. This chart is derived from headcount data for registered nurses. As the 'current policy' scenario assumes that the nurse FTE-to-headcount ratio remains unchanged over time, we use the nurse FTE-to-headcount ratio observed in 2020/21 data to convert headcount numbers to FTE estimates. Numbers are rounded.

Further, sector-specific analysis points to an urgent need for greater policy attention in the relatively neglected areas of nursing in general practice and adult social care. In all three scenarios, nurse supply over the decade to 2030/31 is projected to fall well short of projected increases in demand in these sectors. This becomes even more concerning given the potential increases in unmet need from the pandemic.

4 WHAT'S NEXT FOR THE NURSE SUPPLY MODEL

The REAL Centre recently used the NSM to build on its July 2022 projections in a stakeholder workshop where policy options to address the NHS' longstanding nursing shortages were considered. The Centre plans to continue using the NSM for nurse workforce projections modelling in the future. For instance, there is currently significant policymaking interest in better understanding the relative importance of domestic nurse training and improved retention in terms of boosting nurse supply in specific sectors (such as mental health nursing and learning disability nursing) and regions. The model will facilitate this type of bespoke analysis.

In 2023 and beyond, the data underlying the model will need to be updated to ensure that projections are founded on the latest available numbers in terms of student intakes, student attrition, international recruitment and nurse retention. The REAL Centre may also draw on the rest of its modelling output, including on health care demand, to complement results derived from the NSM.

ACKNOWLEDGMENTS

We would like to acknowledge the contributions made by Rosemary Wildblood (Decision Analysis Services) and Keith Derbyshire (KD Network Analytics) in the development of the Nurse Supply Model.

We are also grateful for insights and input received from a number of REAL Centre colleagues, in particular Anita Charlesworth, Elaine Kelly and James Buchan, in the development of the model.

Finally, Decision Analysis Services Ltd and the REAL Centre would like to thank all those from across the nurse supply system involved over the course of the project.

REFERENCES

Cave S, Woodham E, Derbyshire K, Lewis S, Wildblood R, Shembavnekar N. (2021) *Nurse supply model: overview*. https://www.health.org.uk/publications/nurse-supply-model-overview accessed 10 November 2022

Cave S and Willis, G. (2020). System Dynamics and Workforce Planning. In: Dangerfield, B. (eds) System Dynamics. Encyclopedia of Complexity and Systems Science Series. Springer, New York, NY. https://doi.org/10.1007/978-1-4939-8790-0_659

Derbyshire K, Cave S, Woodham E, Wildblood R, Shembavnekar N. (2021) *Nurse supply model: a review of economic factors affecting nurse supply*. https://www.health.org.uk/publications/reports/nurse-supply-model-a-review-of-economic-factors-affecting-nurse-supply accessed 10 November 2022

Woodham E, Wildblood R, Cave S and Derbyshire K (2021). *Nurse supply model: exploring the potential impact of the first wave of the COVID-19 pandemic on nurse supply*. https://www.health.org.uk/publications/nurse-supply-model-exploring-the-impact-of-covid-19 accessed 10 November 2022

Shembavnekar N, Buchan J, Bazeer N, Kelly E, Beech J, Charlesworth A, McConkey R, Fisher R. NHS workforce projections 2022 (2022). The Health Foundation; (https://doi.org/10.37829/HF-2022-RC01). https://www.health.org.uk/publications/nhs-workforce-projections-2022 accessed 10 November 2022

AUTHOR BIOGRAPHIES

SIÔN CAVE is head of Decision Analysis Service's Analytics + Foresight Hub. The Hub is DAS' centre of excellence for data science, systems modelling and futures techniques. https://uk.linkedin.com/in/sioncave

NIHAR SHEMBAVNEKAR is an Economist at the Health Foundation's REAL Centre.

EMMA WOODHAM is a Lead Consultant at Decision Analysis Services. She has been working as a consultant since graduating with a MMath(Hons) in Mathematics from University of St Andrews in 2011. Her specialisms include Systems Thinking, Data Analysis and simulation modelling. https://uk.linkedin.com/in/emma-woodham-26830a56

SANDRA LEWIS is a principal data scientist at Decision Analysis Services.

Proceedings of the Operational Research Society Simulation Workshop 2023 (SW23)
C. Currie and L. Rhodes-Leader, eds.
https://doi.org/10.36819/SW23.024

THE NEED FOR A SYSTEM DYNAMICS COMMUNITY OF PRACTICE IN THE HEALTH AND SOCIAL CARE SECTOR

Vanessa Perez Perez

Demand and Capacity Programme
NHS England
v.perezperez@nhs.net

Stacey Croft

The Strategy Unit
Midlands and Lancashire CSU
stacey.croft@nhs.net

Hugo Herrera

Analytics Team
East Suffolk and North Essex NHS FT
hugo.herrera@esneft.nhs.uk

Edwin Magombe

Demand and Capacity Programme
NHS England
edwin.magombe@nhs.net

1 BACKGROUND

As a response to the recommendations in the NHS long-term plan, Integrated Care Systems (ICSs) were established as legal entities following the passage of the Health and Care Act 2022. ICSs are partnerships that bring together NHS organisations, the social care and the third sectors to take collective responsibility for planning services, improving health and reducing inequalities across geographical areas. ICSs have also been given the flexibility to develop local arrangements to respond to the needs of their population, in order to help to create local ownership and commitment.

The challenges that the NHS face are increasingly complex, but not unique to any specific region, ICS or provider. Across the country, managers are wondering how long it might take to bring back growing waiting lists to acceptable levels given current capacity constraints, and what the impact of these delays will be on patients' health. Other complex questions that are being asked include:

- What are the capacity and flow requirements to enable effective hospital discharges?
- What are the training and development gaps if the NHS is to have the capacity and skills needed to meet the challenges of changing population health needs into the late 2020s?
- What are the relative costs and benefits of undertaking proactive work with specific population groups over the short and longer term?
- What capacity is required to meet the needs of the population over 1, 2, 5 or even 10 years?

Modelling and simulation approaches such as System Dynamics (SD) can help answer these complex questions.

2 APPROACH

The questions listed in the background section have increasingly led staff in the health and care sector towards using modelling and simulation approaches that have the potential to capture, in relatively simple and transparent ways, the complex behaviours that often emerge. System Dynamics (SD) is a collaborative group model building approach that has been successfully embedded in a range of industries and sectors, including healthcare in several countries.

Using SD models healthcare systems can start to understand how the patients flow across the system, where capacity is located and how it could be used. The benefits of using a SD approach in the NHS and the wider healthcare system could be broken into the three areas: population health, workforce modelling and patient flow and pathway transformation. Using SD as a modelling approach to system level demand capacity planning provides an opportunity to take a system view on population health needs and challenges (e.g., across primary, acute and community care) and its consequences for medium and long-term capacity (including workforce) planning. Having a holistic view of the system can provide insights on opportunities for the re-allocation and reconfiguration of capacity within the system (for example through new ways of care delivery).

Importantly, ICSs could use SD to ask 'what-if' questions to understand the impact of what would happen across the whole system if, for example, patients were diverted to a different service. Being able to test different scenarios in a system model would allow staff to start making some strategic decisions with robust evidence of the potential impacts.

3 IMPACT

There have been attempts to support the improvement in capability alongside a partnership approach in the development of SD models in the NHS. However, the use of SD is still very limited and usually dependent on the use of external consultancies. To reduce the need for external SD expertise, the argument has been strong to improve the capacity and capability of the analytical teams to deliver such models within systems and providers. Seven core domains that are necessary to achieve its effective development and use were identified by NHS England in collaboration with Whole Systems Partnership. However, two of those domains are not yet sufficiently supported by NHS England: SD Modelling technical capabilities and software and infrastructure.

As a collaborative modelling approach, the skills required for SD can span several individuals, so different members of a multi-disciplinary project team can come together to develop models focusing on their area of model development where they could add most value. A Community of Practice would be very useful to consolidate learning and resources and encourage discussion and collaborative working. This document and the below recommendations explore how NHS England could support Systems who are seeking to answer such a complex issue through the application of SD modelling.

The following recommendations to NHS England have been drawn: i. NHS England should advocate for raising even further the profile of modelling and simulation approaches, in particular System Dynamics ii. NHS England to further develop and sustain a System Dynamics Community of Practice iii. In the short and medium term, NHS England to make available the software to develop SD models iv. NHS England to scope and support the development of a SD modelling solution using open-source software.

REFERENCES

Barker M and Zupick N (2016) *A clue, the cash, the commitment, and the courage: the keys to a successful simulation project.* Winter Simulation Conference 2016

Brailsford et al (2014) *Discrete-Event Simulation and System Dynamics for Management Decision making*. Wiley

Charles A (2020, updated in 2022) *Integrated care systems explained: making sense of system, places and neighbourhoods*. Available at: <https://www.kingsfund.org.uk/publications/integrated-care-systems-explained> [Accessed 11 January 2023]

Jahangirian M et al (2017) Key Performance indicators for successful simulation projects. Journal of Operational Research (2017) 68, 747-7

NHS (2019). *The NHS Long Term Plan.* [ebook]. Available at: <https://www.longtermplan.nhs.uk/> [Accessed 11 January 2023]

The Health Services Research Network (HSRN) et al. *Change by design: systems modelling and simulation in healthcare. Tools for health service decision makers*. Available at: <https://mashnet.info/wp-content/files/2016/09/Change-By-Design-Booklet.pdf> [Accessed 11 January 2023].

Proceedings of the Operational Research Society Simulation Workshop 2023 (SW23)
C. Currie and L. Rhodes-Leader, eds.
https://doi.org/10.36819/SW23.025

UNDERSTANDING DEMAND AND CAPACITY OF HOSPITAL CARE USING SYSTEM DYNAMICS

Dr. Hugo Herrera
ESNEFT
hugo.herrera@esneft.nhs.uk

Haeshiya Sivakumaran
DAS
HaeshiyaSivakumaran@das-ltd.co.uk

1 BACKGROUND

The NHS has grouped health care in four broad categories: primary care, secondary care, community care and terciary care (NHS Digital, 2022). Secondary care includes mental health care and 'hospital care'. Hospital care consists of both planned (elective) care such as a cataract operation, or urgent and emergency care such as treatment for a fracture (NHS Providers, 2019).

In recent years the number of patients waiting to receive elective care has increased, lessening patients' experience and quality of life. Between 2019 and 2022, the number of patients waiting for elective care in England increased by 60% and patients are now waiting longer to be treated. For example, the time that patients have to wait to receive elective care in East Suffolk North East Essex NHS Foundation Trust (ESNEFT) has almost doubled between 2019 and 2023 and now many patients are waiting more than 6 months before getting the care they need.

This quick increase on the waiting lists combined with recent challenges seen in the emergency departments has prompted the NHS to reassess the way it looks at its demand and to think on different ways to adjust and configure its capacity. This case study conducted in the two ESNEFT hospitals (Colchester Hosptial and Ipswich Hospital) is part of this new approach driven by the NHS, the Integrated Care Systems and ESNEFT with the aim to alleviate the capacity challenges seen in hospital care.

2 APPROACH

While forecasting of demand and capacity are not new to ESNEFT, the approaches used in the past to estimate demand and required capacity had mainly looked at it from an exogenous perspective (driven by population growth) and have not accounted for the oscilations created by bottlenecks within the system and vicious cycles operating within the system. Hence, in this case study we opted to use system dynamics (see Morecroft, 2015) modelling approach as it is the ideal tool to capture dynamic complexity and the ways system delays influence the system behaviour.

Namely, we developed a system dynamics simulation model that explores the structures driving the increase in the number of patients waiting for hospital care. The model is exogenously driven as a large proportion of the increase can be explained by internal dynamics happening between different specialities and clinical areas. As shown in Figure 1, the model has the RTT waiting list at its core (list of people waiting for treatment) and it covers the elective and non elective pathways of the main seven specialties in the hospital and accounts for important strategic resources, for example the number of beds available and theatres capacity.

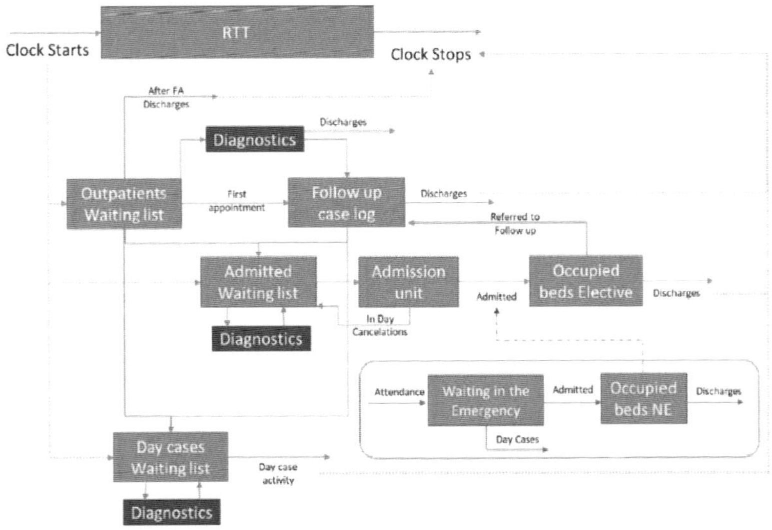

Figure 1: A simplified stock and flow diagram showing the main structure in the demand capacity model

The model was built by ESNEFT analytics team with support from Decision Support Analysis (DAS). ESNEFT analytics team has a deep understanding of patient pathways in the specialities included in the model and work closely with ESNEFT operational teams so we relied on the analysts as subject matter experts of the system. This system dynamics model is part of a wider project led by the ICS and will eventually be linked to other models estimating capacity requirements for primary care, community and social care.

3 IMPACT

While using this model it will be possible to explore potential effects of longer waiting times on emergency attendances, patients' length of stay and changes on patients' need for additional care upon discharge. Once finished, this model will support strategic decisions at an Integrated Care System (ICS) level helping decision makers to develop coherent plans regarding investments in infrastructure and the development of workforce across the system.

While the benefits of this project are only going to be seen a couple of years after its completion, the model building process itself has already delivered substantial insights for ESNEFT. First, it has unconvered inconsistencies in both data and mental models by translating information presented in isolation through regular reports into stock and flows nomenclature. At the same time, the model structure has made the non-linear behaviours of the complex relationships of different specialties and clinical areas throughout the patients treatment pathways explicit. Having these relationships explicitly represented in diagrams has shaped conversations about capacity and perspectives about the root causes of some bottlenecks seen in the system.

REFERENCES

Morecroft, J. D. (2015). Strategic modelling and business dynamics: A feedback systems approach. John Wiley & Sons.

NHS Digital (2022). The healthcare ecosystem. viewed January 2023 (https://digital.nhs.uk/developer/guides-and-documentation/introduction-to-healthcare-technology/the-healthcare-ecosystem).

NHS Providers (2019). The NHS Provider sector. viewed Januay 2023 (https://nhsproviders.org/topics/delivery-and-performance/the-nhs-provider-sector).

Proceedings of the Operational Research Society Simulation Workshop 2023 (SW23)
C. Currie and L. Rhodes-Leader, eds.
https://doi.org/10.36819/SW23.026

DISCRETE-EVENT SIMULATION TO SUPPORT THE MANAGEMENT OF PERISHABLE INVENTORY– A REVIEW

Marta E. Staff

Centre for Simulation, Analytics & Modelling
The Business School
University of Exeter
Exeter, EX4 4ST, UK
Ms640@exeter.ac.uk

Navonil Mustafee

Centre for Simulation, Analytics & Modelling
The Business School
University of Exeter
Exeter, EX4 4ST, UK
N.Mustafee@exeter.ac.uk

ABSTRACT

The importance of inventory management has been widely acknowledged both in practice and in academic research, with perishable products constituting an additional complexity in the field. Discrete Event Simulation (DES) is amongst the most frequently used Operations Research methods that allow the incorporation of stochasticity (reflective of the real-world environment) in modelling the system of interest. Beyond providing a general overview of the publications, the review provides a more detailed analysis of model characteristics (e.g., model objectives and outputs, issuing policy), focusing on supply and demand uncertainty, and identifies possible opportunities for further research. As the first review paper in this area, our work will serve as a key reference on the state-of-the-art in DES for inventory control and its application in the management of supply chains in healthcare, retail and domains that handle perishable inventory. Amongst the important findings is that lifetime is generally assumed to be known a priori, neglecting the product lifetime uncertainty.

Keywords: Discrete Event Simulation, Perishable Inventory, Literature Review

1 INTRODUCTION

Inventory is *"the stock of any item or resource used in an organisation"* (Chase et al., 2007). For any organisation to meet demand, they need to manage and adjust inventory. Holding excess inventory may represent unnecessary costs associated with overbuying and storage. On the other hand, too little inventory may result in stalling of production lines and/or dissatisfaction on the part of customers due to non-availability, or delayed availability, of products. Thus, the optimisation of inventory and the study of associated replenishment strategies are important from both research perspectives and for their practical relevance in organisations. Optimisation aspects include strategies for measuring stock levels, frequency and volume of replenishment (Chase et al., 2007), and associated financial aspects, including the value of inventory and investment trade-offs between overbuying and holding excess inventory versus competing investments for other aspects of production (opportunity costs). Indeed, any inventory control decision environment can be reduced to simple questions of *'when'* and *'how much'* to replenish (Goltsos et al., 2022) to create an appropriate *'buffer'* to deal with the supply and demand uncertainties (Newman et al., 1993). Perishable goods represent an additional challenge when considering the complexity of decision-making in managing inventories. Lower inventory levels for perishable goods are associated with reduced wastage but a higher risk of stock-out probability and vice-versa.

Numerous Operations Research/Management Science (OR/MS) techniques have been previously employed to develop decision support tools for inventory optimisation. Examples of techniques include both deterministic approaches, such as linear mixed integer programming or dynamic programming, as well as stochastic approaches e.g., stochastic optimisation. However as these are analytical approaches, they may offer a limited ability to reach a tractable solution when incorporating the uncertainty and variability of real-life systems. On the other hand, Computer Modelling and Simulation (M&S) presents an OR/MS approach that can handle such uncertainty and variability, and as previously acknowledged by Robinson (2005) *"are normally developed because a system is too complex to be represented in any other way"*. System Dynamics, Agent-based Modelling, Monte-Carlo and Discrete-event Simulation (DES) are examples of M&S techniques that can model stochasticity. Although queueing theory could model stochastic systems, the inherent complexity of an underlying system of interest may often mean that such models may not provide the level of detail that is often necessary for in-depth analysis (Saltzman et al., 2017). M&S methods such as DES are an excellent alternative since they capture the randomness inherent in real-world operations and also provide the ability to model the system at a granular level. Furthermore, they allow for the experimentation of multiple what-if scenarios, that could be co-developed with stakeholders of the real-world system, for better and more informed decision-making. Thus, M&S methods such as DES allow for the experimentation of strategies (e.g., reordering policies, inventory optimisation strategies) before their implementation in practice.

With the increasing complexity of systems being modelled, often multiple M&S techniques need to be combined to develop a *hybrid simulation* (Brailsford et al., 2019). And although hybrid studies are gaining prominence, a recent keyword classification study by Mustafee & Katsaliaki (2020) that surveyed over 82,000 articles published between 1990-2019 in 26 leading OR/MS journals identified conventional DES to be the most popular M&S technique for detailed-level modelling.

The *General Simulation Program (GSP)* was the first specialist DES package developed by Dr K. D. Tocher and his OR team for the UK steel industry during the mid-1950s (Hollocks, 2006). The application of DES has since expanded into fields such as healthcare (Brailsford et al., 2009; Katsaliaki & Mustafee, 2011; Vázquez-Serrano et al., 2021), business (Jahangirian et al., 2010), forestry (Opacic & Sowlati, 2017), logistics and supply chain management (Tako & Robinson, 2012). Beyond incorporating stochasticity and uncertainty, DES is an approach suited for modelling entities through networks of queues and activities/servers (Brailsford & Hilton, 2001). The execution of a DES model could follow a fixed-increment time advance, however as pointed out by Law & Kelton, the most commonly used engines rely on event-driven time advance (1991, p.8)- also referred to as the *ABC of simulation* (Advance simulation time- execute Bound events and execute Conditional events) (Robinson, 2014). As the model executes through simulated time, the system state, which comprises queues, servers, resources and entities, changes at discrete points in time.

Since inventory processes generally consist of moving entities in and out of storage, which is almost always exclusively embedded in the stochastic supply chain environment, DES modelling lends itself naturally to the study of this field. For perishable goods, the age of each item that needs to be tracked adds additional complexity. Currently, there is limited understanding as to the extent DES has been utilised to study perishable inventory systems; hence this paper aims to investigate a sub-set of publications where DES and inventory of perishables converge.

The remainder of this paper is organised as follows. A background and a brief overview of the literature are presented in Section 2. Section 3 describes the search strategy that facilitated the identification of publications used for our analyses. Section 4 presents the analysis and reports on the findings. The paper concludes with Section 5 with a discussion and conclusion.

2 BACKGROUND ON INVENTORY MANAGEMENT

Ford Whitman Harris published the first inventory control model in 1915 - the classic *Economic Order Quantity (EOQ) model* (Harris, 1915). Forty years later, Thomson M. Whitin captured the concept of the deterioration of inventory stocks (Whitin, 1953). Further developments in the field led to classifying inventory models to account for different product characteristics. Goyal and Giri (2001) put forward the three "*meta-categories*" of obsolescence, deterioration, and neither obsolescence nor deterioration, which constitutes the first layer of classifying inventoried goods. Briefly, with obsolescence, the goods will likely become obsolete due to the changing environment, such as related demand, rather than inherent properties. Deterioration implies a reduction of the quantity/quality of the good itself. It could be further divided into so-called perishable and decaying goods, meaning those with maximum usable lifetime and those with no shelf-life, respectively. The final group consists of products that do not exhibit either external or internal changes over time and thus belong to the meta-category of neither obsolescence nor deterioration. The meta-categories cover various inventory problems spanning application areas associated with manufacturing, fresh foods, healthcare products, etc. There is a large body of literature reporting on the use of inventory models that aim to understand the behaviour of inventory systems and attempt to suggest the best and/or feasible solutions for improvement.

In the literature, there are multiple reviews on inventories of deteriorating goods. Amongst the most frequently cited is the review of *Perishable Inventory Theory* (Nahmias, 1982) which predominantly focused on fixed-life perishable inventory literature, while Raafat (1991) limited his survey to mathematical models of continuously deteriorating inventories. Since then, reviews by Goyal and Giri (2001) and Bakker et al. (2012) have included publications of deteriorating inventories that span over two consecutive periods, namely the 1990s and 2000s, respectively. The numbers of publications included in those two reviews amounted to over 300, covering a period of approximately 20 years, indicating the perceived importance and significance of the field. More recent reviews have proposed classification systems focusing on themes such as risk management of inventories through hedging (Svoboda et al., 2021), transportation selection (Engebrethsen & Dauzère-Pérès, 2019) and investigation of models developed across multi-echelon inventories (de Kok et al., 2018).

For the interested reader, the literature reviews mentioned above should provide a good overview of the current state of research employing mathematical approaches. The focus of our review is on the use of DES for tackling perishable inventory problems. As such, we do not consider existing work on mathematical inventory models that aim to find an optimal solution or, indeed, methods utilising heuristics that aim to reach a good but not necessarily optimal solution.

For inventory modelling using DES, referring to Figure 1, in a typical model the entities flow into the system where stock inventory is modelled in the form of queues. Based on the orders received, and on the issuing strategy, they move either to the next echelon of inventory being modelled (in case of multi-echelon system modelling) or exit the system. The replenishment strategy (possibly incorporating lead-time), and product expiry/wastage, are also modelled. Importantly for inventory strategy, each item flowing through the system will have certain attributes associated with it, such as the time it entered and exited the system and associated expiry date when considering perishable goods.

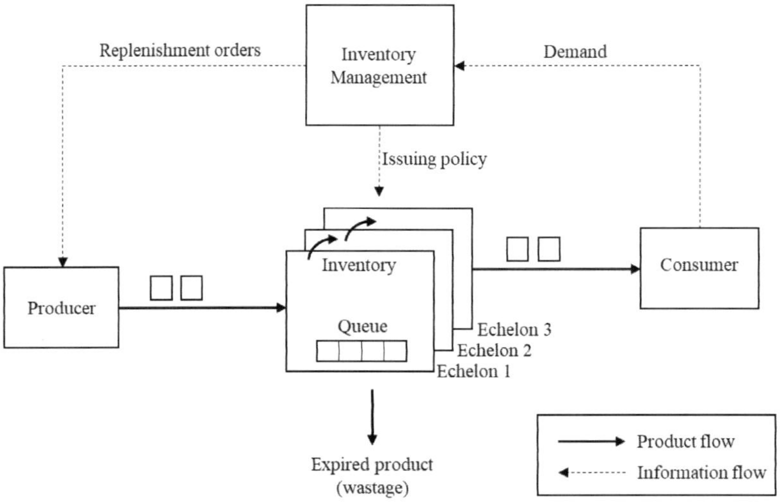

Figure 1 *Information and Product Flows in a DES Model of a Typical Perishable Product*

3 METHODOLOGY AND FRAMEWORK FOR LITERATURE ANALYSIS

The review aims to identify and analyse original research articles on deteriorating inventory models that used DES. We thus selected search terms relating to three themes of *deterioration, inventory/production* and *DES* and developed the following Boolean keyword combination; [(deteriorat* OR perish* OR "shelf life" OR decay*) **AND** ("inventory" OR "operation* research" OR "operation* manage*" OR SCM OR "supply chain manage*" OR "production plan*") **AND** ("discrete event simul*" OR ("discrete event" AND (computer (simulat* OR model*)))]. While the relatively tight control for perishability and DES keywords was informed by Bakker et al. (2012) and Zhang (2018), respectively, a more relaxed approach, informed through multiple literature sources, was taken for the inventory/production search term. This was intentional, as, through this approach, we aimed to capture multiple styles of linguistics that could have referred to the area of our interest.

We used the *Web of Science (WOS) Core Collection* (Clarivate™ Analytics) and conducted a **topic search** on article titles, abstracts, author keywords and Keyword Plus. We performed an unrestricted search in relation to the year of publication and considered all papers published in English. The initial search identified 34 articles. After reading the abstracts, ten articles were removed from our preliminary dataset. After full-text reading, eight papers were removed, resulting in a total of 16 articles considered in this review (our final dataset). Examples of articles excluded include studies by Armbruster et al. (2011) and Chemweno et al. (2016), where the impact of physical hardware/machine breakdown is considered, or where the focus is on decaying military capabilities (Bender et al., 2009), or related to workforce deterioration of productivity (Xu & Hall, 2021), rather than on physical products. Additionally, some of the studies were excluded since they did not provide sufficient detail for our analysis, e.g., the paper by Ma & Meng (2008) did not describe the characteristic of the DES model, beyond stating that DES was used.

The *PPMO framework* for literature synthesis (Mustafee et al., 2021) describes the variables of interest for literature reviews in M&S. The variables focus on profiling research (P), problem definition and context (P), model development & implementation (M), and study outcome (O). Given the relatively small sample of articles and the distinct nature of the current review, the PPMO framework provides an overall structure for reporting the findings from the literature.

4 FINDINGS

For three of the four categories of the PPMO framework (Mustafee et al., 2021), namely profiling research (P) and problem definition and context (P) (considered together in this review) and study outcome (O) we included several variables that fit the context of our study. Regarding the model implementation category (M), we reported whether the DES models had standalone implementations or were combined with other techniques. Additionally, we captured aspects relevant to the management of perishable inventory, such as modelling uncertainty and product flow. We added these variables to category M, thus extending the PPMO framework based on the aim and objectives of the review.

4.1 Research Profiling and Context

4.1.1 Publications Overview

The 16 publications identified in our dataset consist of 15 journal papers and one conference proceeding. The papers appeared between 1998 to 2022. They cover application areas such as blood supply chains, food supply chains, chemical manufacturing, and perishable retail goods (Table 1). Most papers were from OR/MS or Management-related journals, while some were from scholarly sources focussing on computing-related fields or applications (*Transfusion* and the *British Food Journal*).

To report on the subject areas (disciplines) associated with the publications, we have used the Journal Citation Reports™ (JCR, Clarivate Analytics). JCR provides journal-specific citation metrics, such as impact factors and immediacy index. It indexes journals based on subject grouping. Our findings show that the JCR category *OR/MS* and *Industrial Engineering (IE)* were associated with seven and three articles, respectively. This is evidence of the popularity of DES as a quantitative modelling technique in OR/MS and Engineering disciplines. The remaining five journal papers were classified under the following JCR categories: *Management (Man)*, *Health Policy and Services (HPS)*, *Computer Science (SC)*, *Economics (Econ)*, and *Heamatology (Heam)*. Conference papers are not included in JCR; thus, the information presented is for the 15 journal papers included in our dataset.

Table 1 *Fields of Application*

Application Field	Freq	Publications
Blood and blood products	6	Baesler et al. (2014); Abbasi et al. (2017); Osorio et al. (2017); Duong et al. (2020); Ejohwomu et al. (2021); Zhou et al. (2021)
Food	5	Rijpkema et al. (2014); Galal & El-Kilany (2016); Kiil et al. (2018); Xue et al. (2019); Qasem et al. (2021)

Other retail	2	Herbon et al. (2012); Zhang et al. (2021)
Medical supplies	1	Zhou & Olsen (2018)
Chemical manufacturing	1	Sharda & Akiya (2012)
Unspecified/conceptual	1	Alrawabdeh (2021)

4.1.2 Nature of Problem and Stakeholder Engagement

Twelve studies focussed on real-world problems and had access to some form of empirical data, although the evidence of stakeholder engagement was somewhat limited (Table 2). While the importance of stakeholder engagement in the realisation of a successful simulation study has been widely accepted (Robinson & Pidd, 1998), according to Jahangirian et al. (2010), among simulation techniques, DES had relatively lower stakeholder engagement. In light of this, it is worth noting that, even though not observed within our sample, the participatory modelling approaches, such as PartiSim (Tako & Kotiadis, 2015) or SimLean (Robinson et al., 2012) were developed as a way to address that issue. In the future if more studies engage in such approaches the stakeholder engagement might improve.

Within our dataset, Rijpkema et al. (2014) and Kiil et al. (2018) conducted interviews to understand the system under investigation (food-related in both cases). Sharda & Akiya (2012) allude to the management's involvement in the problem specification stage of their chemical plant modelling study, although no further stakeholder involvement was apparent. Qasem et al. (2021) discussed the results of their DES study on milk production with the management; however, there is no mention of stakeholder engagement in the earlier stages of the study. An exception is a work by Osorio et al. (2017). The authors report the involvement of problem owners in multiple stages of simulation study of their blood supply chain, which extended as far as to the organisation agreeing to use the model for a pilot study (*ibid.*). Four papers studied problems of a hypothetical nature (Table 2).

Table 2 *Nature of Problem and Stakeholder Engagement*

Nature of problem	Real (empirical data) (n=12)	Sharda & Akiya (2012); Baesler et al. (2014); Rijpkema et al. (2014); Galal & El-Kilany (2016); Abbasi et al. (2017); Osorio et al. (2017); Kiil et al. (2018); Zhou & Olsen (2018); Xue et al. (2019); Ejohwomu et al. (2021); Qasem et al. (2021); Zhou et al. (2021)
	Hypothesised (n=4)	Herbon et al. (2012); Duong et al. (2020); Alrawabdeh (2021); Zhang et al. (2021)
Stakeholder involvement	Yes (n=6)	Sharda & Akiya (2012); Rijpkema et al. (2014); Osorio et al. (2017); Kiil et al. (2018); Ejohwomu et al. (2021); Qasem et al. (2021)
	No (n=8)	Herbon et al. (2012); Baesler et al. (2014); Abbasi et al. (2017); Zhou & Olsen (2018); Xue et al. (2019); Duong et al. (2020); Zhang et al. (2021); Zhou et al. (2021)
	N/A (n=2)	Galal & El-Kilany (2016); Alrawabdeh (2021)

4.2 Model Characteristics

4.2.1 Integrated Approaches with DES (Modelling Methodology)

Simulation models are often run in tandem with optimisation methods to facilitate better decisions. Beyond four studies combing DES with optimisation (Alrawabdeh, 2021; Osorio et al., 2017; Xue et al., 2019; Zhou & Olsen, 2018) (see Table 3), a single publication (Ejohwomu et al., 2021) described a hybrid simulation, which combined DES with ABS. Table 3 also lists two studies employing a hybrid modelling approach using DES with multi-criteria ranking techniques, which assists in the interpretation of the results of the DES model, to either evaluate the Key Performance Indicators (KPIs) (Duong et al., 2020) or to evaluate different policies (Zhou et al., 2021).

Table 3 *Modelling Methodology*

Publication	Integrated Approach with DES	DES Objectives	DES Model Outputs
Herbon et al. (2012)	No	Effect of price discounting policy	Profit
Sharda & Akiya (2012)	No	To select best manufacturing/ inventory management policy	Order fulfilment, cost of lost materials, inventory cost

Baesler et al. (2014)	No	To test adjustment of reorder point; optimum inventory; extra donations	Total production for each product; unsatisfied demand; expired units
Rijpkema et al. (2014)	No	To calculate the cost for different inventory policies	Costs (incl. waste, inventory, stock-outs)
Galal & El-Kilany (2016)	No	Inventory replenishment, effect of changing order quantity	Costs and emissions within supply chain; service level
Abbasi et al. (2017)	No	Understand system behaviours with reduced shelf-life	Outdates, cost, sufficiency of supply
Osorio et al. (2017)	Yes - Optimisation via ILP	To generate KPIs and ILP inputs	Stockouts, discards, cost, inventory levels, donors required
Kiil et al. (2018)	No	To test replenishment policies as a function of shelf life	Fill rate, waste, number of deliveries, inventory level
Zhou & Olsen (2018)	Yes - Optimisation	Evaluation of stock rotation policy	Cost
Xue et al. (2019);	Yes - Optimisation	Replenishment policy to prevent stock-outs and reduce waste	Number of units consumed; number of units remaining
Duong et al. (2020)	Yes - AHP & DEA to rank KPIs of DES results	Replenishment policy with DES as a part of wider decision making	Fill rate, average inventory, order rate variance ratio
Alrawabdeh (2021)	Yes - Optimisation	To find optimal order quantity with age-based demand	Outdates, shortages, age-related mismatch
Ejohwomu et al. (2021)	Yes – Hybrid simulation (DES-ABS)	To evaluate "pull system" reliant on hospital demand. DES for internal SHU operations	Stock level
Qasem et al. (2021)	No	To test different inventory policies	Cost (for stock holding, deterioration, shortage, ordering), customer service level
Zhang et al. (2021)	No	Determine optimal profit as a function of discounting policy	Profit, waste rate, average selling price
Zhou et al. (2021)	Yes - MADM to rank inventory policies	Assessment of four different inventory policies	Shortages, outdated stock, fairness index

*AHP: Analytical Hierarchy Process; SHU: Stock Holding Unit; DEA: Data Envelop Analysis; ILP: Integer Linear Programming; MADM: Multiple Attribute Decision Making

4.2.2 Product Variants and Issuing Policy

Eight papers model multiple product variants; however, many of these consider the blood supply chain and relate to a single blood product (e.g., red blood cells or platelets) with multiple variants corresponding to different blood types, rather than heterogeneity of product characteristics such as shelf-life, processing requirements etc., see Table 4. For example, the study by Ejohwomu et al. (2021) considered 24 platelet types. Nevertheless, this does not take away from the complexity when the substitution of blood products is modelled, for instance, by Abbasi et al. (2017).

Two studies included multiple products displaying more distinct characteristics. Xue et al. (2019) modelled multiple products in food retailing with differing shelf lives and Sharda & Akiya (2012) modelled the manufacturing of multiple chemical products with different packaging sizes. It should be noted that although some studies consider only single products, the existence of discounting mechanisms, e.g. Qasem et al. (2021), or multiple demands for age-differentiated products, e.g., Alrawabdeh (2021) & Zhou et al. (2021), created multiple streams which were modelled through DES.

Depending on the problem, different systems might be considered in terms of the issuing policy of perishable inventories. For example, in the supermarket environment, customers are likely to select products based on expiry date/use-by date (Zhang et al., 2021). In blood banking, the quality of product to be issued is likely to be considered based on pragmatic objectives that aim to limit waste and possible shortages. It has been widely assumed that the issuing model of *first-in-first-out* (FIFO) policy (the oldest product leaves the system first) is the most efficient way to manage a perishable inventory. However, in the field of blood banking, for some medical conditions, it might be beneficial to aim for a fresher blood product (Koch et al., 2008); this might necessitate a combination of FIFO and *last-in-first-out* (LIFO).

Table 4 lists the articles describing the issuing policy. Six studies include models that implement FIFO logic, with three considering both FIFO and LIFO. Qasem et al. (2021) modelled the *first-expired-first-out* (FEFO) policy, and Alrawabdeh (2021) implemented a bespoke model based on age. Aspects of customer behaviour when selecting products were incorporated by Rijpkema et al. (2014) and Kiil et al. (2018), which employed mixed FIFO/LIFO logic, to represent the alternative consumer decisions of whether to take the freshest product, or select the product in front of the shelf. However, the policy mix was expressed as a simple fraction of occurrence between the two prioritisation logics.

4.2.3 Modelling Uncertainty

It is widely acknowledged that supply chains exhibit uncertainty at multiple levels. As simulation methods are better suited than mathematical models in terms of their ability to incorporate uncertainty, we assessed the extent to which the authors incorporated uncertainty in their models, specifically whether stochastic models are employed to represent supply/demand uncertainties.

When considering the inventory models, the focus classically has been on uncertainty on the demand side and how to mitigate the associated risk. However, according to Schmitt et al. (2010), since the mid-2000s, the focus on supply disruption when studying the inventory and supply chain models has led to an "*explosion of research*" in the field. However, in our sample, such evidence was lacking (Table 5). Most studies assumed unconstrained supply, ignoring possible yield uncertainty concerning the supply level (Silver, 1976). Only three studies modelled constrained supply: Osorio et al. (2017), in which supply is pre-determined by an integer linear programming optimisation module before reaching DES, and Baesler et al. (2014) and Abbasi et al. (2017), both of which model the supply of red blood cells as being constrained according to historical fitted supply data. However, in Beasler et al. (2014), the model allows this constraint to be relaxed through public calls for additional blood donations, allowing the necessary supply to be realised.

Additional considerations around lead time and supply disruption leading to supplier uncertainty have also been highlighted (Fang & Shou, 2015). The lead time for most of the studies was set to a fixed value (with some equal to zero) (Table 5). Rijpkema et al. (2014) differentiate between two lead times (corresponding to regular and expedited orders) with static characteristics. Galal & El-Kilany (2016) have considered a level of uncertainty expressed as a stochastic lead time for otherwise unconstrained yield from the supplier.

In our sampled literature, the uncertainty around demand is acknowledged and incorporated in the modelling, with all the publications considering non-constant demand (Table 5). Random demand models are considered in 13 out of the 16 publications, of which seven (Abbasi et al., 2017; Baesler et al., 2014; Ejohwomu et al., 2021; Galal & El-Kilany, 2016; Kiil et al., 2018; Qasem et al., 2021; Xue et al., 2019) are based on fitting to historical data. In contrast, studies by Sharda & Akiya (2012) and Osorio et al. (2017) relied on historical data without distribution fitting. Sharda & Akiya (2012) applied a random sampling technique; Osorio et al. (2017) directly used historical data.

Models for the lifetime of perishable products assumed a fixed shelf-life; with the exception of the studies by Duong et al. (2020) and Rijpkema et al. (2014). The study by Duong et al. (2020) relate to platelets, for which, although they have a fixed shelf-life at the time of collection, at the moment they are received at the distribution centre, the remaining shelf-life is modelled as an exponential distribution. Rijpkema et al. (2014) model a case study of strawberries with a random shelf-life parameterised by environmental factors related to storage conditions in the supply chain. With only the single study (Rijpkema et al., 2014) factoring in exogenous factors that influence the lifetime of perishable products, we suggest that far more knowledge could be gained through employing DES for scenarios exhibiting more elaborate characteristics, such as non-deterministic lifetime models.

Table 4 *Multi-product and FIFO/LIFO/FEFO Characteristics*

Publication	Multi-product	FIFO/LIFO/FEFO	Product
Herbon et al. (2012)	No	N/A	Retailer goods
Sharda & Akiya (2012)	Yes (60 products)	Unclear	Chemicals
Baesler et al. (2014)	Yes	Unclear	Blood
Rijpkema et al. (2014)	No	Combination 60:40 FIFO:LIFO at retailer, unclear at distributer	Food-strawberries
Galal & El-Kilany (2016)	No	FIFO	Food-oranges
Abbasi et al. (2017)	Yes (different blood types)	FIFO	Blood
Osorio et al. (2017)	Yes	FIFO	Blood
Kiil et al. (2018)	No	Combination 90:10 FIFO:LIFO	Food products

Zhou & Olsen (2018)	No	FIFO at the reserve, at hospital unclear	Unspecified medical supplies
Xue et al. (2019);	Yes	Unclear	In shop production of sandwiches
Duong et al. (2020)	Yes	FIFO	Unspecified perishable health supplies
Alrawabdeh (2021)	No (single product but demand differentiated by age)	Bespoke, based on age	Not specified
Ejohwomu et al. (2021)	Yes (24 platelet types)	FIFO	Blood- Platelets
Qasem et al. (2021)	No	FEFO	Milk
Zhang et al. (2021)	Yes (two products)	LIFO	Retailer goods
Zhou et al. (2021)	No (but demand differentiated by shelf-life)	FIFO and LIFO considered as part of different policy options	Blood

Table 5 *Uncertainty Characteristics*

Publication	**Supply side, as input to the system modelled**	**Demand side**	**Modelling of product lifetime**
Herbon et al. (2012)	Unconstrained, no lead-time	Random uniform for time distribution between events, and for customer preference (freshness vs price)	Fixed lifetime
Sharda & Akiya (2012)	Unconstrained, no lead-time	Random sampling of historical data	Fixed lifetime
Baesler et al. (2014)	Constrained, using historical fitted distributions; mitigated via calls for donation	Empirical and standard distributions fitted to historical data, differentiated by product type and distribution site	Fixed lifetime
Rijpkema et al. (2014)	Unconstrained, fixed lead-time (regular and expedited)	Poisson distribution at the retailer	Random shelf-life based on storage environment
Galal & El-Kilany (2016)	Unconstrained, stochastic lead-time	Distribution fitted to historical data	Fixed lifetime after original quality check
Abbasi et al. (2017)	Constrained, using historical fitted distributions	Distribution fitted to historical (1 year) data, daily, by location	Fixed lifetime, varied for different product
Osorio et al. (2017)	Constrained, using historical data directly or fitted to distribution, based on scenario	Historical data	Fixed lifetime
Kiil et al. (2018);	Unconstrained, fixed lead-time	Distribution fitted to historical data, differentiated by weekday and subgroup of stores	Fixed lifetime
Zhou & Olsen (2018)	Unconstrained supply but lead time is a parameter	Fixed size, random (exponential) demand rate, gamma distribution for modelling emergency occurrence	Fixed lifetime for reserve
Xue et al. (2019);	Unconstrained	Erlang fitted to historical data	Fixed, differentiated by product type
Duong et al. (2020)	Unconstrained, fixed lead-time	Poisson	Exponential distribution of lifetime
Alrawabdeh (2021)	Unconstrained, fixed lead-time	Poisson with pre-determined mean items/day	Fixed lifetime
Ejohwomu et al. (2021)	Unconstrained, fixed lead-time	Normal distribution fitted to historical data	Fixed lifetime
Qasem et al. (2021)	Unconstrained, lead-time unclear	Distribution fitted to historical data, interarrival-exponential, quantity per arrival-normal	Fixed lifetime
Zhang et al. (2021)	Unconstrained replenishment once a day	Poisson customer arrival, "bespoke" for number if items (Monte-Carlo)	Fixed lifetime with threshold for discounting
Zhou et al. (2021)	Normal distribution fitted to data per day of week	Historical data provided, however unclear how data was used to generate stochastic demand events	Fixed lifetime

4.3 Study Outcome

4.3.1 Model Outputs

Regarding model outputs, it is not surprising that when assessing the inventory, the most commonly used KPIs are around metrics associated with *fill rates*, *inventory/product outdates* and *finance* (Table 3- column "DES Model Outputs"). However, Galal & El-Kilany (2016), in their study of the supply chain of oranges, simulated an inventory replenishment policy that went beyond the economic considerations. They considered environmental sustainability aspects, whereby adjustment of the order quantity presents an opportunity to improve costs and emission levels without compromising the customer service levels.

Most papers surveyed in our review used DES to compare competing policy options associated with inventory optimisation. Four notable exceptions are described briefly. Zhou & Olsen (2018) assessed the potential benefits of stock rotation between an emergency reserve and regular hospital use. Ejohwomu et al. (2021) considered the potential benefits of introducing a "*pull-based*" system between a stock holding unit and a hospital for blood platelets based on hospital demand (rather than the practised solution of having fixed stock-level targets); Herbon et al. (2012) and Zhang et al. (2021) considered the retail environment and the effect of applying an age-based discounting policy to maximise profits.

4.3.2 Implementation of the Results of the DES study

None of the studies reported real-world implementation or having influenced policy (Table 6). Thus, following Brailsford et al. (2009), the papers were classified into the two remaining categories in the three-level scale of model implementation: "*suggested*" (theoretical application) and "*conceptualised*" (if discussion with a client organisation had taken place). The low level of implementation is broadly in line with the findings of previous larger-scale surveys of simulation modelling in healthcare by Brailsford et al. (2009) and Katsaliaki & Mustafee (2011), as well as a more recent review of hybrid simulation by Brailsford et al. (2019).

Table 6 *Result Implementation*

Implemented (n=0)	None
Conceptualised (n=2)	Osorio et al. (2017); Qasem et al. (2021)
Suggested (n=14)	Herbon et al. (2012); Sharda & Akiya (2012); Baesler et al. (2014); Rijpkema et al. (2014); Galal & El-Kilany (2016); Abbasi et al. (2017); Kiil et al. (2018); Zhou & Olsen (2018); Xue et al. (2019); Duong et al. (2020); Alrawabdeh (2021); Ejohwomu et al. (2021); Zhang et al. (2021); Zhou et al. (2021)

5 CONCLUSION

Given the importance of the volatility of perishable products in inventory management and the long-standing application of DES to study product flows, we identified the need to conduct a methodological review of the literature to find a sample of publications wherein the two fields converge. Our findings show that the management of blood products, with its associated complexities of preservation and storage, received significant attention from the DES community. This is not surprising considering the decades of research in quantitative modelling to manage the inventory of blood products for transfusion. On the other hand, the lack of modelling of inventories related to, for instance, pharmaceutical supplies, was rather unexpected since pharmaceutical supplies constitute an important and growing economic sector, and associated expiry dates need careful attention to manage patient safety.

As pointed out by Robinson et al. (2012), DES and lean have similar motivations for achieving improvements in processes and service delivery, hence considering them together in healthcare was recommended to be an appropriate trajectory. Similarly, we argue that the overlap of (perishable) inventory and the DES methodology allows for a deeper understanding of system behaviour and, through experimentation, identification of specific areas for improving the system. As such, more research needs to be devoted to the study of DES specific to perishable inventory.

To conclude, DES presents an exciting opportunity for researchers to investigate yet unexplored and underrepresented problems concerning inventory management for perishable goods. A deeper understanding of the system's behaviour, especially by incorporating further aspects of uncertainty (e.g., for the product lifetime) in the supply chains to better mimic the real world, will likely facilitate addressing significant problems related to inventory management issues such as waste, storage and overall resilience. As such, the authors are actively employing DES to study the human donor milk system, and associated perishable inventories, with the intent of addressing some of the identified gaps.

While a relatively small sample was selected through our search which constitutes a limitation, and does not at this stage allow for generalization, a more extensive search of the literature (e.g., using literature snowballing and grey literature search) could provide further evidence for some of the findings presented in this review; this is an avenue for future research.

REFERENCES

Abbasi, B., Vakili, G., & Chesneau, S. (2017). Impacts of Reducing the Shelf Life of Red Blood Cells: A View from Down Under. *Interfaces,* **47(4)**, 336–351.

Alrawabdeh, W. (2021). Multi- period age-discriminated perishable inventory. *Management Systems in Production Engineering,* **29(2)**, 97–105.

Armbruster, D., Gottlicht, S., & Herty, M. (2011). A scalar conservation law with discontinuous flux for sypply chains with finite buffers. *SIAM Journal on Applied Mathematics,* **71(4)**, 1070–1087.

Baesler, F., Nemeth, M., Martínez, C., & Bastías, A. (2014). Analysis of inventory strategies for blood components in a regional blood center using process simulation. *Transfusion,* **54(2)**, 323–330.

Bakker, M., Riezebos, J., & Teunter, R. H. (2012). Review of inventory systems with deterioration since 2001. European *Journal of Operational Research,* **221(2)**, 275–284.

Bender, A., Pincombe, A. H., & Sherman, G. D. (2009). Effects of decay uncertainty in the prediction of life-cycle costing for large scale military capability projects. *18th World IMACS Congress and MODSIM 2009 - International Congress on Modelling and Simulation: Interfacing Modelling and Simulation with Mathematical and Computational Sciences, Proceedings, July*, 1573–1579.

Brailsford, S. C., Eldabi, T., Kunc, M., Mustafee, N., & Osorio, A. F. (2019). Hybrid simulation modelling in operational research: A state-of-the-art review. *European Journal of Operational Research,* **278(3)**, 721–737.

Brailsford, S., Harper, P. R., Patel, B., & Pitt, M. (2009). An analysis of the academic literature on simulation and modelling in health care. *Journal of Simulation,* **3(3)**, 130–140.

Brailsford, S., & Hilton, N. (2001). A comparison of discrete event simulation and system dynamics for modelling health care systems. Proceedings from ORAHS 2000, 18–39. http://eprints.soton.ac.uk/35689/1/glasgow_paper.pdf

Chase, R., Jacobs, F., & Aquilano, N. (2007). Operations management for competitive advantage. McGraw-Hill: Boston.

Chemweno, P., Pintelon, L., & Muchiri, P. (2016). Simulating the Impact of Deferred Equipment Maintenance. *Proceedings of the 10th World Congress on Engineering Asset Management (WCEAM 2015)*, 133–140.

de Kok, T., Grob, C., Laumanns, M., Minner, S., Rambau, J., & Schade, K. (2018). A typology and literature review on stochastic multi-echelon inventory models. *European Journal of Operational Research,* **269(3)**, 955–983.

Duong, L. N. K., Wood, L. C., & Wang, W. Y. C. (2020). Inventory management of perishable health products: a decision framework with non-financial measures. *Industrial Management & Data Systems,* **120(5)**, 987–1002.

Ejohwomu, O. A., Too, J., & Edwards, D. J. (2021). A resilient approach to modelling the supply and demand of platelets in the United Kingdom blood supply chain. *International Journal of Management Science and Engineering Management,* **16(2)**, 143–150.

Engebrethsen, E., & Dauzère-Pérès, S. (2019). Transportation mode selection in inventory models: A literature review. *European Journal of Operational Research,* **279(1)**, 1–25.

Fang, Y., & Shou, B. (2015). Managing supply uncertainty under supply chain Cournot competition. *European Journal of Operational Research,* **243(1)**, 156–176.

Galal, N. M., & El-Kilany, K. S. (2016). Sustainable agri-food supply chain with uncertain demand and lead time. *International Journal of Simulation Modelling,* **15(3)**, 485–496.

Goltsos, T. E., Syntetos, A. A., Glock, C. H., & Ioannou, G. (2022). Inventory – forecasting: Mind the gap. *European Journal of Operational Research,* **299(2)**, 397–419.

Goyal, S. K., & Giri, B. C. (2001). Recent trends in modeling of deteriorating inventory. *European Journal of Operational Research,* **134(1)**, 1–16.

Harris, F. W. (1915). What Quantity to Make at Once. In The Library of Factory Management, Operation and Costs. (pp. 47–52). A. W. Shaw Company: Chicago.

Herbon, A., Spiegel, U., & Templeman, J. (2012). Simulation study of the price differentiation effect in a stochastic deteriorating inventory with heterogeneous consumers - freshness sensitivity. *Applied Economics,* **44(24)**, 3101–3119.

Hollocks, B. W. (2006). Forty years of discrete-event simulation—a personal reflection. *Journal of the Operational Research Society,* **57(12)**, 1383–1399.

Jahangirian, M., Eldabi, T., Naseer, A., Stergioulas, L. K., & Young, T. (2010). Simulation in manufacturing and business : A review. *European Journal of Operational Research,* **203(1)**, 1–13.

Katsaliaki, K., & Mustafee, N. (2011). Applications of simulation within the healthcare context. *Journal of the Operational Research Society,* **62(8)**, 1431–1451.

Kiil, K., Hvolby, H. H., Fraser, K., Dreyer, H., & Strandhagen, J. O. (2018). Automatic replenishment of perishables in grocery retailing: The value of utilizing remaining shelf life information. *British Food Journal,* **120(9)**, 2033–2046.

Koch, C., Li, L., Sessler, D., Figueroa, P., Hoeltge, G., Mihaljevic, T., & Blackstone, E. (2008). Duration of Red-Cell Storage and Complications After Cardiac Surgery. *New England Journal of Medicine,* **358**, 1229–1239.

Law, A. M., & Kelton, W. D. (1991). Simulation Modeling & Analysis. 2nd ed. McGraw-Hill, Inc: New York.

Ma, Q. G., & Meng, L. J. (2008). Simulation study about perishable products inventory system with resalable product return. *Proceedings of the International Conference on Information ManagementProceedings of the International Conference on Information Management, Innovation Management and Industrial Engineering,* ICIII 2008, 2, 214–217.

Mustafee, N., & Katsaliaki, K. (2020). Classification of the Existing Knowledge Base of OR/MS Research and Practice (1990-2019) using a Proposed Classification Scheme. *Computers and Operations Research*, **118**, 104920.

Mustafee, N., Katsaliaki, K., & Taylor, S. J. E. (2021). Distributed Approaches to Supply Chain Simulation. *ACM Transactions on Modeling and Computer Simulation,* **31(4)**.

Nahmias, S. (1982). Perishable Inventory Theory: a Review. *Operations Research,* ***30*(4)**, 680–708.

Newman, W., Hanna, M., & Jo Maffei, M. (1993). Dealing with the Uncertainties of Manufacturing: Flexibility, Buffers and Integration. *International Journal of Operations & Production Management,* **13(1)**, 19–34.

Opacic, L., & Sowlati, T. (2017). Applications of discrete-event simulation in the forest products sector: A review. *Forest Products Journal,* **67**, 219–229.

Osorio, A. F., Brailsford, S. C., Smith, H. K., Forero-Matiz, S. P., & Camacho-Rodríguez, B. A. (2017). Simulation-optimization model for production planning in the blood supply chain. *Health Care Management Science,* **20(4)**, 548–564.

Qasem, A. G., Aqlan, F., Shamsan, A., Alhendi, M., Gailan Qasem, A., Aqlan, F., Shamsan, A., Alhendi, M., Qasem, A. G., Aqlan, F., Shamsan, A., & Alhendi, M. (2021). A simulation-optimisation approach for production control strategies in perishable food supply chains. Journal of Simulation. https://doi.org/10.1080/17477778.2021.1991850

Raafat, F. (1991). Survey of Literature on Continuously Deteriorating Inventory Models. *The Journal of the Operational Research Society,* **42(1),** 27–37.

Rijpkema, W. A., Rossi, R., & van der Vorst, J. G. A. J. (2014). Effective sourcing strategies for perishable product supply chains. *International Journal of Physical Distribution and Logistics Management*, **44(6)**, 494–510.

Robinson, S. (2005). Discrete-event simulation: From the pioneers to the present, what next? *Journal of the Operational Research Society*, **56(6)**, 619–629.

Robinson, S. (2014). Simulation: The Practice of Model Development and Use, 2nd ed. palgrave macmillan: New York.

Robinson, S., & Pidd, M. (1998). Provider and customer expectations of successful simulation projects. *Journal of the Operational Research Society,* **49(3)**, 200–209.

Robinson, S., Radnor, Z. J., Burgess, N., & Worthington, C. (2012). SimLean: Utilising simulation in the implementation of lean in healthcare. *European Journal of Operational Research*, **219(1)**, 188–197.

Saltzman, R., Roeder, T., Lambton, J., Param, L., Frost, B., & Fernandes, R. (2017). The impact of a discharge holding area on the throughput of a pediatric unit. *Service Science*, **9(2)**, 121–135.

Schmitt, A. J., Snyder, L. V., & Shen, Z. J. M. (2010). Inventory systems with stochastic demand and supply: Properties and approximations. *European Journal of Operational Research,* **206(2)**, 313–328.

Sharda, B., & Akiya, N. (2012). Selecting make-to-stock and postponement policies for different products in a chemical plant: A case study using discrete event simulation. *International Journal of Production Economics,* **136(1)**, 161–171.

Silver, E. (1976). Establishing the order quantity when the amount received is uncertain. *INFOR,* **14(1)**, 32–39.

Svoboda, J., Minner, S., & Yao, M. (2021). Typology and literature review on multiple supplier inventory control models. *European Journal of Operational Research,* **293(1)**, 1–23.

Tako, A. A., & Kotiadis, K. (2015) PartiSim: A multi-methodology framework to support facilitated simulation modelling in healthcare. *European Journal of Operational Research,* **244(2)**, 555–564.

Tako, A. A., & Robinson, S. (2012). The application of discrete event simulation and system dynamics in the logistics and supply chain context. *Decision Support Systems,* **52(4)**, 802–815.

Vázquez-Serrano, J. I., Peimbert-García, R. E., & Cárdenas-Barrón, L. E. (2021). Discrete-event simulation modeling in healthcare: A comprehensive review. *International Journal of Environmental Research and Public Health,* **18(22)**, 12262.

Whitin, T. M. (1953). The Theory of Inventory Management. Princeton University Press.

Xu, S., & Hall, N. G. (2021). Fatigue, personnel scheduling and operations: Review and research opportunities. *European Journal of Operational Research,* **295(3)**, 807–822.

Xue, N., Landa-Silva, D., Figueredo, G. P., & Triguero, I. (2019). A simulation-based optimisation approach for inventory management of highly perishable food. *ICORES 2019 - Proceedings of the 8th International Conference on Operations Research and Enterprise Systems, Icores*, 406–413.

Zhang, X. (2018). Application of discrete event simulation in health care: A systematic review. *BMC Health Services Research,* **18(1)**, 1–11.

Zhang, Y., Lu, H., Zhou, Z., Yang, Z., & Xu, S. (2021). Analysis and optimisation of perishable inventory with stocks-sensitive stochastic demand and two-stage pricing: A discrete-event simulation study. *Journal of Simulation,* **15(4)**, 326–337.

Zhou, Q., & Olsen, T. (2018). Rotating the medical supplies for emergency response: A simulation based approach. *International Journal of Production Economics,* **196(July 2017)**, 1–11.

Zhou, Y., Zou, T., Liu, C., Yu, H., Chen, L., & Su, J. (2021). Blood supply chain operation considering lifetime and transshipment under uncertain environment. *Applied Soft Computing,* **106**, 107364.

AUTHOR BIOGRAPHIES

MARTA STAFF received a BSc (Hons) in Biology from the University of London, and an MBA from the University of Exeter. After being awarded a scholarship from the UK Economic and Social Research Council, she successfully completed an MRes and is currently studying for a PhD within the University of Exeter Business School, UK, with a focus on Operations management in healthcare. Her email address is ms670@exeter.ac.uk.

NAVONIL MUSTAFEE is a Professor of Analytics and Operations Management at the University of Exeter Business School, UK. His research focuses on M&S methodologies and Hybrid Modelling and their application in healthcare, supply chain management, circular economy and resilience and adaptation due to climate change. He is a Joint Editor-in-Chief of the Journal of Simulation (UK OR Society journal) and Vice-President of Publications at The Society of Modeling and Simulation International (SCS). His email address is n.mustafee@exeter.ac.uk.

Proceedings of the Operational Research Society Simulation Workshop 2023 (SW23)
C. Currie and L. Rhodes-Leader, eds.
https://doi.org/10.36819/SW23.027

EXTENDING AN "OUT OF THE BOX" SIMULATION AND OPTIMISATION SUPPLY CHAIN TOOL TO INCLUDE BESPOKE FEATURES DESIGNED FOR FAST-MOVING CONSUMER GOODS

Adam Coleman,
Simulation Consultant
Decision Lab
adam.coleman@decisionlab.co.uk

Jacob Whyte,
Simulation Consultant
Decision Lab
jacob.whyte@decisionlab.co.uk

Aanand Davé,
Chief Revenue Officer
Decision Lab
aanand.dave@decisionlab.co.uk

Peter Riley,
Principal Simulation Consultant
Decision Lab
peter.riley@decisionlab.co.uk

Gavin Wilkinson,
Head of Marketing
Decision Lab
gavin.wilkinson@decisionlab.co.uk

ABSTRACT

Managing a supply chain is a complex and difficult task in Fast Moving Consumer Goods (FMCG). Failing to keep up with technology can make management even harder and cause organisations to fall behind their competitors. Supply chain management technologies can help identify bottlenecks and support informed decision-making, often resulting in greater efficiencies. Modelling tools allow companies to simulate, optimise, and analyse different scenarios for inventory management, logistics, and demand forecasting.

This paper discusses the process of extending an out-of-the-box simulation and optimisation modelling tool to include bespoke 'data-driven' features. These features help identify potential problems and risks before they occur in a client's supply chain network. The extensions were developed as part of the client's digital transformation.

1 INTRODUCTION TO THE CLIENT

The client is a UK-based FMCG manufacturer that produces a wide range of baked goods. Their products have a short two-to-three-day shelf-life and are sold to consumers through retail outlets. The manufacturer is also responsible for supplying bread and other baked goods to food and catering services through a secondary distribution network. The client will remain nameless within this paper due to ongoing work.

The operations of this supply chain are complex but still follow common standard practices within manufacturing and logistics (Hopp and Spearman, 2011). They have several production facilities where they produce the baked goods before sending them to their distribution centres; from here they are sent to depots that are responsible for transporting the goods to the retailers such as supermarkets and grocery stores who then sell the products to consumers.

One of the main challenges for this company is the consistent production of products in a short timeframe in response to high demand; in this case, a retailer expects their order to arrive within 2 days of placing it. They must also consider factors such as the cost of ingredients, transportation, and storage, to satisfy consumers and remain competitive in the market.

2 IDENTIFYING THE PROBLEM, CHALLENGES, AND THE SOFTWARE

The fast-moving consumer goods sector faces many challenges, such as changes in consumption trends and large fluctuations in demand, that are difficult for the food and drinks industry to adapt to quickly (Adebanjo and Mann, 2000). The client identified inefficiencies and bottlenecks in their day-to-day operations that were challenging their ability to adapt, slowing down their growth and development (Büyüközkan and Göçer, 2018) within the FMCG industry.

Underlying these challenges was a lack of operational and network insight and understanding, such as for stock capacity, production capacity, and demand fluctuations. And, with few digital systems and technologies in their network, they would react to demand (Zhengping Li et al., 2008) rather than predict it and make plans. The result was slow progress in improving areas such as efficiency, bottleneck identification, cost reduction, and agility within a rapidly developing market.

Overall, they lacked digital platforms and tools that would aid them in analysing, coordinating, and optimising their supply chain. With them, they could improve goods-flow visibility and transportation, reduce costs and supply chain risks, better respond to demand fluctuations, and optimise facility locations.

2.1 Defining a Goal: Achieving Improved Efficiency and Effectiveness with Technology

The client identified that they were falling behind in using modern technology and systems for defining organisational goals. By moving towards a more digital approach, by implementing a digital transformation, they could improve their data analytics and make more informed strategic decisions in areas such as the movement of goods and the allocation of storage. Leveraging modern supply chain technologies would make them more responsive, flexible, and resilient – allowing them to adapt to dynamic market conditions and customer demand more quickly and efficiently.

Due to the dynamic nature of FMCG and the difficulty that creates when making predictions for them, the company would need to overhaul the way they solved their business problems in conjunction with implementing a new digital tool. The overall aim was to achieve a complete digital transformation (Albukhitan, 2020) and implement a supply chain modelling tool that allows them to design and analyse their supply chain within a virtual environment, enabling the easy identification of failure points and inefficiencies, and to test solutions within a risk-free environment. Simulation and optimisation would allow for this, as it gives them the ability to predict, optimise, and implement solutions.

2.2 Defining the Business Questions and Identifying the Key Challenges

Through a series of formal workshops involving the client and a separate UK-based consultancy, user stories were collaboratively constructed based on the problems that were addressed when identifying the supply chain's issues.

The workshop process involved multiple members and key stakeholders from different teams within the organisation such as warehousing, transportation and logistics, production planning, information technology, and the board of directors. Team members were given the chance to put forward their business problems or concerns related to their area within the supply chain to develop an extensive list of questions.

Identifying business problems was useful for producing functional requirements which can provide guidance during the development of solutions and aid in defining a model architecture. The list of business questions was categorised to form distinct user stories that focused on specific business areas as seen below:

Production:

- Improve insight into production capacity and location of bakeries for optimal new site placement

Warehousing:

- Improve insight into stock capacity and location of warehouses for optimal new site placement, consolidate impact, lease expiration decisions, and cross-docking optimization

Products:

- Understand spare capacity for third-party products and impact of demand changes

Logistics:

- Understand fleet size and reverse logistics impact

Policies:

- Understand impact of current inventory and lead time policies on total cost and cost impact of changes made

Costs:

- Optimize total network to reduce costs and understand cost impact of changes on the network.

2.3 Choosing Software to Address Supply Chain Design Questions

anyLogistix provides an "out of the box" solution suited to answer most, if not all, of the related supply chain questions by providing functionality for both simulation and optimisation.

anyLogistix is a user-friendly supply chain simulation and optimization tool for various industries, such as manufacturing, retail, and logistics services (Ivanov, 2020). It enables users to design, analyze, and optimize their supply chain and logistics networks through a visual interface without the need for programming languages. It simulates and optimizes facility locations such as a centre of gravity (Sanjaya et al., 2019), product flows, and transportation methods, providing insights into key performance indicators such as delivery times, costs, warehouse capacity, and transportation costs.

Using the information the tool provides, the user can then identify bottlenecks, potential risks, and inefficiencies within their logistics network before carrying out what-if experiments and testing various scenarios to find the most cost-effective and efficient solutions.

3 SOLUTION

It was agreed with the client that certain features would need to be added that were not included in the "out of the box" version of anyLogistix and that they would require the use Anylogic Extensions to implement. An Anylogic Extension allows for the customisation of processes within the four walls of a supply chain object.

Anylogic is the underlying simulation engine for anyLogistix. It combines the three main simulation modelling methods: system dynamics, discrete event, and agent-based modelling. It allows users to choose from any one of the methods or to combine them in one model (Borshchev, 2014). anyLogistix makes use of a cut-down version of this engine and implements it via extensions. The key features added by the solution are:

- Resource utilization: Staff as a resource for bay tasks, custom logic for loading/unloading with queues and priorities.
- Production logic: Custom bakery logic for using production line queues and capacity constraints.
- Custom SQL outputs: Implementation of custom statistics by writing directly to a database.
- Product sourcing logic: Customizing sourcing algorithms to align with the client's supply chain.
- Cross-docking: Altering handling to avoid issues with standard functionality in the client's supply chain.

The standard anyLogistix model makes use of custom Anylogic blocks for which there is currently no application programming interface (API), and some internal logic is hidden from developers. This makes it difficult to make substantial changes to the model logic because the core functionality still needs to be preserved for the simulation to run. The anyLogistix Support Team provided versions of the model that replaced some of the custom blocks with blocks from the Anylogic Process Modelling Library and code that performed previously unseen core functionality. This provides more flexibility to how the standard implementation can be adapted to create new functionalities.

3.1 Supply Chain Network Structure

The structure of the supply chain is comprised of multiple tiers and types of facilities. The bakery is at the topmost tier and is responsible for producing products. The primary is a warehouse that is responsible for distributing products to other warehouses and cross-docking at other primaries if required. Lastly, there is a depot that distributes the goods to the retail stores, the simulation model focuses on the bakery, primary, and depot, so all demand is aggregated up to the depot level which represents customers.

All bakeries have an attached primary warehouse in the same location, so products can be temporarily stored in them after being produced before they are sent elsewhere. Most primaries have an attached depot (customer), which products are sent to as soon as they are available. However, not all primaries are attached to a bakery, meaning some warehouses stand alone. This increases the complexity of the structure of the network, making it more difficult when modelling the different sourcing, shipping, and inventory policies using standard anyLogistix functionality.

A primary is a customer or bakery-attached warehouse that shares the same code. The customer can only source from it if attached, or the bakery can only ship to it if attached. The vehicle functionality was altered to reflect this. In standard anyLogistix, vehicles move products from the bakery to the attached primary. Dummy vehicles were used to represent product movement within the same location. These dummy vehicles do not occupy bays and there is no restriction on their number. The same concept applies to products moving from a primary warehouse to the attached customer.

3.2 Resource Utilisation

The client's requirements specified that staff utilisation should be modelled during the loading/unloading of vehicles. The custom anyLogistix blocks that handled bay assignments, warehouse queues, and loading/unloading process times were replaced by a flowchart that handles queueing for bays, queueing for staff, and variable loading/unloading processing times depending on the number of workers assigned to the task. The flowchart also contains logic for prioritising which vehicle types are allowed into certain bay types, the seizing/releasing of staff partway through a task, and the recalculation of processing times to reflect this.

The solution has the flexibility to determine which individual vehicles go to which bay and generates new outputs related to staff utilisation and queueing for staff that the client can use to determine how many staff they should assign to each site.

3.3 Production Logic and Bakery Behaviour

The client's production process utilises production lines and production is planned based on demand across the UK which determines where specific types of products should be made. When selecting a bakery to produce the products, the number of production lines is considered as this determines the overall production capacity each day. Different sites have different numbers of production lines, but each line equates to 24 hours of production meaning that: depending on the number of production lines a site has, they might have 48 hours or more of production within one day (Figure 1.). Products have a production time per unit associated with them, therefore restricting the amount that can be produced in a day. Products are produced in batches and categorised by type; for example, all types of bread are in the same group. This is important as certain production lines can only produce certain categories of products. The client agreed that the additional logic should be added to determine if it is possible to source from a bakery based on its daily production limit. This limit is the maximum time the bakery's production lines can spend producing orders (in total) and is refreshed at the start of each day. Whenever a bakery accepts an order, the remaining daily production capacity is decremented by how long it would take the bakery to produce that order. When checking if an order can be sourced from a bakery, the bakery checks if there is enough production capacity left to produce that order and rejects the order if there is not. To consider the restrictions surrounding product categories and their production lines, it was agreed upon that each bakery would be split into multiple objects to represent the categories. Each object would have a specified number of production lines associated with it.

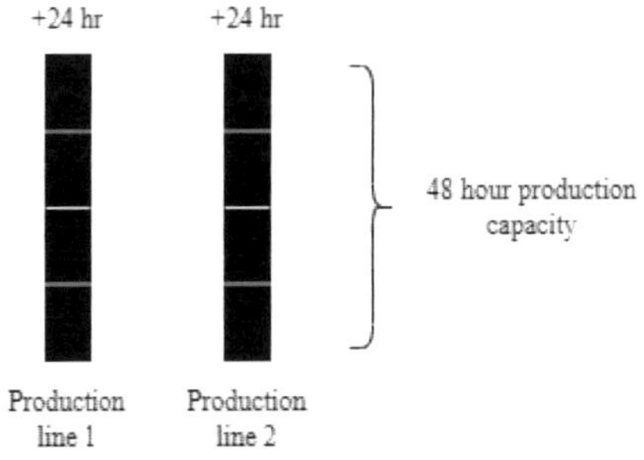

Figure 1 *Production lines*

3.4 Limiting The Bakery's Production Output

A rule was requested by the client when checking if a bakery could produce an order, to help address an issue where bakeries were continuing to produce even when the warehouse was full. This rule is a maximum queue length per production line that is allowed, which is measured as the time it would take to produce the orders currently queueing for a production line and the remaining time to produce any orders currently in production. If there were 2 production lines and the limit per production line was 24, the maximum queue length would be 48 hours.

When the customer checks if a bakery can produce the order, the bakery checks to see if the current production queue length plus the amount of time to produce the new order is less than the maximum. If it is, the bakery accepts the order and it is added to the queue, otherwise, the order is rejected, and the customer will have to check the next closest bakery. By imposing this limit, the bakery production line queue is no longer continually added to, even if batches are currently queueing at the end of the production lines.

A feature was added to check inventory levels of the bakery's attached primary warehouse or the next location. If the warehouse was near capacity, production would pause or drop orders. A 99% threshold was applied to all sites so that the bakery could check if the warehouse's inventory level was above this. Before accepting an order, the bakery checks the warehouse's inventory. If it exceeds the threshold, the order is dropped. The model backtracks through all primaries that the customer visited and removes the order from ***waitingOrders***. The client wanted to set a daily production capacity for the bakeries but also consider if production had to be paused due to warehouse queues. This gives the client greater control over how many orders the bakeries can accept and how they deal with backlogs caused by warehouse queues, to match their real-life supply chain.

3.5 Custom SQL Database Output

It was determined that new outputs should be created to give information on the new features such as staff queue times. anyLogistix provides a detailed set of outputs from each run and displays them in an internal dashboard; however, the system for creating custom outputs is too restrictive for what the client needed. Custom tables are possible, but they only allow for selection from a small number of pre-set column names. The client wanted a straightforward way to get the outputs into their SQL database in as few steps as possible and preferred not to use an interim step like outputting to Excel first. The JDBC library was selected to output directly to the client's SQL database at the end of each model run (Reese, 2000). The extension now uses custom collections and data classes to store outputs for each bakery, customer, and primary. The outputs are formatted in a way that is more readable for the client and output to SQL at the end of each run. For example, shipment actions (loading, unloading etc.) are split by product and quantity for easy tracking of individual product types. Orders also include the selected sourcing path, allowing the client to quickly identify any unusual paths.

3.6 Product Sourcing Logic

The Client's supply chain network is complex and required an extended version of the default sourcing logic to be implemented. The primary warehouses are designed such that in most cases, a primary that can source from another primary is also able to ship to that same primary and there is no restriction on what products a primary can ship/source. The standard out-of-the-box implementation is too simple to handle this network in a way that would reflect the real-life sourcing choices that would be made by our client. anyLogistix support provided a solution using a custom subclass of order called ***MyOrder,*** which can have additional variables and functions added to it. They also provided insight into how the ***waitingOrders*** and ***orderSource*** (the two blocks handling the order arriving at a primary) so that when a new order was created, it could be passed information about the original order. The default logic creates an entirely new order each time an order arrives at a primary and does not preserve the knowledge of what the original order was. The new implementation passes the ID of the original order every time a new order is created, allowing the primaries to keep track of what orders they have seen already and allowing shipments that arrive to match their orders to the corresponding waiting order that created them. This allowed the orders to flow more naturally through our client's complex network of primaries and was better equipped to handle multiple primaries that could source from each other.

The client's supply chain model also differs from the standard implementation in terms of how products are produced, shipped, and sourced. In the standard implementation, when entering a primary or bakery, orders would check to see if there are enough products already in the inventory to satisfy the order. When products enter the inventory, either after arriving as part of a shipment or after being produced, each of the waiting orders is checked to see if there are now enough products to fulfil the order. This is done because the standard implementation can have products which are produced even when there is no order for them currently. For the client's scenario, products are only produced when demand is received from a customer. The production order has been adapted so that it knows which of the waiting orders it corresponds to using the order ID. When the order has finished being produced, the corresponding order is released and seizes the products that were produced for it.

Similarly, when an order has finished being unloaded at a primary, the corresponding order is found using the knowledge of the original order ID and is released to seize those products. This had to be adapted slightly to work with anyLogistix's 'splitting' logic, which is embedded within the custom ***shipmentGenerator*** block. When placing orders into a shipment, the logic can sometimes choose to split an order into two or more, so that some go onto a shipment immediately and some are placed onto another shipment. In the splitting logic, it can tell the new split orders what the original order was to preserve the information. When an order arrives at a primary or the customer, the logic checks to see if the full expected quantity has arrived yet. If it has not, then it waits until the rest arrives before marking the order as fulfilled or releasing the order from waiting orders, to be shipped to the next location.

Upon testing this iteration of the model and analyzing the outputs, the client identified that using the closest source policy was leading to sourcing paths that were suboptimal and unrealistic, even with the new method of tracking the original order ID. As Figure 3 shows, Customer A is trying to source from a Bakery. In this example, the optimal path is to source from Primary D and then Primary D will source from Bakery A. However, because Primary A is the closest source to Customer A and there is no way for Customer A to see Bakery A, it would choose the path Customer A, Primary A, Primary B, Primary C, Primary D and finally Bakery A.

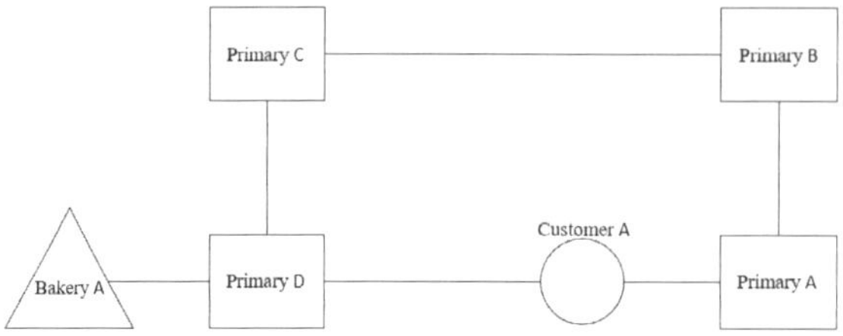

Figure 2 *Customer sourcing example*

The simulation in anyLogistix handles sourcing by finding the next best primary/bakery to source from based on the sourcing policy used. However, it does not determine the best path. For a complex network like our client's, it would be preferable to have a system where the customer determines which bakery to source from first and then finds the best path to that bakery. The client agreed to an approach where the customer, on model startup, determines the best path to each bakery and sorts those paths by total distance. The model first checks for a "direct" path from the bakery to the customer, and if not, uses the shortest path. A direct path is one where the customer can source directly from the bakery, the customer's attached primary can source directly from the bakery, the customer can source directly from the bakery's attached primary, or the customer's attached primary can source directly from a bakery's attached primary. If there is no direct path, the model uses the shortest path.

3.6.1 Dijkstra's Algorithm for Shortest Paths

To get the shortest path, it made sense to use Dijkstra's algorithm (Zhan and Noon, 1998) as the supply chain network is already set up with nodes (the customer, bakeries, and primaries) and the distances between them. Below is the outline of the process for finding the shortest path within the simulation.

1. The algorithm starts at the customer and identifies all sites it can source from.
2. It updates the current distance to that site to be the distance from the customer to that site and sets the previous node to be the customer.
3. It then identifies the primary that has the current shortest distance and checks its sources.
4. If the distance from the customer to the identified primary plus the distance from the identified primary to the source is less than the current distance recorded for that source (or there is not already a distance), then it is updated to the new distance and set its previous node to be the identified primary.
5. This process is repeated until all primaries have been exhausted.

This process is not performed on the bakeries as these do not have any sources and their minimum distance and the previous node will be identified by checking the primaries. Now the customer has identified the shortest distance to each bakery and can identify the shortest path by starting at a bakery and checking its previous node and then that node's previous node until it reaches the customer.

3.7 Cross-Docking

The client wanted to consider how the different primaries are used within the supply chain network. Some primaries are designed to handle many shipments moving through via cross-docking (Liao et al., 2010) by having a larger warehouse capacity and more bays. Others are much smaller and are not meant to be used for cross-docking to other primaries, but rather are meant to be used by a small number of bakeries or customers. To do this, the client added two tables to their SQL database that could be read in on model startup using JDBC.

The first table has a bakery, a customer, and then 0-5 primaries per row. This specifies a preferred path that the order should be shipped along between the specified customer and bakery. If the customer and/or the bakery has an attached primary, it is not specified in the table but is still identified and added as part of the sourcing path. This is used instead of the shortest path identified using Dijkstra's, however, we still use a direct path instead if one exists. If a customer-bakery pair does not appear in this table, then the shortest path is used. This allows the customer to route orders along a specific path that it would take in real life.

The second table specifies a list of primaries that we do not want to consider for cross-docking, and therefore should only be used if it is the only viable way to create a path from a customer to a bakery. To enforce this, when performing Dijkstra's, the algorithm applies an arbitrarily large value to the distance when a non-cross-dock primary is identified as a source. This means it would never not be selected as part of a shortest path unless there was no other path to one of the bakeries. For selecting the order to check the bakeries, it still uses the actual length of the path and not the distance with a large value added to it.

Now when checking where to source from, the customer checks the bakeries starting from the one with the shortest path up to the one with the longest path. If a bakery accepts the order, the customer provides the custom ***MyOrder*** agent with the sourcing path (customer to bakery). When an order arrives at a primary, rather than trying to determine a source, the next primary (or the bakery) is obtained from the sourcing path and the order is sent there.

4 RESULTS

The results of the simulation are visualized within a Microsoft Power BI dashboard. This is a business intelligence tool used for transforming, cleaning, and combining data sets into a model, to allow users to create charts or graphs to provide a visualisation of the data within a central dashboard. It is often useful for collaboratively analysing results as it allows stakeholders to share and view findings with others, without them having to run any simulation or optimisation models themselves (Ferrari and Russo, 2016).

It is possible to upload data sets as Excel files or via SQL databases. The client did not want the extra step of uploading a file to the dashboard but instead, wanted the process to be streamlined. The outputs of the simulation model are written directly into their SQL database, and from there they are automatically pulled into the Power BI dashboard and visualised through graphs. This is important for the stakeholders and management team who are not users of anyLogistix, as it allows them to refresh the dashboard's data set to see the latest results and scenarios where they can analyse the results and compare them.

The simulation outputs are broken down into five categories and can be viewed within the client's dashboard. A requirement as part of the data analysis is the ability to compare scenario results. Each time the user runs a simulation, the outputs are stored in the database with a scenario name and ID and can be selected from a drop-down menu to view the associated outputs. The dashboard overlays scenario results so they can easily be compared on the same graph (shown in Figure 4). The figures shown will remain hidden to respect client agreements.

4.1 Warehouse Capacity & Product Flows

The client relies on inventory-level data to plan the movement of products and inventory management. They need to know which warehouses have spare capacity or the potential to store more products to avoid others becoming too full as this often results in bottlenecks and can result in failure to meet the service level of the retailers. The dashboard includes a visualisation of the inventory level at each warehouse across the period of a week (Figure 4). It displays the current date and time, and the current inventory level so that any user can focus on a specific point in time to understand how many products were incoming and outgoing at a particular site and how that affected the inventory level. This is broken down into to include the average percentage of warehouse space by product so they can understand how much of each one is currently there. Using this, they can decide if they are storing a specific product in the right location based on where the demand is.

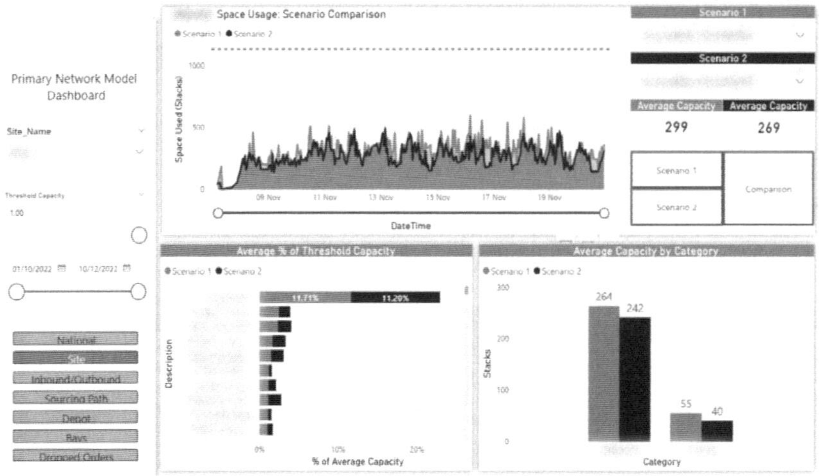

Figure 3 *Warehouse Inventory Levels Dashboard View*

Inventory level statistics aid in planning product storages however, it is often useful to understand the inbound and outbound flows for specific warehouses to gain insight into why certain products are moving through this site. The inbound and outbound section of the dashboard (Figure 5) shows the number of stacks that were inbound vs outbound per site, this is viewable per scenario allowing them to be compared. Viewing the product flows in terms of stacks, allows the inventory planning team to understand how this affects physical space in a warehouse as each stack occupies a certain area of space. The graph (shown in Figure 5) also shows the inventory levels so that scenarios can be compared. The bottom of this view shows the location for the outbound products, and how many are to be sent to that destination in terms of stacks and individual units.

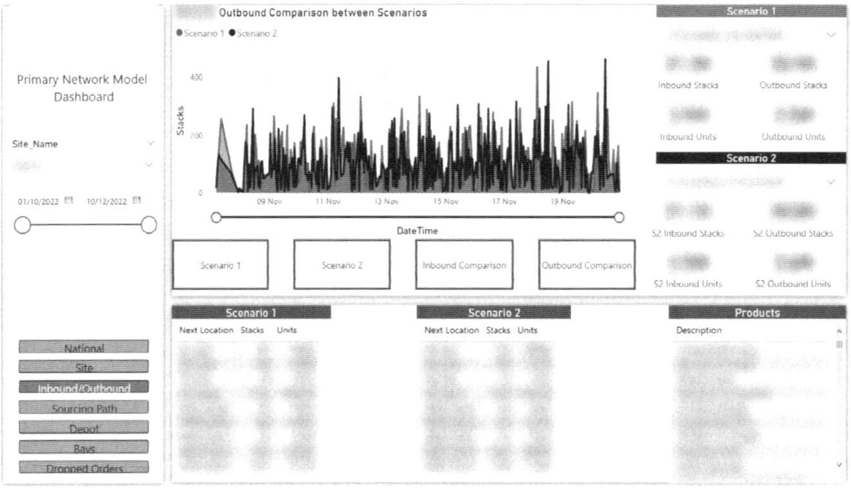

Figure 4 *Inbound & Outbound Flows Dashboard View*

4.2 Sourcing Paths & Product Flows

The sourcing logic within the simulation determines which bakery the customers will source their orders from and which sourcing path it will take. These sourcing paths a determined at model startup as either a direct path, predefined path or shortest path. The sourcing path selected is included within the dashboard to aid in transportation planning. Viewing the paths predicted by a simulation allows the transportation planning team to understand the movement of products based on customer-specified sourcing paths. This also gives the team insight into how they should move products around the network. It provides the full sourcing path for a specific order, starting from the bakery where it was produced, through the different warehouses, where it may have been cross-docked, all the way to the customer. The tables (shown in Figure 6) break down these paths including detail on which orders were fulfilled and which orders were cross-docked. Sourcing outputs from multiple scenarios can be compared to understand which sourcing paths are most likely.

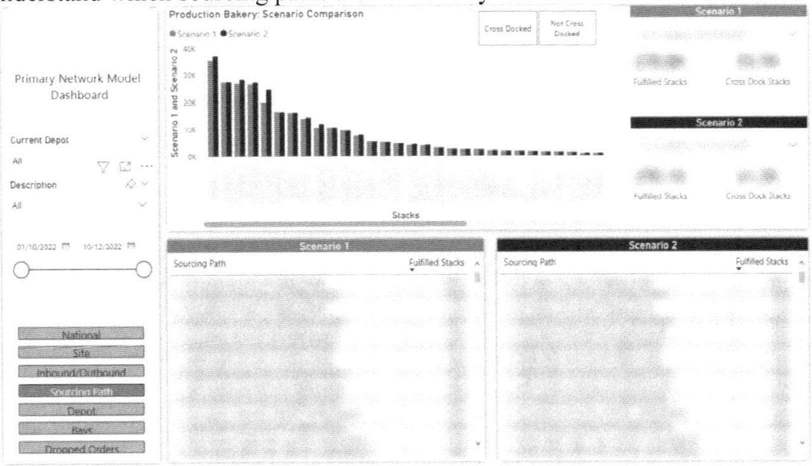

Figure 5 *Sourcing Paths Dashboard View*

4.3 Dropped Orders

When the customer is attempting to source products, they may be unable to find a bakery that can produce and ship the products right away. This can happen because, the bakery does not have enough remaining production capacity, the bakery production line queue time is too long or because the attached primary (or first primary visited on route to a customer) is above the 99% warehouse threshold. This results in their order being dropped. Similarly, the order can be dropped when it reaches a production queue, and the attached/first primary warehouse is above the threshold.

When an order is dropped a message is logged within the database to specify the order they requested and the reason it was dropped. This is visualised through the dropped orders dashboard (Figure 7). The total number of dropped products in individual units, for the network, is displayed per scenario so that if changes were applied to the simulation such as production capacity or sourcing routes then this would be reflected to help understand which scenario may be better for meeting demand. It also includes a breakdown by customer so the user can view the number of individual products they demanded and how many were dropped, this is represented by the total of units ordered as well as by product type.

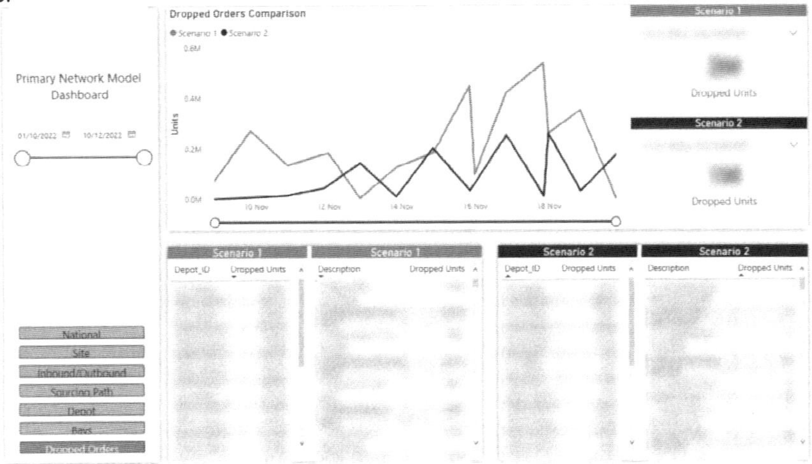

Figure 6 *Dropped Orders Dashboard View*

5 CONCLUSION

The features that were added to the anyLogistix simulation model enabled us to extend the functionality of the default behaviour of supply chain objects by modelling withinside the four walls. This allowed for more complex processes to be added to the standard anyLogistix factory, warehouse, and customer objects, specifically tailored to the client's requirements.

The addition of SQL outputs and the production of a Power BI dashboard, allowed for greater data granularity and visualisation when gathering statistics from the simulation. anyLogistix was integrated into the client's existing IT infrastructure and helped to streamline the process of supply chain modelling as they now have a way of taking existing data and automating the development of simulation models. This was one of the main goals as part of the wider digital transformation. With a better understanding of the data associated with their supply chain, the client gained greater insight into where certain areas of the network may be failing due to inefficiencies or bottlenecks. This information helps the various planning teams and stakeholders within the client's organisation make data-driven and informed decisions to mitigate risks to improve efficiency and performance.

REFERENCES

Adebanjo, D. and Mann, R. (2000) "Identifying problems in forecasting consumer demand in the fast moving consumer goods sector," Benchmarking: An International Journal, 7(3), pp. 223–230. Available at: https://doi.org/10.1108/14635770010331397.

Albukhitan, S. (2020) "Developing digital transformation strategy for manufacturing," Procedia Computer Science, 170, pp. 664–671. Available at: https://doi.org/10.1016/j.procs.2020.03.173.

Borshchev, A. (2014) "Multi-method modelling: Anylogic," Discrete-Event Simulation and System Dynamics for Management Decision Making, pp. 248–279. Available at: https://doi.org/10.1002/9781118762745.ch12.

Büyüközkan, G., & Göçer, F. (2018). Digital Supply Chain: Literature review and a proposed framework for future research. Computers in Industry, 97, 157-177.

Ferrari, A., & Russo, M. (2016). Introducing Microsoft Power BI. Microsoft Press.

Hopp, W. J., Spearman, M. L. (2011). Factory Physics: Third Edition. United States: Waveland Press.

Ivanov, D. (2020) "Predicting the impacts of epidemic outbreaks on global supply chains: A simulation-based analysis on the coronavirus outbreak (COVID-19/SARS-COV-2) case," Transportation Research Part E: Logistics and Transportation Review, 136, p. 101922. Available at: https://doi.org/10.1016/j.tre.2020.101922.

Liao, C.-J., Lin, Y. and Shih, S.C. (2010) "Vehicle routing with cross-docking in the supply chain," Expert Systems with Applications, 37(10), pp. 6868–6873. Available at: https://doi.org/10.1016/j.eswa.2010.03.035.

Reese, G. (2000). Database Programming with JDBC and JAVA. " O'Reilly Media, Inc.".

Sanjaya, A., Sembiring, A.C. and Willyanto, W. (2019) "Determination of the optimal distribution centre location with Gravity Location Model," Journal of Physics: Conference Series, 1402(2), p. 022041. Available at: https://doi.org/10.1088/1742-6596/1402/2/022041.

Zhan, F.B. and Noon, C.E. (1998) "Shortest path algorithms: An evaluation using Real Road Networks," Transportation Science, 32(1), pp. 65–73. Available at: https://doi.org/10.1287/trsc.32.1.65.

Z. Li, Cheng Hwee Sim, M. Y. H. Low and Y. G. Lim (2008) "Optimal product allocation for crossdocking and warehousing operations in FMCG Supply Chain," 2008 IEEE International Conference on Service Operations and Logistics, and Informatics [Preprint]. Available at: https://doi.org/10.1109/soli.2008.4683042.

AUTHOR BIOGRAPHIES

Adam Coleman is a simulation consultant at Decision Lab. He is the technical product lead of anyLogistix and leads our model development, training, and mentoring activities relating to anyLogistix. He is also the head of training for Anylogic and a member of the simulation team where he works on various simulation projects.

Jacob Whyte is a simulation consultant who joined Decision Lab as part of the kickstart program and has since become a full-time member of the simulation team as a Junior Simulation Consultant. He has experience working with Simulations in AnyLogic and helped build models for multiple clients.

Aanand Davé leads a simulation team at Decision Lab, delivering Anylogic and anyLogistix products and support services to clients in the UK and Italy. He guides senior managers in making sustainable business decisions through modeling in various industries, using simulation and AI technologies. Aanand has an academic background in design, architecture, and engineering, and is a leader in the field of modeling science, combining systems thinking, sustainability and simulation to add value through services and high-quality products.

Peter Riley is the technical lead of the simulation team at Decision Lab, with experience in developing various types of models for multiple industries. He leads validation and verification efforts and has worked on projects in several other fields, including deep reinforcement learning, optimization, and data science. He has a BSc in Mathematics and has worked as an operational research analyst prior to joining Decision Lab. He is currently an Applied AI Engineer with Microsoft Bonsai and has applied reinforcement learning in Anylogic.

Gavin Wilkinson is a technical marketing lead with a background in computer networks, software, and education. He spent several years in remote communications for Guardian News & Media, and five years in marketing for The Anylogic Company. He has a BSc in Computing Networks, is a certified Anylogic simulation modeller, and holds a certificate in AI from the University of Helsinki.

Proceedings of the Operational Research Society Simulation Workshop 2023 (SW23)
C. Currie and L. Rhodes-Leader, eds.
https://doi.org/10.36819/SW23.028

SIMULATING REVERSE LOGISTICS IN THE FASHION INDUSTRY: A CASE STUDY FOR A UK FOOTWEAR COMPANY

Miss Trung My Tran

Kent Business School, University of Kent
Canterbury, Kent, CT2 7NZ
trungmytran.tmt@gmail.com

Professor Kathy Kotiadis

Kent Business School, University of Kent
Canterbury, Kent, CT2 7NZ
K.Kotiadis@kent.ac.uk

Dr. Antuela Tako

Loughborough Business School
Epinal Way, Leicestershire, LE11 3TU
a.takou@lboro.ac.uk

Dr. Virginia Spiegler

Kent Business School, University of Kent
Canterbury, Kent, CT2 7NZ
V.L.Spiegler@kent.ac.uk

ABSTRACT

We report on a case study of a footwear retailer in the UK, experiencing a higher rate of returns during the pandemic and needing support with the operational planning of its reverse logistics processes. We use semi-structured interviews to derive an understanding of the problem addressed with simulation modelling. Operational concerns about the costs related to product returns were raised in the interviews. Hence the simulation model focuses on this aspect by exploring the number of staff to returns ratio to achieve the targeted percentage of returned items processed within a certain number of working days. We conclude that other fashion companies might benefit from reviewing their reverse logistics operations especially in anticipation of escalating product returns.

Keywords: Reverse Logistics; Fashion Industry; Discrete Event Simulation;

1 INTRODUCTION

The online retail channel has seen a huge growth in the last decade (Cullinane et al 2019), estimated to account for approximately one quarter of the 2022 overall UK retail market according to the Mintel marketing intelligence agency. This trend was accelerated even more during the pandemic (McKinsey 2021). The increase in online retail has resulted in high product returns, which according to Cullinane et al (2019), ranges globally between 20-60%, depending on specific product characteristics. High return rates related to reverse logistics affects the operational logistics processes of retail companies, and ultimately their financial ability to cope with the additional stock in the product sale cycle (McKinsey, 2021).

The aim of this study is to examine reverse logistics practices in the fashion industry through a UK footwear retailer case study. The studied company is one of the major players in the footwear retail industry in the UK, with about 1,300 employees and annual revenue of £130 million in 2019-2020. This study will contribute towards our understanding of the UK fashion industry management of reverse logistics, and their perspective toward tackling environmental issues. Reverse logistics was defined by Rogers and Tibben-Lembke (1999, p.2) as "the process of planning, implementing, and controlling the efficient, cost-effective flow of raw materials, in-process inventory, finished goods and related information from the point of consumption to the point of origin for the purpose of recapturing value or proper disposal". Reverse logistics has gained importance due to political, environmental, and economical concerns (Sasikumar and Kannan, 2008a, 2008b, 2009; Difrancesco et al, 2018; Janeiro et al., 2020). Environmental considerations are of particular interest in reverse logistics (Agrawal et al, 2015).

We use a combination of qualitative (semi-structured interviews) and quantitative analytics (discrete event simulation) to understand the management and role of reverse logistics activities in the fashion retail industry. As a contribution to literature, this combined approach offers a more systematic means for the identification of factors that affect the management of reverse logistics. Interview results offered a better understanding of the context of study and provided a focus for the simulation model which are the internal operations of reverse logistics. The simulation model explores the number of staff to returns ratio needed to achieve the targeted percentage of returned items processed within a certain number of working days, which was an operational decision identified as important during the interviews. This type of modelling has not been reported in the fashion industry.

The remainder of this paper is structured as follows. Section 2 provides a brief review of the literature of simulation in reverse logistics. This is followed by a brief description of the methodology in Section 3. Section 4 is dedicated to findings from the qualitative research and simulation analysis, respectively. The final section discusses the study findings and limitations as well as directions for future research.

2 SIMULATION LITERATURE IN REVERSE LOGISTICS

Simulation modelling is an efficient tool used to gain deeper understanding in the operations of the supply chain as well as to identify and solve several problems as it can deal with the variability, interconnectedness, and complexity of systems (Tako and Robinson 2012; Persson and Saccani, 2009). A recent review of literature on the use of quantitative models of green supply chain models found that simulation is the second most widely used tool (Becerra et al, 2021).

There are several simulation techniques such as Monte Carlo Simulation, Agent-based Simulation, Continuous Simulation, Hybrid Simulation, among which, System Dynamics (SD) and Discrete Event Simulation (DES) are two simulation modelling approaches that are commonly used in logistics and supply chain management to help in decision making (Tako and Robinson, 2012). The authors indicate that DES is more frequently used for strategic to operational/tactical level supply chain issues as compared to SD among 127 published journal articles between 1996 to 2006. Agnusdei et al (2019) reviewed 78 papers using simulation tools in studying closed-loop supply chains. The authors comment that DES is suitable to model logistics processes at operational level including more detailed analysis and individual characteristics of the parts of the system, while SD is more suitable to carry out analysis at "higher abstraction levels" due to its ability to capture the complex dynamics resulting from the interaction of the multiple actors in the system. Becerra et al (2021) identify that DES models are used more frequently to study sustainable inventory management systems in green supply chains.

Within reverse logistics, DES combined with other methods (heuristics, genetic algorithms, tabu search, neural network, etc.) has been noted to be the most widely used simulation approach to tackle multiple problems related to uncertainty, network design, inventory control, production planning and control (Abid and Mhada, 2021). The same authors state that DES has an advantage over other analytical methods, in that it facilitates the development of stochastic models, where parameters can be easily altered to evaluate the impact of changes on system performance (Abid and Mhada, 2021).

3 METHODOLOGY

We use a mixed methods approach combining qualitative and quantitative research methods applied to a case study of a fashion footwear company. Combining qualitative and quantitative research methods is beneficial to ensure that analysis of the problem from different perspectives takes place and better more relevant outcomes can be achieved for decision making (Kotiadis and Mingers 2006; Howick and Ackerman, 2011; Callaghan et al 2021). The study was undertaken over a period of two months and led by the first author, an MSc student. The actual company's name is not revealed for confidentiality purposes.

The qualitative study undertakes analysis based on secondary and primary data and it aims to gain a general understanding of reverse logistics practices and issues faced in the fashion retail sector and the case study company. Secondary data from academic journals, books, industry reports, and company's website is gathered to provide a background. Semi-structured interviews were the main method used for primary data collection. The aim was to understand the reverse logistics process, key drivers and strategic priorities placed by the case study company in managing it. We carried out 6 interviews with employees from different positions within the company, ranging from the CEO, logistics and distribution manager, e-commerce manager to operatives and warehouse staff. Besides interviews, observation of the reverse logistics operation of the company also took place to gain understanding of the process in order to model it. The qualitative aspect of the study supported the process of simulation modelling and in particular supporting the data collection needs, which will be described in greater detail in the next section.

In terms of quantitative research, this study uses discrete event simulation (DES) to model the company's reverse logistics process specifically at their Distribution Centre (DC) to assist the company in planning its operations by evaluating several scenarios of product returns. We chose to use DES to build the model as it is considered a suitable tool for modelling the behaviour of operational logistics supply chain systems, where operations are represented as a series of events (Abid et al, 2019). SIMUL8 is the software used.

4 STUDY FINDINGS

4.1 The Key Findings from the Semi-structured Interviews

The studied company is a major player in the footwear retail industry in the UK and operates various channels for both the domestic and international markets. They own UK standalone stores, concessions in major global department stores such as Topshop, House of Fraser, and John Lewis, and franchised stores and concessions in 10 countries around the world. Beside outlets, they run their own website with shipping options to over 130 different countries. They also supply products to major e-tailers and sell to other retailers on a wholesale basis.

One of the main reverse logistics drivers is regulatory pressure. The regulation on returns and refund is part of the consumer protection law, which states that businesses must allow customers to return the goods purchasing online, via mail and telephone within 28 days from the date of receiving (UK Government, n.d.). Reverse logistics is a requisite activity for organisations such as this, where the company recovers the value of the returned products. They operate a high level of quality control in both forward and reverse logistics when inspecting and sorting returned shoes so as to maximise the number of products that can be resold at full price.

The company's average return rate on an item basis is about 30% across different categories and across the year, which is slightly higher in comparison with 25% return rate of the fashion industry in general and similar to the return rate (30%) of the footwear sector (Interactive Media in Retail Group, 2020). One reason for the high return rate in the footwear sector is due to the challenge of standardising the sizes of products. There are various forms and styles of shoes such as trainers, high heels, and boots, as well as various shoes manufacturers, which makes size standardisation in manufacturing products difficult. The return figure of the company is different across countries, with Germany being the highest of 70% to 80% due to high customer standards. The online channel generates higher returns rate than the offline channel, with the rate being 35%. The obvious reason is that customers do not have a chance to try on products as they do in store. Secondly, it becomes a learned behaviour of customers, in which they are getting familiar and confident in returning products purchased online. There is a minimal fluctuation in the return rate throughout the year, in which autumn-winter season experiences higher return rate than spring-summer season. The return rate at the time of the study was about 32% in spring-summer and 36% in autumn-winter. It is believed to be affected by the pandemic.

The fashion retailer places great emphasis on the speed of the returns process influenced by third-party logistics providers as well as technology and infrastructure. The fashion company experiences the efficient-responsive trade-off in working with carriers. From their perspective, it is essential to have a smooth flow of returns in order to avoid inconsistency in returns volume between days and to quickly process refunds to customers. The company accepts all returns from customers provided that returns are made within the return time window. The reason behind this is to maximise customer satisfaction and avoid disputable situations. The returns to stores are collected by the company's trucks that deliver new orders to stores, whereas others are consolidated by a third-party logistics and dispatched directly to the company's Distribution Centre (DC).

At the time of the study, summer 2021, the reverse logistics process at the DC was paper-based. When returned items arrive at the DC, there was no electronic system to scan the product and automatically trigger a notification of receipt to customers. Additionally, checking returns forms and capturing data are also done manually. This in turn limits the transparency in the reverse supply chain and creates extra efforts in the customer service and logistics department to handle customer enquiries about their returns and refunds. Another obstacle involves sales partners such as Zalando and Asos, where they have very strict KPIs on returns management. For example, customers should receive their refund within 48 hours from when the parcels is scanned back into their return network. In some circumstances such as long-distance travel between other countries and the UK, the policy does not apply for the fashion retailer.

The DC receives on average around 3,500 returned items per week. Delivery of returns regularly arrives at the warehouse after 5 to 7 days, sometimes 10 to 15 days. The refund window offered by the company is 5 working days from receiving the returned items. Currently, the company finds its reverse logistics operations efficient enough to meet the refund window. However, there are days that they receive a large quantity of returns, up to 8,000 returned items, due to returns consolidation of carriers or general delays in the logistics network, which affects the normal operation in the distribution facility. Although this is a rare situation, the scheduling and allocation of staff needs to be planned in advance to address the issue effectively.

Inspection and sorting is the main activity of the fashion retailer in operating reverse logistics, which is performed by the quality control team at the centralised DC. Parcels are opened and checked individually. Depending on the quality of the returned items, they are sorted into three categories: prime stock, which will be resold at full price; sub-standard stock, which will be resold at discount price; and those which cannot be resold at stores will be sold to third-party market traders or donated to charity. If there is no issue with the shoes, they will be cleaned, repacked, and restocked. If there are some problems such as labelling, remedial action will be carried out and shoes will be restocked after this work. After that, the team inspect the returns paperwork, highlight the item being returned and the reasons for returns, and pass the paperwork to customer service department to process the refund. From the activities outlined above, sorting and disposition activities are managed in-house and centralised at the retailer's Distribution Centre (DC), while collection activity is managed partly in-house and partly outsourced to third-party logistics.

The manual processing of returns makes reverse logistics labour intensive and time-consuming.

> *"Despite the full automation in delivery process, or forward logistics, when it comes to reverse logistics, there are still someone sitting there on a packing bench opening in, looking at the condition of the item and deciding where it will go. There is no way of getting around that manual element." – Warehouse Operative*

The company is flexible in allocating human resources for processing returns.

> *"It is very often for small operation like we have here to move your staff to where you need them. There is not fixed number of people in quality control team just doing the web returns because we can have heavy in-take days (deliveries from suppliers), then we put more people on doing the in-take. That is the biggest priority for the business." - Head of Logistics*

4.2 The Simulation

The data collected from the qualitative analysis assisted in developing the simulation model, with further inputs provided by the Head of Logistics, who helped us to identify the simulation model objective. The model focused on determining the number of staff required at each stage in the reverse logistics process in order to ensure 95% of returned parcels to the DC are processed within 3 working days. The 3-day time is set as a reverse logistics KPI to finish inspection, sorting, and disposition so as to spare time for paperwork and refund activities. Table 1 presents the model inputs and outputs as indicators to achieve the objective.

Table 1 *Model Key Inputs and Performance Indicators (KPIs)*

Key inputs	The number of returned items arriving at the DC (range = 3,500; 8,000) The number of staff performing each reverse logistics activity (handling staff range = 3-5; quality control staff range = 4-12)
Outputs (Responses)	To determine the achievement of objectives: • The percentage of items processed within 3 working days (= items completed within 3 working days/total number of items received) • The mean total processing time (from unloading to restocking) To identify reasons for failure of achieving objectives: • Queuing time at each activity: mean, minimum and maximum • Staff utilisation • Time-series of hourly throughput • Histogram of hourly throughput: mean, standard deviation, minimum and maximum • Time-series of total processing time • Histogram of total processing time: mean, standard deviation, minimum and maximum

Figure 1 presents the simulation model of the retailers reverse logistics process.

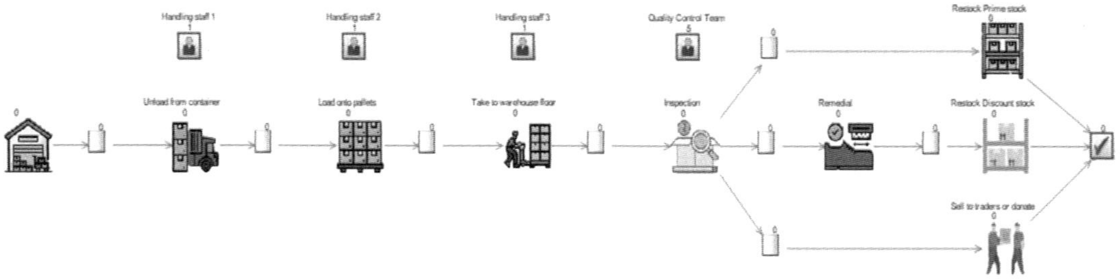

Figure 1 *Simul8 model of reverse logistics for the retailer.*

We introduced the following assumptions and simplifications to our model.

Assumptions:

- Only one batch of returned items arrives at the DC on a Monday.
- Physical operation staff will take 50 returned items to warehouse floor at a time. The same will be applied to items being restocked after checking.
- The ability and productivity of all staff are equal, and all staff have the same working schedule.

Simplifications:

- Travel time of staff within the DC, except for moving stock to warehouse floor and restocking, is not included.
- Uncertainty in operations such as equipment failure or break time are not considered.
- Instead of modelling the operations through a long period with several intakes of returns, only one intake of return is examined and becomes the basis for other scenarios.

Referring to the data used to build the model, the main issue faced was that data regarding the time to complete activities were not recorded by the company, these were estimated during on-site visits and relevant distributions were fitted using the StatFit software. Table 2 presents the main distributions and key data inputs included in the model. Activities in the DC are numbered from 1 to 8. The data used for the model were collected through semi-structured interviews, on-site visits, and email communications with the company's management. The conceptual model's content, assumptions, and simplifications were checked with the CEO and head of logistics and found to be sensible.

Table 2 *Model Distributions and other inputs. All times are in minutes unless otherwise stated.*

Component	***Attribute***	
Simulation run time	5 days (Monday to Friday) 8 hours per day from 8am to 4pm	
Arrival of returned items		
Quantity	Batch arrival (3500)	
Arrival time	8:30 am Monday	
Activities (time to complete)		
(1) Unloading items from container	Erlang (mean=0.05, k=5)	1 physical operation staff
(2) Stacking items onto pallets	Erlang (mean=0.1, k=5)	1 physical operation staff
(3) Taking items to warehouse floor	Beta (min=3, max=5, alpha$_1$=1, alpha$_2$=4) Perform task after collecting 50 items	1 physical operation staff
(4) Inspection and Sorting (10 workstations)	Triangular (min=1, mode=1.5, max=2.5)	4 quality control staff
(5) Remedial (3 workstations)	Beta (min=5, max=10, alpha$_1$=1, alpha$_2$=1)	
(6) Restocking as prime stock	Beta (min=10, max=15, alpha$_1$=1, alpha$_2$=3) Perform task after collecting 50 items	3 physical operation staff. Each staff has to finish activities (1), (2), (3) before starting activities (6), (7).
(7) Restocking as discount stock	Beta (min=10, max=15, alpha$_1$=1, alpha$_2$=3) Perform task after collecting 50 items	
(8) Selling to traders or donation	Fixed (0)	
Number of resources		
Physical operation staff	3	
Quality control staff	4	

Four scenarios were considered important for the model based on the interviews with company staff. The results of the experiments are presented in Table 3 below. The results show that the main model objective can be achieved with 3 physical operation staff and 5 or 6 quality control staff for 3,500 returned items (scenarios 1 or 2); 4 or 5 physical operation staff and 12 quality control staff for 8,000 returned items (scenarios 3 or 4). We statistically compared the pairs of scenarios (scenario 1 vs 2 and scenario 3 vs 4) using the standard T-test and the paired t-test where appropriate to determine which scenario is better than another. The comparisons show that the percentage of items processed within 3 working days in scenario 1 is less than that in scenario 2, whereas the mean total processing time in scenario 1 is greater than that in scenario 2. We found no significant difference in the results between scenarios 3 and 4, as shown in Table 3. Based on the experiments and findings, we can draw the following conclusions:

- Increasing the number of quality control staff improves the percentage of items processed within 3 working days and the mean total processing time significantly.
- For normal quantity of returns (3,500 returned items), 95% of items processed within 3 working days can be achieved with 3 physical operation staff and 5 quality control staff.
- For high quantity of returns (8,000 returned items), 95% of items processed within 3 working days can be achieved with 4 physical operation staff and 12 quality control staff.

Table 3 *Experimentation (10 Replications)*

	Scenario			
	1	**2**	**3**	**4**
Experimental Factors				
Number of returned items	3,500	3,500	8,000	8,000
Total handling staff	3	3	4	5
Activity (1)	1	1	1	1
Activity (2)	1	1	1	1
Activity (3)	1	1	2	3
Activity (6), (7)	Pooled resource	Pooled resource	Pooled resource	Pooled resource
Total quality control staff	5	6	12	12
Responses				
% Items processed within 3 working days - Mean	94.48	100	96.12	96.05
95% CI	(93.77, 95.20)	(100, 100)	(95.32, 96.93)	(95.35, 96.74)
Mean total processing time	826.56	699.87	797.42	800.94
95% CI	(819.44, 833.69)	(691.76, 707.98)	(792.73, 802.12)	(797.10, 804.78)

5 DISCUSSION AND CONCLUSIONS

This paper provides a mixed methods, qualitative and quantitative (simulation) study of the reverse logistics process of a fashion retailer. We believe this is the first modelling study to focus on the reverse logistics process of the fashion retail industry. We study the inhouse operational process of the company's reverse logistics with the view to identifying areas for improvement. The computer model and the process of developing it provide an opportunity to extend our understanding of this context. Below we discuss the empirical findings of our study and position it within the literature around the context area.

Five main activities constitute the reverse logistics process in the fashion retail industry, which include: mitigation, gatekeeping, collection, sorting, and disposition (de Leeuw et al., 2016; Cullinane and Cullinane, 2021). This is roughly similar to the reverse logistics process of the studied company. Mitigation is not mentioned as an important activity in the literature; however, the fashion retailer focuses on this step to reduce product return rates. In particular, they aim to provide production information as detailed as possible on website, collect online customer feedback about the products through their website, and analyse reasons for product returns to acknowledge the problems and find suitable solutions. In contrast, literature considers gatekeeping as the most critical stage as it could reduce the number of returns by choosing specific item that are allowed into the reverse flow (Cullinane and Cullinane, 2021). Meanwhile, the studied company accepts all returns from customers. This could affect the manageability and profitability of the reverse logistics according to theory, but as the company emphasises decent customer service, gatekeeping is not an option they consider. Future studies for fashion retailers could consider including mitigation and gatekeeping as behaviours affecting sales. This could be more suitable to model using an agent based modelling approach or considering a hybrid simulation model.

Collection of returned items can be performed in three ways: directly from customers, by the company itself, or via third-party logistics companies (Lambert et al, 2011). Our empirical findings show that most of the collection activities are handled by third-party logistics providers, including international returns and online national returns. The company collects and transports the returned items only by its own truck fleet for in-store returns. This is a more economical solution for the company when they do not have enough resources and it is not part of their core competencies. They offer customers several collection options, which has been proved by an exploratory study by de Leeuw et al. (2016), to increase customer satisfaction and boost repurchase intention.

Sorting is the next step in the process, which can be performed at centralised or decentralised facilities (de Leeuw et al., 2016). Decentralised sorting at stores or drop-off locations could increase the speed of processing returns; however, it requires skilled staff at each location for inspection and assessment of returned items. The studied company has a quality control team based in the DC managing the inspection and sorting activities. Having a centralised facility to process returns can help business in minimising cost of processing returns (Loomba and Nakashima, 2012), standardising the operations and reducing labour costs for inspection (de Leeuw et al., 2016).

The last activity is disposition, which can take various forms such as reuse, repair, refurbish, recycling, and disposal (Agrawal et al, 2015). Theory stated that this stage involves the most important goal of organisations, which is recapturing product value (Lambert et al, 2011; Mollenkopf et al, 2011). This corresponds well to the fashion retailer. The company restock the items that are qualified for sale at either full price or discount price, and their objective is to maximise the percentage of returned items that can be resold at full price. They do not deal with recycling but sell unwanted items to an expert in the field.

Our simulation model represents the main activities of the company's reverse logistics operations, sorting and disposition activities,. The model simulates the operations from the delivery of returns at the DC to the restocking of returned and refurbished items before being sold to the market again, providing recommendations to the company about the number of return items that can be processed and the handling and quality control staff, in order to ensure that the returns can be processed within the set time targets (3 working days). The simulation model could be improved by incorporating collection activities to evaluate the retailer's whole process of reverse logistics, however, more data, which are not available, would be needed such as quantity of returns managed by third-party logistics, transportation time to deliver returns to DC, distance between stores and DC, etc, hence these were not included.

This study has been useful for the company to help understand and inform their plans in further developing their reverse logistics process. The qualitative study identified the current practices and areas of uncertainty around which reverse logistics process could be improved. We then homed in with a more detailed simulation model that focused on the staff resources needed on achieving return time targets.

ACKNOWLEDGMENTS

The authors acknowledge the support of the footwear company and their staff for the inputs in completing this study. This study has received ethical approval.

REFERENCES

Abid, S., Radji, S. and Mhada, F. Z. (2019) 'Simulation Techniques Applied in Reverse Logistic: A Review', in International Colloquium on Logistics and Supply Chain Management, LOGISTIQUA 2019. doi: 10.1109/LOGISTIQUA.2019.8907293

Abid, S. & Mhada, F.Z. (2021) 'Simulation optimisation methods applied in reverse logistics: a systematic review', *International Journal of Sustainable Engineering*, 14(6), pp. 1463-1483, DOI: 10.1080/19397038.2021.2003470.

Agnusdei, G. P., Gnoni, M. G. and Tornese, F. (2019) 'Modelling and simulation tools for integrating forward and reverse logistics: A literature review', in 31st European Modeling and Simulation Symposium, EMSS 2019. doi: 10.46354/i3m.2019.emss.045.

Agrawal, S., Singh, R. K. and Murtaza, Q. (2015) 'A literature review and perspectives in reverse logistics', Resources, Conservation and Recycling. doi: 10.1016/j.resconrec.2015.02.009.

Bastan, M., Zarei, M., Tavakkoli-Moghaddam, R. and G., H.S. (2022), "A new technology acceptance model: a mixed-method of grounded theory and system dynamics", Kybernetes, Vol. 51 No. 1, pp. 1-30. R

Becerra, P., Mula, J., Sanchis, R. (2021) 'Green supply chain quantitative models for sustainable inventory management: A review', *Journal of Cleaner Production*, 328. doi: 10.1016/j.jclepro.2021.129544.

Callaghan, H, Tako, A, Jackson, L, Dunnett, S (2021) Developing a discrete event simulation model using multiple data sources. In Fakhimi, M, Boness, T, Robertson, D (ed) SW21 The OR Society

Simulation Workshop; Proceedings of SW21 The OR Society Simulation Workshop, pp 192-199. March 2021. Online Conference.

Cullinane, S., Browne, M., Karlsson, E., Wang, Y. (2019). Retail Clothing Returns: A Review of Key Issues. In: Wells, P. (eds) Contemporary Operations and Logistics. Palgrave Macmillan, Cham. https://doi.org/10.1007/978-3-030-14493-7_16.

Cullinane, S. and Cullinane, K. (2021) 'The Logistics of Online Clothing Returns in Sweden and How to Reduce its Environmental Impact', Journal of Service Science and Management, 14(01). doi: 10.4236/jssm.2021.141006.

de Leeuw, S. et al. (2016) 'Trade-offs in managing commercial consumer returns for online apparel retail', International Journal of Operations and Production Management, 36(6). doi: 10.1108/IJOPM-01-2015-0010.

Difrancesco, R. M., Huchzermeier, A. and Schröder, D. (2018) 'Optimizing the return window for online fashion retailers with closed-loop refurbishment', Omega, 78. doi: 10.1016/j.omega.2017.07.001.

Howick, S., & Ackermann, F. (2011). Mixing OR methods in practice: Past, present and future directions. European Journal of Operational Research, 215(3), 503–511.

Janeiro, R. et al. (2020) 'New conceptual model of reverse logistics of a worldwide fashion company', Procedia Manufacturing, Vol 53. doi: 10.1016/j.promfg.2020.10.232.

Kotiadis, K., Mingers, J. Combining PSMs with hard OR methods: the philosophical and practical challenges. *J Oper Res Soc* **57**, 856–867 (2006). https://doi.org/10.1057/palgrave.jors.2602147

Loomba, Arvinder P.S.; Nakashima, K. (2012) 'Enhancing value in reverse supply chains by sorting before product recovery', *Production Planning & Control*, 23(2-3), pp. 205–215. doi:10.1080/09537287.2011.591652.

Lambert, S., Riopel, D. and Abdul-Kader, W. (2011) 'A reverse logistics decisions conceptual framework', Computers and Industrial Engineering, 61(3). doi: 10.1016/j.cie.2011.04.012.

McKinsey (2021) The State of fashion 2021. McKinsey Company Report, p. 37, Available online: https://www.mckinsey.com/~/media/McKinsey/Industries/Retail/Our%20Insights/State%20of%20fashion/2021/The-State-of-Fashion-2021- vF.pdf (accessed on 11 April 2021).

Mollenkopf, D. A., Frankel, R. and Russo, I. (2011) 'Creating value through returns management: Exploring the marketing-operations interface', Journal of Operations Management, 29(5). doi: 10.1016/j.jom.2010.11.004.

Persson, F., Saccani, N. (2009) 'Managing the after-sales logistic network–a simulation study', *Production Planning & Control*, 20(2), pp. 125–134. doi:10.1080/09537280802707530.

Rogers, D. and Tibben-Lembke, R. (1999) Going backwards: Reverse Logistics Trends and Practices, RLEC Press, Pittsburgh, PA

Sasikumar, P. and Kannan, G. (2008a) 'Issues in reverse supply chains, part I: End-of-life product recovery and inventory management - an overview', International Journal of Sustainable Engineering. doi: 10.1080/19397030802433860.

Sasikumar, P. and Kannan, G. (2008b) 'Issues in reverse supply chains, part II: Reverse distribution issues - an overview', International Journal of Sustainable Engineering. doi: 10.1080/19397030802509974.

Sasikumar, P. and Kannan, G. (2009) 'Issues in reverse supply chain, part III: Classification and simple analysis', International Journal of Sustainable Engineering, 2(1). doi: 10.1080/19397030802673374.

Tako, A. A. and Robinson, S. (2012) 'The application of discrete event simulation and system dynamics in the logistics and supply chain context', *Decision Support Systems*, 52(4). doi: 10.1016/j.dss.2011.11.015.

AUTHOR BIOGRAPHIES

TRUNG MY TRAN received a MSc (Hons) Logistics and Supply Chain Management from the University of Kent in 2021. Her major focuses on Planning, Supply Chain Modelling and Analytics, Simulation Modelling, Operations Management and Digital Transformation, Sustainable Procurement, Warehouse and Global Transportation Management. She works for Mastercard as a Product Operations

Specialist, specialising in demand planning, product distribution and inventory management. Her email address is trungmytran.tmt@gmail.com.

KATHY KOTIADIS is a is a Professor in Operational Research (OR) at the Kent Business School, University of Kent. Her main research interests include facilitative modelling in health care, simulation modelling in health and transport, conceptual modelling, Soft Systems Methodology and problem structuring methods. She is the co-founder of the PartiSim approach which supports the involvement of stakeholders in the modelling process. In 2009 she was awarded the KD Tocher Medal by the UK Operational Research Society. She is a founding member of Women in OR and Analytics (WORAN) of the Operational Research society (UK), a member of the UK Engineering and Physical Sciences Research (EPSRC) Peer Review college and the co-Editor in Chief of Health Systems (journal). Her email address is K.Kotiadis@kent.ac.uk

ANTUELA TAKO is a Reader in Operational Research at Loughborough Business School, Loughborough University. She holds a PhD in Simulation and an MSc in Management Science and Operational Research from the University of Warwick. Her research interests include the comparison of simulation approaches, facilitated and participative simulation modelling, conceptual modelling, and health care modelling. She is a member of the UK Engineering and Physical Sciences Research (EPSRC) Peer Review college and Associate Editor of the Journal of the Operational Research Society, Journal of Simulation and Health Systems. Her email address is a.a.takou@lboro.ac.uk.

VIRGINIA SPIEGLER is Reader in Operations and Supply Chain Management. She obtained a PhD degree in the area of Supply Chain Dynamics from Cardiff Business School. Virginia has worked in the areas of material, production planning and traceability in the automotive and aircraft industries. Her research spans the areas of supply chain resilience, production and inventory control, system dynamics and nonlinear control theory. Her email address is V.L.Spiegler@kent.ac.uk

Proceedings of the Operational Research Society Simulation Workshop 2023 (SW23)
C. Currie and L. Rhodes-Leader, eds.
https://doi.org/10.36819/SW23.029

A DISCRETE EVENT SIMULATION MODEL OF A HOSPITAL FOR PREDICTION OF THE IMPACT OF DELAYED DISCHARGE

Laura M. Boyle

Mathematical Sciences Research Centre
Queen's University Belfast
Belfast, BT7 1NN, UK
laura.boyle@qub.ac.uk

Christine S.M. Currie
Carlos Lamas Fernandez
Ly Nguyen

University of Southampton
Highfield Campus
Southampton, SO17 1BJ, UK
christine.currie@soton.ac.uk

Clare Halpenny

Advanced Care Research Centre
University of Edinburgh
Old College, South Bridge
Edinburgh, EH8 9YL, UK
clare.halpenny@ed.ac.uk

ABSTRACT

Patients who have additional care needs after a stay in hospital can often experience delays in being discharged. These delayed discharges and their impact on the smooth running of hospitals has been well publicised in the UK in recent years. In this preliminary work, we build a discrete event simulation model to describe the process of a patient leaving the hospital. Based on a proposed plan for Southampton, we investigate the impact of providing intermediate care in the form of Discharge to Assess places on the number of patients who remain in hospital longer than necessary. We describe the model, a sensitivity analysis and preliminary results.

Keywords:

Discrete Event Simulation, Delayed discharge, Social care, Hospital

1 INTRODUCTION

Recent work from the Nuffield Trust shows that delayed discharges in the UK increased by 57% between April 2021 and April 2022 (Flinders and Scobie 2022). In Scotland, the media have highlighted that one in six patients should not be in hospital, and consequently Accident & Emergency Waiting times are at a record high (BBC News 2022). A delayed discharge or 'delayed transfer of care' (DTOC) is defined to be one that occurs after a patient is medically optimised for discharge (The King's Fund 2018). Such delays are bad for patient outcomes as longer stays in hospital tend to result in longer recovery times or even an irrecoverable deterioration in health (Rojas-García et al. 2018). But they are also problematic for hospitals by limiting the number of beds they have available for new patients. The reasons for the delays appear to be largely due to problems in the social care sector which have led to a lack of availability of the care needed. We do not consider here how to improve availability in the social care sector but instead focus on modelling the use of what is described as a Discharge to Assess (D2A) process (NHS Improvement 2022), a form of intermediate care between the hospital and social care which we describe below. This allows us to investigate how varying the number of places available in this intermediate process will affect the number of patients with a delayed discharge and the number of hospital bed days being used for patients who are medically optimised for discharge but unable to leave.

We consider the specific case of University Hospital Southampton (UHS) in this preliminary work which wrote and published a plan for a discharge process in Schofield (2021). This describes two options for D2A: using private care home beds outside of the hospital and a ward in a smaller community hospital for what we describe as inpatient D2A, or alternatively the Home First scheme whereby patients are sent home with some care provision. In both cases patients are assessed for their longer term care needs during their time in D2A with one of the stated aims being the provision of more long term care taking place at home. We use the simulation model to determine the impact of changing the number of available D2A places on the key measures of the number of delayed discharges and the number of extra bed days being used by patients who are medically optimised for discharge.

This work aims to make contributions by: (i) modelling the flow of patients to both intermediate and long term care in the UK NHS, and (ii) modelling delayed discharge using an acceptance probability to represent the situations where a delayed transfer of care occurs due to reasons other than capacity constraints (e.g., a disagreement with the patient's family about the care placement). This paper is applied to the Southampton context, but future iterations of the work will consider a reusable model that can be applied to multiple settings.

We discuss previous work in this area in Section 2 before providing a more detailed problem description and details of the DES model in Section 3. Preliminary results are given in Section 4 before we conclude and describe our plans for future work.

2 LITERATURE REVIEW

Discrete event simulation (DES) models the operation of a system as a sequence of events in discrete time. It has been used extensively to model patient flow through components of healthcare systems internationally (Mustafee et al. 2010). There are many models of individual hospital units, for example emergency departments and intensive care units, however it is recognised that models of individual units developed in isolation of the wider healthcare system may not appropriately account for blockages further downstream (Salmon et al. 2018). For example, overcrowded emergency departments and queues of ambulances outside hospitals are often the most visible symptom of delays in discharging patients from in-patient hospital to the community care (Kelen et al. 2021). This section presents a literature review in respect of papers which use simulation modelling to consider the problem of delayed discharge from in-patient hospital care to community care facilities.

Tobail et al. (2013) used value stream mapping and discrete event simulation (DES) to show that length of stay and bed availability could be improved by reducing the rate of late discharges from hospital. Khanna et al. (2016) developed a DES of an Australian hospital to experiment with the impact of various hospital discharge policies on length of stay, hospital occupancy, and emergency department overcrowding. The study highlighted that improvements to hospital patient flow could be achieved by discharging patients early in the day and by spreading discharges across the day. Busby and Carter (2017) built a generic DES to assess hospital-wide bed management strategies, which include expedited discharge from hospital. Qin et al. (2017) developed a comprehensive DES of an Australian hospital which captured patient flow from emergency department arrival to hospital discharge. The authors used scenario experimentation to determine the impact of various hospital discharge strategies (e.g., patients can be discharged at any time of the day or length of stay can be reduced by 12 hours) on overall hospital occupancy. Although each of these four studies experimented with changes to hospital discharge policies, they did not model the link between hospitals to community care, so the logistics of discharging patients and the capacity constraints within downstream community care facilities were not considered.

The number of studies which have used simulation to model the flow of patients between in-patient hospital and community care are limited as highlighted in Patrick et al. (2015) and Harper et al. (2021). Zhang et al. (2012) combined simulation, optimisation, and survival analysis to develop a model for planning long-term care capacity in British Columbia, Canada. The model was used to find the minimum capacity level needed each year to satisfy a waiting time threshold. Patrick et al. (2015) developed a DES of patient flow between hospital and the community to long-term care homes in a region of Canada. The model included patient preferences and was treated as an inventory management problem. Both studies modelled capacity as the total number of beds in community care facilities.

Focusing on modelling the link between hospital and 'step down' home-based intermediate care, Harper et al. (2021) used DES to model the patient pathway from hospital discharge to community visits in the UK. The model was designed such that patients could transition from hospital to step-down community care if there was sufficient capacity (defined as the number of community visits). If no community visits were available, the patient would wait in hospital until capacity became available, representing a delayed discharge. The authors used the DES model to determine an optimal capacity of step-down care, which minimised cost of delayed discharge and the cost of under-utilised community care services. Onen-Dumlu et al. (2022) extended this work to model the flow between hospital and three types of step-down intermediate community care in the UK NHS. The objective of this study was to determine an appropriate level of intermediate community care capacity in the context of the COVID-19 pandemic. In both of these papers, capacity was modelled as the total number of community care visits.

Some of the papers reviewed in this section have experimented with improving patient flow by changing hospital discharge policies; these studies make an assumption that downstream community care facilities are not capacity constrained. Other papers have modelled the capacity of community care as the number of available beds or visits. The assumption that patients can be discharged if a bed becomes available in community care is a limitation of these models. Patients cannot always be moved even if there is an available bed in community care, due to reasons such as (i) disagreements with patients and families about the care placement, (ii) funding problems, (iii) housing issues, and (iv) waits for equipment to be installed (The King's Fund 2018).

This paper addresses a gap in the literature by viewing the problem of delayed hospital discharge as multifaceted, rather than directly related to capacity, where the availability of community care placements are modelled using an acceptance probability. This study is also novel in that it considers flow of patients to both intermediate and long term care in the UK NHS.

3 METHODOLOGY

3.1 Problem Description

Our focus is on modelling the impact of different resourcing on the number of patients who are medically optimised for discharge but remain in hospital. As a result, we do not explicitly model the hospital stay or difficulties accessing hospital care. Only patients who need care after a stay in hospital enter the model and patients leave the model when they are directed to a care provider. We use the proposal in Schofield (2021) as a guide to the pathways through the hospital and care systems, which are shown in Figure 1. We discuss these in more detail below.

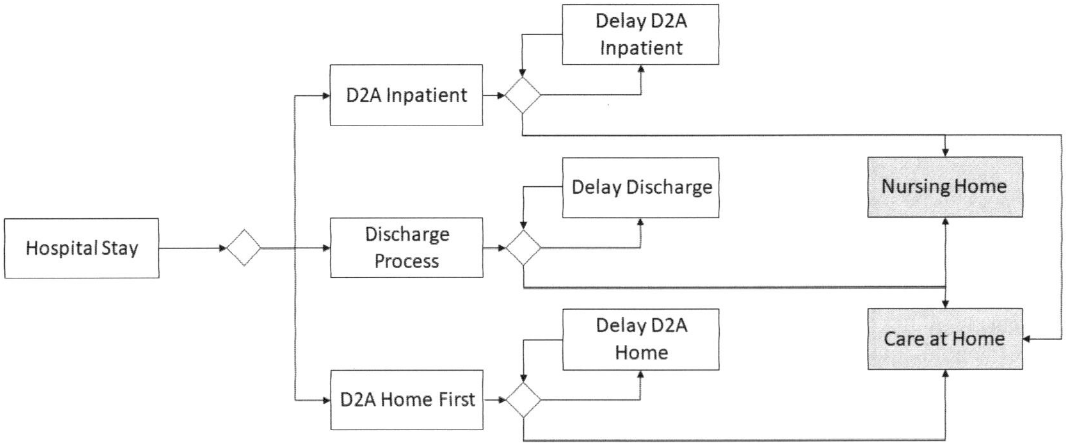

Figure 1: A conceptual model of the delayed discharge simulation.

Following completion of hospital treatment, patients will start one of three processes:

1. Move to a *Discharge to Assess* ward run by the hospital (D2A Inpatient) in which their needs will be assessed and from which they will be sent to either a nursing home or to their own home with some care, labelled as care at home. Within Southampton, these D2A beds are community rehabilitation beds at a neighbouring hospital (the Royal South Hants) and interim beds that have been contracted from the nursing home market.
2. Move home and follow a *Discharge to Assess* process at home (D2A Home) before being set up with care at home.
3. Follow a discharge process while remaining in the hospital. This is the least preferred option and patients will only follow this path if there is no availability in the relevant D2A activity (either inpatient or at home).

Following completion of the D2A process either as an inpatient or at home or following completion of the hospital discharge process, patients will try to access the care that they need. Rather than explicitly modelling capacity of care places, we instead assume that there is a probability that a patient will be given a care place. As Figure 1 shows, if patients are unable to access a nursing home place or care at home they enter a delay state. Delay states are assumed to have a fixed duration of 1 day, assuming that staff will not repeatedly make enquiries of the same care providers during one day for a particular patient.

3.2 Simulation Model

The simulation model is built in Simul8 and follows the process displayed in Figure 1 where Nursing Home and Care at Home are included as exit points. Patients enter the model when they are assumed medically fit for discharge rather than when they enter the hospital.

There are two key metrics for the model: the number of days that patients are delayed before accessing care and the number of patients who are medically fit for discharge and are still in hospital. We may also be interested in the utilisation of the D2A activities and whether patients are being blocked from leaving them because of an inability to access nursing home or care at home places. The number of days a patient spends in either a D2A bed or a hospital bed after being declared ready for discharge can be measured by counting the number of times they enter the delay state. We can also count the number of patients in the Delay Discharge state to provide an estimate of the number of patients in hospital who are medically fit for discharge. Similar results can be collected for the two Delay D2A states.

In order to limit the capacity in D2A Inpatient and D2A Home we use two resources: D2A Inpatient Resource and D2A Home Resource that are needed in both the actual state and the delay state. This ensures that patients will only enter D2A states when there are spaces available. If there are no spaces available when they enter the model they start the discharge process in hospital and approach the nursing home or care at home directly following a discharge process. We assume an unlimited capacity for the Delay Discharge state.

Patients are given a label when they enter the model which indicates whether they will need either a nursing home bed or care at home. This affects their routing: patients who are labelled as needing a care home bed will only be allowed to enter D2A Inpatient or the Discharge Process. Patients labelled as needing care at home are currently only able to enter D2A Home and the Discharge Process but in future versions of the model we will experiment with allowing these patients to also enter D2A Inpatient.

As the model is in its early stages we have not made it publicly available but subsequent work will translate the Simul8 model into SimPy to allow for more straightforward experimentation and easier sharing.

3.3 Parameterisation

A full set of parameters is given in Table 1. The scenario we include here is based on University Hospital Southampton (UHS) and estimated from Schofield (2021). We provide further details below.

UHS received 154,350 admissions in 2021-22 in the most recently published Hospital Episode Statistics for England (NHS Digital 2022). Of these, the percentage needing some form of care is estimated as 27% (Schofield 2021), 5% needing a nursing home and 22% needing care at home.

Table 1: Simulation model parameters.

Parameter	Distribution
Arrival rate (number of patients becoming ready for care per day)	Poisson(114)
Percentage given nursing home label	18.52%
Percentage given care at home label	81.48%
Duration in D2A Inpatient	1 day (fixed)
Duration in D2A Home	1 day (fixed)
Duration in hospital discharge process	1 day (fixed)
Probability of acceptance at a nursing home (per day)	1/7
Probability of acceptance for care at home (per day)	1/3
Number of beds available in D2A Inpatient	75
Number of spaces available in D2A Home	Varied in Table 2
Number of spaces available in Delay Discharge	Unlimited

Assuming that arrivals are spread evenly over the year, this gives an overall arrival rate of 114 per day with 18.52% of patients entering the model requiring a nursing home place and the remainder requiring care at home.

The duration of time spent in the D2A states and in the hospital discharge state before being ready for discharge is currently set to 1 day. We are aware that this may not be the correct duration to use and future work will investigate different public datasets to enable a better estimate of this variable.

We can derive a probability of acceptance p from the expected time spent waiting for a place using the result that

$$\text{Expected time spent waiting} = p\sum_{i=1}^{\infty} i(1-p)^i = 1/p.$$

Therefore, using an expected time spent waiting for a nursing home bed of 7 days and for care at home of 3 days, we use $p = 1/7$ and $p = 1/3$ respectively.

Based on Schofield (2021) we assume 75 beds are available in the D2A inpatient wards. No details are given on the capacity of D2A home and consequently we experiment with different values in the results section to determine an appropriate value to use in order to reproduce published statistics for UHS. We do not put a limit on the number of spaces available in Discharge Process and Delay Discharge and instead experiment with how high this number gets under different scenarios.

3.4 Model Verification and Sensitivity Analysis

Model verification was performed to ensure that the conceptual model in Figure 1 was correctly translated into the Simul8 model shown in Figure 2. This was achieved by maintaining documentation, performing sensitivity analysis to check the model behaved as expected, and using Simul8's graphical interface to check the flow of entities through the model.

The model outputs of interest are the *number of days that patients are delayed before accessing care* and *the number of patients who are medically fit for discharge but remain in hospital*. An initial validation was conducted to compare the model output with information on delayed transfers of care at UHS using data from NHS England (2022) on the number of patients in hospital who no longer meet the criteria to reside and the number of additional bed days for patients with a length of stay of 7+, 14+ and 21+ days.

Preliminary results are presented in Section 4. The model was run 5 times for each of the scenarios presented. All of the results are presented with 95% confidence intervals. The results were calculated using 1 year of simulated time with a warm-up period of 8 weeks. This was determined to be appropriate by visual inspection of the model outputs.

Sensitivity analysis was conducted to (i) determine an appropriate level of capacity in the D2A Home resource, (ii) experiment with the probability of acceptance at a nursing home, and (iii) experiment with the probability of acceptance for care at home. Scenario analysis was carried out to investigate the impact of introducing additional D2A Inpatient beds.

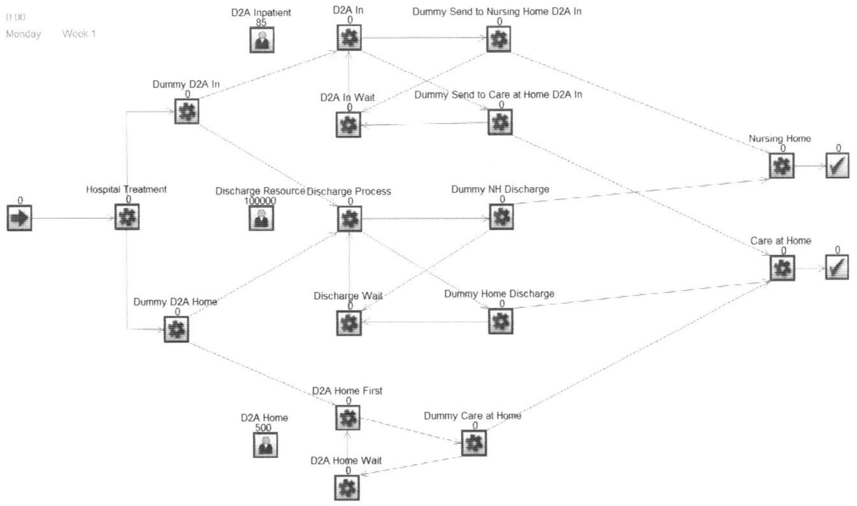

Figure 2: Screenshot of the SIMUL8 Model.

4 PRELIMINARY RESULTS

4.1 Determining an Appropriate Capacity of the D2A Home Resource

The first set of experiments presented in Table 2 were conducted to determine an appropriate capacity of the D2A Home resource. The simulation model was run with the parameter settings listed in Table 1 with D2A Home varied from a capacity of 350 to 650. The model output shows that the average number of patients delayed in hospital decreased as the capacity of the D2A resource was increased. The model output was closest to the NHS data when a capacity of 500 was used. The NHS data for 'average number of patients delayed in hospital' and 'average delayed bed days for patients waiting 7+ days' was closely matched by the simulation output, and the metrics 'average delayed bed days for patients waiting 14+ days' and 'average delayed bed days for patients waiting 21+ days' were underestimated by the model. The underestimation of these metrics is due to the current modelling assumption of homogeneity in patient characteristics. Future work will consider introducing variables that affect a patient's discharge from hospital e.g., medical complications and conditions which make it difficult to find a care placement.

Table 2: Baseline model output with varied D2A Home Capacity compared with NHS England (2022). In each case the point estimated is reported with 95% Confidence Intervals in brackets.

Experiment	Average no. patients delayed in hospital	Average delayed bed days for patients waiting 7+ days	Average delayed bed days for patients waiting 14+ days	Average delayed bed days for patients waiting 21+ days
NHS Data	208	2543	2357	2089
350 D2A Home	320 (315,324)	2837 (2790,2884)	2086 (2046,2125)	1479 (1441,1517)
400 D2A Home	273 (269,276)	2751 (2682,2821)	2090 (2025,2154)	1516 (1456,1575)
450 D2A Home	230 (227,233)	2605 (2530,2680)	2036 (1963,2109)	1498 (1430,1566)
500 D2A Home	205 (203,207)	2493 (2417,2570)	1977 (1908,2047)	1466 (1406,1526)
550 D2A Home	202 (199,205)	2514 (2442,2586)	2000 (1931,2068)	1468 (1422,1551)
650 D2A Home	202 (199,205)	2514 (2442,2586)	2000 (1931,2068)	1468 (1422,1551)

4.2 Sensitivity Analysis

The D2A Home capacity was therefore fixed at 500 and used for the remaining experiments. Sensitivity analysis was performed on the probability of acceptance at a nursing home (per day). The baseline assumption is an acceptance probability of 1/7 (where patients wait an expected time of 7 days). Table 3 shows the model output when this parameter is varied between 1/3 and 1/10. The model behaves as expected, where a lower acceptance probability corresponds to higher average numbers of patients delayed in hospital and higher average delayed bed days across all categories. A similar sensitivity analysis is presented in Table 4 for the probability of acceptance for care at home, where the baseline assumption is 1/3 and the parameter is varied from 1/2 to 1/5. Similarly, a decrease in this acceptance probability corresponds to higher numbers of delayed patients and delayed bed days. The results in Tables 3 and 4 demonstrate that the baseline acceptance probabilities of 1/7 for nursing home and 1/3 for care at home provide the best match to the data.

Table 3: Sensitivity analysis for Nursing home acceptance probability, with care at home acceptance probability fixed as 1/3 and D2A Home capacity set as 500. In each case the point estimated is reported with 95% Confidence Intervals in brackets.

Nursing home acceptance probability	Average no. patients delayed in hospital	Average delayed bed days for patients waiting 7+ days	Average delayed bed days for patients waiting 14+ days	Average delayed bed days for patients waiting 21+ days
1/5	119 (118,121)	923 (897,949)	609 (585,632)	368 (349,386)
1/6	162 (160,164)	1624 (1588,1660)	1200 (1164,1236)	822 (784,859)
1/7	205 (203,207)	2493 (2417,2570)	1977 (1908,2047)	1466 (1406,1526)
1/8	246 (243,250)	3475 (3386,3564)	2872 (2787,2958)	2226 (2142,2310)
1/9	288 (284,291)	4688 (4595,4780)	4019 (3928,4110)	3265 (3187,3343)
1/10	329 (326,332)	6063 (5925,6200)	5340 (5205,5474)	4488 (4359,4616)

Table 4: Sensitivity analysis for care at home acceptance probability, with nursing home acceptance probability fixed as 1/7 and D2A Home capacity set as 500. In each case the point estimated is reported with 95% Confidence Intervals in brackets.

Care at home acceptance probability	Average no. patients waiting in hospital	Average delayed bed days for patients waiting 7+ days	Average delayed bed days for patients waiting 14+ days	Average delayed bed days for patients waiting 21+ days
1/2	202 (199,204)	2514 (2442,2586)	2000 (1931,2068)	1486 (1422,1551)
1/3	205 (203,207)	2493 (2417,2570)	1977 (1908,2047)	1466 (1406,1526)
1/4	356 (352,360)	3376 (3316,3436)	2477 (2419,2534)	1722 (1668,1776)
1/5	539 (535,543)	5166 (5095,5237)	3757 (3694,3821)	2544 (2489,2599)

4.3 Scenario Analysis

Table 5: Scenario analysis for the capacity of D2A Inpatient resource, where D2A Home capacity is set at 500 and acceptance probabilities 1/7 and 1/3 for nursing home and care at home respectively. In each case the point estimated is reported with 95% Confidence Intervals in brackets.

D2A Inpatient Capacity	Average no. patients waiting in hospital	Average delayed bed days for patients waiting 7+ days	Average delayed bed days for patients waiting 14+ days	Average delayed bed days for patients waiting 21+ days
75 (baseline)	205 (203,207)	2493 (2417,2570)	1977 (1908,2047)	1466 (1406,1526)
85	195 (192,197)	2419 (2393,2444)	1927 (1909,1945)	1438 (1424,1453)
95	185 (182,187)	2237 (2170,2304)	1755 (1686,1823)	1289 (1226,1343)

Scenario analysis was performed to investigate the effect of increasing the number of available D2A Inpatient resources from 75 to 85 and 95. Table 5 shows a reduction of approximately 10 in the average number of patients waiting in hospital for an increase of 10 D2A Inpatient beds. There is also a marked reduction in the number of delayed bed days for patients in each of the 7+, 14+ and 21+ categories.

5 CONCLUSION

This paper presented a conceptual model and discrete event simulation of the discharge process from inpatient hospitals in the UK. The model specifically considered delayed discharge at the University Hospital Southampton (UHS). Although this is only preliminary work, the results showed a close match to the most recent data from UHS, where 208 patients were delayed in hospital and there were 2543 excess bed days for patients waiting more than 7 days. The number of excess bed days for patients waiting over 14 and 21 days was underestimated by the model. Sensitivity analysis showed that decreasing the probability of acceptance to a nursing home and care at home led to an increased number of patients delayed in hospital and an increase in excess bed days. Scenario analysis showed that increasing the capacity of the inpatient discharge to assess process could lead to a reduction in the number of patients delayed in hospital and associated excess bed days.

Future work will consider how to improve the model so that the output better represents the distribution of excess bed days. One of the key parameters in the simulation model is the probability of acceptance in care, which we estimate differently for nursing home care and care at home but do not vary between patients. In reality the probability of a patient being accepted into care varies significantly between different patient types. For example, Flinders and Scobie (2022) suggests that patients who have an initial hospital stay of 3 weeks or more are more likely to suffer long delays to discharge. These patients tend to be elderly and frail and have more complex needs, hence requiring a higher level of care. Taking into account the differences in probability of acceptance into care between different patient groups rather than assuming the population is homogeneous will improve the ability of the model to reproduce observed behaviour and will also help to better understand the relative benefits of increasing provision for hard-to-place patients versus improving the system more generally. Future development of the model will also consider adding variation in the probability of acceptance to community care facilities by day of the week and obtaining more data to better inform estimates of the duration of time patients spend in D2A processes. The model will be translated into SimPy to make it open source and to give more control over running experiments.

ACKNOWLEDGEMENTS

The authors would like to thank the Isaac Newton Institute for Mathematical Sciences for support during the Virtual Forum for Knowledge Exchange in Mathematical Sciences (V-KEMS) study group on Communities for an Aging Society where preliminary work on this project was undertaken; in particular Dawn Wasley who was instrumental in managing the logistics for the workshop. This work was supported by EPSRC grant number EP/V521929/1. Clare Halpenny's role in this research was independent of her PhD research project which is funded by the Legal & General Group (research grant to establish the independent Advanced Care Research Centre at University of Edinburgh). The funders had no role in conduct of the study, interpretation, or the decision to submit for publication. The views expressed are those of the authors and not necessarily those of Legal & General.

REFERENCES

BBC News 2022. "One in six patients should not be in hospital". Available at https://www.bbc.co.uk/news/uk-scotland-63455657. Accessed November 2022.

Busby, C. R., and M. W. Carter. 2017. "Data-driven generic discrete event simulation model of hospital patient flow considering surge". In *Procedings of the 2017 Winter Simulation Conference*, edited by W. K. V. Chan, A. D'Ambrogio, G. Zacharewicz, N. Mustafee, G. Wainer, and E. Page, 3006–3017. Piscataway, New Jersey: Institute of Electrical and Electronics Engineers, Inc.

S. Flinders and S. Scobie 2022. "Hospitals at capacity: understanding delays in patient discharge?". Available at https://www.nuffieldtrust.org.uk/news-item/hospitals-at-capacity-understanding-delays-in-patient-discharge. QualityWatch: Nuffield Trust and Health Foundation. Accessed on 1 December 2022.

Harper, A., M. Pitt, M. D. Prez, Z. Onen-Dumlu, C. Vasilakis, P. Forte, and R. Wood. 2021. "A Demand and Capacity Model For Home-Based Intermediate Care: Optimizing The'Step Down'Pathway". In *2021 Winter Simulation Conference*, edited by B. Kim, K. Feng, S. Masoud, Z. Zheng, C. Szabo, and M. Loper, 1–12. Arizona: Institute of Electrical and Electronics Engineers, Inc.

Kelen, G. D., R. Wolfe, G. D'Onofrio, A. M. Mills, D. Diercks, S. A. Stern, M. C. Wadman, and P. E. Sokolove. 2021. "Emergency department crowding: the canary in the health care system". *NEJM Catalyst Innovations in Care Delivery* 2 (5).

Khanna, S., D. Sier, J. Boyle, and K. Zeitz. 2016. "Discharge timeliness and its impact on hospital crowding and emergency department flow performance". *Emergency Medicine Australasia* 28 (2): 164–170.

Mustafee, N., K. Katsaliaki, and S. J. Taylor. 2010. "Profiling literature in healthcare simulation". *Simulation* 86 (8-9): 543–558.

NHS Digital 2022. "Hospital Admitted Patient Care Activity, 2021-22". Available at https://digital.nhs.uk/data-and-information/publications/statistical/hospital-admitted-patient-care-activity/2021-22#resources. Accessed November 2022.

NHS England 2022. "Discharge delays (acute)". Available at https://www.england.nhs.uk/statistics/statistical-work-areas/discharge-delays-acute-data/. Accessed November 2022.

NHS Improvement 2022. "Quick Guide: Discharge to Assess". Available at https://www.nhs.uk/NHSEngland/keogh-review/Documents/quick-guides/Quick-Guide-discharge-to-access.pdf. NHS England Publications Gateway Reference 05871. Accessed November 2022.

Onen-Dumlu, Z., A. L. Harper, P. G. Forte, A. L. Powell, M. Pitt, C. Vasilakis, and R. M. Wood. 2022. "Optimising the balance of acute and intermediate care capacity for the complex discharge pathway: computer modelling study during COVID-19 recovery in England". *PLOS ONE* 17 (6): 1–16.

Patrick, J., K. Nelson, and D. D. Lane. 2015. "A simulation model for capacity planning in community care". *Journal of Simulation* 9 (2): 111–120.

Qin, S., C. Thompson, T. Bogomolov, D. Ward, and P. Hakendorf. 2017. "Hospital occupancy and discharge strategies: A simulation-based study". *Internal Medicine Journal* 47 (8): 894–899.

Rojas-García, A., S. Turner, E. Pizzo, E. Hudson, J. Thomas, and R. Raine. 2018. "Impact and experiences of delayed discharge: A mixed-studies systematic review". *Health Expectations* 21 (1): 41–56.

Salmon, A., S. Rachuba, S. Briscoe, and M. Pitt. 2018. "A structured literature review of simulation modelling applied to Emergency Departments: Current patterns and emerging trends". *Operations Research for Health Care* 19:1–13.

Jamie Schofield 2021. "Hospital Discharge Operational model and Home First Discharge to Assess (D2A)". Available at https://www.southampton.gov.uk/moderngov/documents/s52758/Enc.20220for20Hospital20Discharge20Operational20and20Urgent20Community20Response20Models.pdf. Accessed November 2022.

The King's Fund 2018. "Delayed transfers of care: a quick guide". Available at https://www.kingsfund.org.uk/publications/delayed-transfers-care-quick-guide. Accessed November 2022.

Tobail, A., P. Egan, W. Abo-Hamad, and A. Arisha. 2013. "Application of lean thinking using simulation modeling in a private hospital". In *Fifth International Conference on Advances in System Simulation, SIMUL 2013*, 22–28. Venice, Italy.

Zhang, Y., M. L. Puterman, M. Nelson, and D. Atkins. 2012. "A simulation optimization approach to long-term care capacity planning". *Operations Research* 60 (2): 249–261.

AUTHOR BIOGRAPHIES

LAURA BOYLE is a Lecturer in Data Analytics at the Mathematical Sciences Research Centre, Queen's University Belfast. Prior to this she worked as a Research Fellow of the Australian Research Council Centre of Excellence for Mathematical and Statistical Frontiers at the University of Adelaide. Her research interests are in simulation modelling and data anlaytics with applications in healthcare. Her email address is laura.boyle@qub.ac.uk and her website is https://pure.qub.ac.uk/en/persons/laura-boyle.

CHRISTINE CURRIE is a Professor of Operational Research in Mathematical Sciences at the University of Southampton and a member of the Centre for Operational Research, Management Sciences and Information Systems (CORMSIS). She is Editor-in-Chief for the Journal of Simulation and has previously co-chaired the UK SimulationWorkshop. Her research interests include simulation optimization, applications of simulation in health care, optimal pricing and disaster management. Her email address is christine.currie@soton.ac.uk and her website is http://www.southampton.ac.uk/maths/about/staff/ccurrie.page.

CARLOS LAMAS FERNANDEZ is a Lecturer in Business Analytics/Management Science in Southampton Business School. He completed his PhD in Operational Research in the Southampton Business School in 2018. Prior to that, he had worked as a research fellow in NIHR ARC Wessex and as a scientific developer at ETH Zurich. In his research, Carlos is interested in using optimisation techniques in areas such as sports, health care, transportation and logistics. He has developed both heuristic and exact methods for cutting and packing, vehicle routing and facility location problems. His email address is C.Lamas-Fernandez@soton.ac.uk and his website is https://www.southampton.ac.uk/people/5xjdpm/doctor-carlos-lamas-fernandez.

CLARE HALPENNY is a PhD student at the Advanced Care Research Centre (ACRC), University of Edinburgh. Her research interests include care trajectories, transitions and decision making processes for older adults. She has worked as an occupational therapist for the NHS in both England, and Scotland. Her e-mail address is clare.halpenny@ed.ac.uk.

LY NGUYEN is a research fellow supported by the NIHR at the University of Southampton. Ly is interested in using optimisation techniques in health care, transportation and logistics applications. She has worked with mathematical programming and developed heuristic methods for applications such as container packing, vehicle routing, inventory routing, and planning and logistic in health care. Her e-mail address is L.Nguyen@soton.ac.uk and her website is https://www.southampton.ac.uk/people/5zbys6/doctor-ly-nguyen.

Proceedings of the Operational Research Society Simulation Workshop 2023 (SW23)
C. Currie and L. Rhodes-Leader, eds.
https://doi.org/10.36819/SW23.030

A FRAMEWORK TO SHARE HEALTHCARE SIMULATIONS ON THE WEB USING FREE AND OPEN SOURCE TOOLS AND PYTHON

Dr. Alison Harper

University of Exeter
a.l.harper@exeter.ac.uk

Dr. Thomas Monks

University of Exeter
t.m.w.monks@exeter.ac.uk

ABSTRACT

We present a simple framework to support sharing of executable healthcare discrete-event simulation models in python over the web. Our sharing framework is based on combining remote version control repositories with free and open source software - Jupyter Notebooks, Jupyter Books, and streamlit - along with free digital infrastructure provided by Binder, streamlit.io, GitHub pages, and open science repositories such as Zenodo. The framework enables executable models to be shared with users of different technical abilities: from coders to software literate users. We provide an applied example including a full web application of an executable model. Our framework aims to support NHS organisations to preview, validate, and use models. Academic research teams can also benefit from enhanced scrutiny of their work and long term archiving of models.

1 INTRODUCTION

In the UK, health and social care research is often funded using public money. This might be from a funder such as the National Institute for Health and Care Research (NIHR). This money is part of the overall NHS budget. Modelling and simulation within health and social care research often tackles very human problems as well. The systems being modelled affect hundreds, thousands, and potentially hundreds of thousands of people (infants, children, adults, carers, friends, family and potentially yourself) per year. As such research using these funds must maximise transparency, re-usability, adaptability (by the NHS and other researchers), and reduce barriers to uptake by health systems, and scrutiny by research teams. These barriers can include cost (such as the license fees for purchasing a commercial off the shelf simulation package), technical skills (such as coding skills or use of commercial software), security (such as installation on NHS machines) and software dependency management.

One possible route to enable sharing/reuse/adaptation and eliminate licensing costs is to adopt a Free and Open-Source Software (FOSS) for your modelling and simulation. FOSS is best defined using Stallman's four freedoms:

0. The freedom to run the program as you wish, for any purpose.
1. The freedom to study how the program works, and change it so it does your computing as you wish.
2. The freedom to redistribute copies so you can help your neighbour.
3. The freedom to distribute copies of your modified versions to others. By doing this you can give the whole community a chance to benefit from your changes.

Note that FOSS here is more than simply open source code. It grants the rights for users to adapt and distribute copies however they choose. FOSS software within a data science context is often provided with a *permissive* type of license such as the MIT or BSD-2 licenses. These grant the user the four freedoms (within commercial or non-commercial work), waive liability, and require crediting (citation) of that authors and software in subsequent work.

Many FOSS simulation tools exist. For a full review of open source Discrete-Event Simulation (DES) software readers should refer to Dagkakis and Heavey (2016). Here we limit our FOSS focus to tools in Python. We choose Python as it is consistently ranked in the top 4 of Stack OverFlows's most used programming languages (4th in 2022) and top 5 most loved programming languages by developers. Python is also synonymous with *modern data science* and is used extensively in research and development of computational methods and applications.

FOSS and Python have gained significant popularity and use in the UK's NHS in recent years. This transformation has been inspired via communities such as NHS-Python (and NHS-R) as well as NHS-England promoting the use of FOSS tools (NHS England 2022). However, widespread access to Python still remains challenging in the security conscious health care setting. A big challenge is *model installation* on locked down and controlled NHS machines. A possible solution to the installation problem is web based technology and in particular *simulation on the web*.

Here we adopt a straightforward definition of simulation on the web as a simulation model that has been deployed to a remote server, is accessed via a public web URL, and can be executed, in some manner with varying processing power, without local installation of any software or components. The contribution of our paper is to propose a simple python based framework for sharing executable simulation models on the web via Binder, Streamlit and Github pages.

2 AIMS

1. Briefly review related work in FOSS simulation packages implemented in Python, and sharing simulations on the web;
2. Outline a straightforward framework for deploying a simulation developed in Python on the web for users of varying technical skills;
3. Provide an applied simulation example implementing our framework;
4. Provide guidance for modellers to begin sharing models via the web.

3 RELATED WORK

3.1 Python Discrete-Event Simulation Tools

Dagkakis and Heavey (2016) reviewed open source software for discrete-event simulation (DES). Their review identified three FOSS Python packages: Simpy (Team SimPy 2020), Pysimulator (Pfeiffer, Hellerer, Hartweg, Otter, and Reiner 2012) and ScipySim (McInnes and Thorne 2011). PySimulator, and ScipySim were last updated in 2014, and 2010 respectively. Simpy has a process based simulation worldview and has continued to be maintained (last update Apr 2020). It has been used in several Operations Research relevant publications (Bovim, Gullhav, Andersson, Dale, and Karlsen 2021, Allen, Bhanji, Willemsen, Dudfield, Logan, and Monks 2020, Monks, Harper, Anagnostou, and Taylor 2022).

An updated list of Python DES packages now includes Salabim (van der Ham 2018, MIT licensed, last updated Apr 2022), Ciw (?? , MIT licensed; last updated Oct 2022), and de-sim (Goldberg and Karr 2020, MIT licensed, last updated Nov, 2020). Salabim is a fork of Simpy2 and includes many simulation tools including automatic results collection and animation. Ciw (the Welsh word for queue) was developed at Cardiff University and allows users to very quickly build complex multi-class queuing networks. An advanced feature of Ciw is network deadlock detection. Palmer and Tian (2021b) provide two reproducible, open source implementations of Ciw for hybrid modelling, archived here (Palmer and Tian 2021a). De-sim is an object orientated (OO) framework for developing complex interacting DES models. The de-sim authors argue that their OO framework is an advancement over Simpy's process-based worldview; although we note that Simpy itself is highly flexible and can easily be used within an OO framework; for example see (Allen, Bhanji, Willemsen, Dudfield, Logan, and Monks 2020).

3.2 Sharing Models on the Web

The idea of simple sharing and executing simulations via the web is not a new one. Fishwick (1996) provides an early example of an executable queuing simulation model accessed via a public URL. Modern takes on simulation on the web include full Open Science Gateways that provide substantive e-infrastructure for controlled model access, parameterisation and execution. Anagnostou, Taylor, Groen, Suleimenova, Anokye, Bruno, and Barbera (2019) provide an impressive example of open science gateways in action using their PALMS model.

Simpler recent examples include a generic hospital ward model developed and made available online (Penn, Monks, Kazmierska, and Alkoheji 2020), which aimed to address the issues raised in this paper. The Simul8 model is made available through Zenodo - an open science repository (Penn and Monks 2018). The authors note that a weakness of their approach is that NHS users must purchase a COTS package license in order to reuse/adapt the model. Tyler, Murch, Vasilakis, and Wood (2022) addressed the need to improve uptake and routine use of simulation by NHS analysts by developing an open source model using *R* which supports user-defined configurations of patient pathways. The *R Shiny* user interface improves usability and functionality, but the model needs to be downloaded and executed locally on the user's machine. Similarly, an open, verifiable model to enable the organisation of dialysis outpatient services during the pandemic was developed by Allen, Bhanji, Willemsen, Dudfield, Logan, and Monks (2020). The code and documentation are available to download and a limited user interface to execute the model is provided via Binder. There is no support for experimentation. Monks, Harper, Anagnostou, and Taylor (2022) proposed a 6-level framework for supporting M&S open working, with a case study demonstrating an approach to achieving openness, including a simple, reproducible pipeline for sharing simulation experimentation and results. The focus of the paper is the practical issues to be considered with different levels of openness.

4 A FRAMEWORK FOR SHARING PYTHON MODELS

We developed a framework to allow an executable simulation model developed in Python to easily be shared to different classes of users with varying knowledge of coding, computer simulation, and software. Our first class considers more technical simulation users that have coding experience; for example NHS analysts or simulation researchers. Our second class of user assumes no experience of coding, but a general familiarity of using software. Figure 1 provides an overview of our framework. In summary, we provide two approaches to sharing executable models:

- FOSS licensed, interactive, and executable scientific notebooks using BinderHub;
- A web app (browser based) model front end built via StreamLit and deployed via Streamlit.io

In addition to the sharing of executable python models our framework also supports enhanced model documentation and open working using Github pages and an (interactive) Jupyter Book.

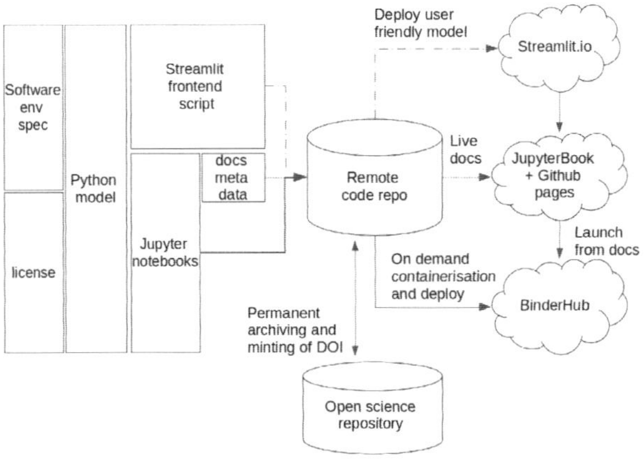

Figure 1: Proposed sharing framework for Python models

4.1 Software Environment Specification

For any type of software deployment problem it is necessary to adopt a language that allows model developers to specify the software libraries, version numbers and sources that are required in order for the model to run. Three options of environment file exist for python: pip requirements, conda environments, and poetry. Our research often makes use of conda virtual environments that documents dependencies in the *Yet Another Mark-up Language (.yml)* format.

4.2 Licensing of Research Artefacts

Before sharing (or making any Python code public), it is essential that authors select an appropriate FOSS license for the code and other research artefacts. This, for example, ensures appropriate use of the model and removes liability of the authors. A common approach in data science is to adopt a permissive license such as the MIT. See Monks, Harper, Anagnostou, and Taylor (2022) for more details on licence types.

4.3 Python Model

Our framework is general to any Python model and Python simulation package. For an application exemplar, we chose to use Simpy as our FOSS simulation package. Simpy provides a simple process based worldview simulation engine. This simplicity makes the package highly flexible and usable in many simulation applications in healthcare. Our research group also has extensive experience of using it in real world simulation applications. One limitation of Simpy relative to commercial off the shelf simulation packages is the lack of a user friendly front end to control simulation execution, scenarios and results analysis. The Python model should be independent of any *front-end* and executable standalone.

4.4 Remote Code Repository

Central to Figure 1 is a remote code repository. The most popular solutions are GitHub, GitLab, and BitBucket. Most developers opt for GitHub, but we recommend modellers consider the unique benefits and cons of each solution. All code artefacts (including licence, meta-data files, and readme) should be committed to the remote repo to provide long lasting version control. Some developers store data here too; although other specialist solutions exist for data version control.

4.5 Jupyter Notebooks

Project Jupyter is a FOSS project that supports scientific coding in Python, *R* and Julia in a classical notebook format. A simple way to conceptualise Jupyter notebooks are as browser based interface and an execution kernel. The interface provides a set of markdown and code sells that a user can control. Markdown is simply a way to present formatted text, mathematics, and code together. As such it provides an excellent way to explain, document and then execute simulation experiments. For example, an analyst may have created a set of Jupyter notebooks for model validation, determining the number of replications, warm-up analysis, defined experiments, or interactive experimentation.

Jupyter's kernel executes all of the code. Typically the kernel is located on the users local machine. But given the server nature of Jupyter, the kernel could be located on a remote (possibly powerful) machine. In these remote cases the user need not have Python, Simpy, or Jupyter installed on their own machine.

4.6 Binderhub

Jupyter's use of the browser and ability to access a remote kernel allows for easy deployment of simulation notebooks to a technical user-base. For example, well trained NHS analysts from the NHS-python / NHS-R community or other researchers familiar with reading and executing code.

BinderHub provides a Jupyter environment in the cloud that others can use to run your simulation model. One simple requirement is that the Python simulation model code, and Jupyter notebooks are available in a public respository hosted in the cloud; for example in GitHub or GitLab.

4.7 Streamlit and streamlit.io

A web app interface to the model is suitable for less technical simulation users such as NHS analysts with no coding experience, NHS managers, researchers not familiar with python, or the general public. A web app will provide simple ways to setup and execute the model. This might include parameter fields, logic diagrams, basic animation, and buttons.

Streamlit is a simple modern Python library that provides a way to script web applications. Streamlit has many built in controls and display functions, for buttons, sliders, text fields, tables, and charts. Using Streamlit, it is possible to create a very simple simulation app with only a few lines of code; for example, basic parameter settings, an execute button, and results displayed as a table in the browser. Streamlit.io is a free hosting service for Streamlit apps. When used together they provide a robust and easy way to share simulation models to healthcare users. *These models can then be accessed from any device*. For example, models can be executed from a phone or a tablet instead of traditional laptop or desktop environment.

4.7.1 Documentation via Jupyter Book and Github Pages

While not strictly required for sharing the simulation model our framework advocates that any simulation used for health decision making should provide open and high quality documentation on how the model is implemented and works. In our example, we provide a link to another web based technology: a Jupyter Book deployed in GitHub pages https://tommonks.github.io/treatment-centre-sim. Jupyter Book is a high level tool to create a structured website based on meta-data (a table of contents and config file), markdown, and Jupyter Notebooks. GitHub Pages is a free web hosting service provided by GitHub (owned by Microsoft). As such it can publish cleanly annotated Python code describing the working of each component within the model. In our example, we have supplemented this with discrete sections for the Strengthening the Reporting of Empirical Simulation Studies (STRESS) for DES models (Monks, Currie, Onggo, Robinson, Kunc, and Taylor 2019).

One benefit of a Jupyter Book deployed on the web in this manner is that it automatically provides links to BinderHub. This means that different parts of the model and model process (such as model testing) can be viewed on a static website and then opened live to run via a link in the book if desired. A second benefit of online documentation managed with Jupyter book is that we have provided a simple mechanism for users/readers to provided feedback (such as queries, clarification requests, and bug reports). Within our Jupyter Book we have provided *contribution* guidelines and implemented the technology using GitHub Issues. The latter allows a conversation between authors and contributors to facilitate improvement in the documentation and model web app.

4.8 Open Science Repository

Lastly, for long term archiving (and preservation) of simulation models our framework links code in the remote repository to an open science repository. Examples, include Figshare, Zenodo, and the Open Science Framework. Each of these has guarantees on storage and mints DOIs to enable citation and improve discoverability.

5 APPLIED EXAMPLE

This section describes a text-book simulator implemented in Python and shared via our framework. In addition to the code we provide, we include some basic instructions for use of Binder and required modifications to sharing desktop based simulations.

5.1 Case Study Model

We adapt a textbook example from Nelson (2013, p.170): a discrete-event simulation model of a U.S based treatment centre. In the model, patients arrive to the health centre between 6am and 12am following a non-stationary Poisson process. On arrival, all patients sign-in and are triaged into two classes: trauma and non-trauma. Trauma patients include impact injuries, broken bones, strains or cuts etc. Non-trauma include acute sickness, pain, and general feelings of being unwell etc. Trauma patients must first be stabilised in a trauma room. These patients then undergo treatment in a cubicle before being discharged. Non-trauma patients go through registration and examination activities. A proportion of non-trauma patients require treatment in a cubicle before being discharged. The model predicts waiting time and resource utilisation statistics for the treatment centre. The model allows managers to ask questions about the physical design and layout of the treatment centre, the order in which patients are seen, the diagnostic equipment needed by patients, and the speed of treatments. For example: "what if we converted a doctors examination room into a room where nurses assess the urgency of the patients needs."; or "what if the number of patients we treat in the afternoon doubled".

5.2 Code and Software

We have shared our code via GitHub. Binder examples can be found at https://github.com/TomMonks/treatment-centre-sim while the Streamlit example can be found at https://github.com/TomMonks/treat_sim_streamlit. We used Conda to manage our dependencies. In summary, we used Python 3.8. Data manipulation, simulation, and general mathematical modeling were done using Simpy v4.0.1 (Team SimPy 2020), NumPy v1.19.2 (van der Walt, Colbert, and Varoquaux 2011) and Pandas v1.2.3 (McKinney 2011). Charts were produced with MatPlotLib v3.3.4 (Hunter 2007). All analyses were conducted in Jupyter-Lab v2.4.3 (JupyterLab 2022). The web app was produced using StreamLit v1.13.0 (streamlit.io 2022). Documentation was created using Jupyter Book v0.13.1 (Executable Books Community 2020).

5.3 Binder

For this example, we have created a simulation model and committed the code to a public GitHub repository https://github.com/TomMonks/treatment-centre-sim. Binder expects that a conda virtual `environment.yml` file is placed into a sub-directory `binder/`. Once this has been committed it is a simple case of navigating to https://mybinder.org/ and copy and pasting the URL into the build and launch text box. Binder will then create a remote instance that contains both your code and dependencies specified in the `binder/environment.yml` file. The first time the instance is built it will likely take several minutes. This process may be need to be repeated if the instance is not used regularly. The setup form is illustrated by Figure 2

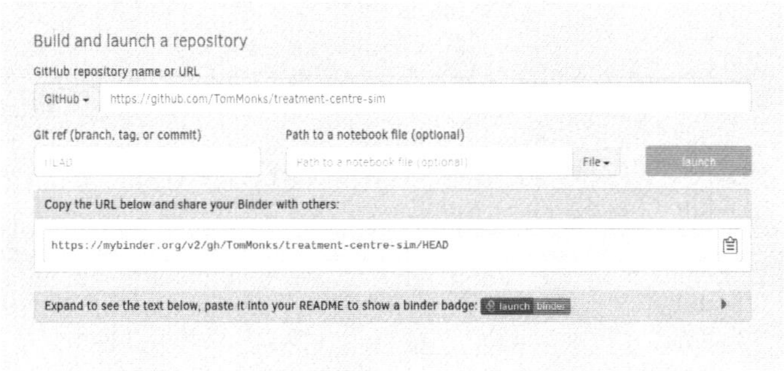

Figure 2: Binderhub setup

We invite interested readers to launch our treatment sim instance on Binder via the following link: https://bit.ly/treat_sim_binder. The model notebook can be found `src/full_model.ipynb`. You should be presented with a notebook in Figure 3.

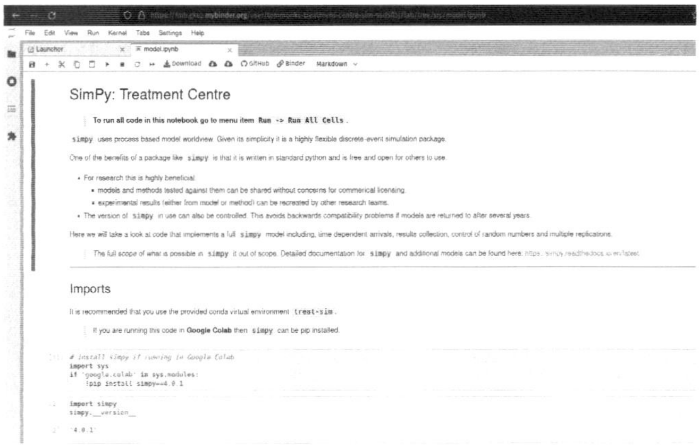

Figure 3: Simulation model running in binderhub

5.4 StreamLit.io App

Our Streamlit implementation of the case study model can be accessed via https://treat-sim.streamlitapp.com. To create the Streamlit app we logged into streamlit.io, created a new app, and specified the GitHub repo plus app launch file (Overview.py). Streamlit.io requires information on app dependencies. We used a pip `requirements.txt`. The initial build is automatic and takes 5-10 minutes. Figure 4 illustrates the interactive simulation page of our app.

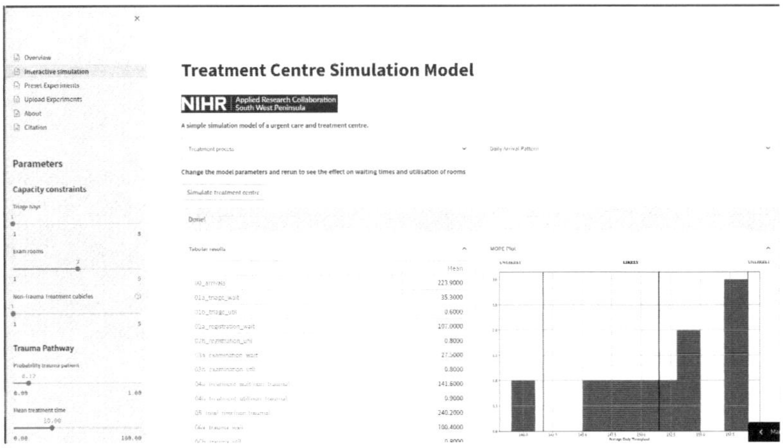

Figure 4: Treatment centre user interactive simulation

6 GUIDELINES FOR SHARING MODELS VIA STREAMLIT

When deployed, a Streamlit application must be thought of as a remote instance of a simulation model in the cloud. The model is accessed via a URL. Each user that accesses the URL is provided with a unique instance of the web app (i.e. it is not shared between users). To be successful, a **slight change** in a developers way of thinking is needed as you are working with an instance of a web application as opposed to a model on a local machine.

6.1 Parameterisation

A simulation model that will be executed by clients and people other than those that developed it will typically provide a way to parameterise a model for one or more experiments. Two such ways are:

- Default parameters, interactive controls, and fields built into the simulation interface;
- An external file that contains parameter configurations for a single or multiple experiments.

The first of these options is perhaps best used for setting up a singular experimental run of the model. For example, one or multiple replications of the model to test if changing a parameter has an impact on results. For interactive simulation of this type a web app requires no modification from classical desktop simulation. For example, a set of sliders representing model parameters, each with a default value, could be provided for users to manipulate and rerun the model.

The second of these options in its simplest form might use a format such as a Comma Separated Values (.CSV), containing a simple row and column format for experiment and parameters. The user would populate (or more likely modify) these parameters in some way before a simulation model loads the parameters on each run. If a simulation model is deployed via a web-app the model may be setup with an option for a user to *upload* the CSV file to the instance (e.g. by click and dragging a file into a web form). It is recommended that an example file is provided. The format might vary depending on the application, for example, a format might be rows of named experiments and columns of parameter values (or vice versa). One option is for the web-app to auto-generate an exemplar file that a user can download, modify, and then upload. A secondary option for parameterisation is to specify a URL where parameter data are stored. This would still require a user to ensure the data are in the correct location and it may be simpler to upload directly. Alternatively an application might allow a user to specify a Digital Object Identifier (DOI) where data are stored. This latter option might be useful if the web app is part of a publication and a permanent record of data used in experiments is required.

6.2 Results Visualisation

One significant advantage of developing a front end to the simulation model in Python is the opportunity to produce publication quality charts and outputs directly from each run of the model and then allow a user to export them. For example, Python's Matplotlib library could be used to create a Measure of Risk and Error (MORE) plot after a user specified experiment. These packages offer superior plotting options and output resolution than utilities found in a typical spreadsheet package, or specialist simulation tool. Once a model run is complete, a high resolution image, meeting printing specifications such as DPI > 300, can then be downloaded by the user. Of course, like desktop simulation packages the data underlying plots might be of more interest to users; particularly if they wish to create a custom set of plots. In these cases web-apps need to provide the functionality to *download* the data in a suitable file format (for example, a CSV file). Another option is to allow a user to copy data from the web-app to the local machine's clipboard. The data can then be pasted into whatever software the user desires; for example, a spreadsheet or statistical analysis package. In these instances, the developer needs to remember that the data reside on a remote instance of the model not the local machine. For external plotting web-apps may also contain links to (or generate) Python code templates that users can use to quickly produce plots and customise.

Many interactive visualisation technologies have been developed in Python for web applications. Two such powerful and popular visualisation libraries are bokeh and plotly. Interactive options might include 'hover and see more detail', such as data point values, or zooming in on different parts of a plot, for example in the case of a time series plot.

6.3 Content and Organisation

The digital content that is included with a simulation model deployed online will vary depending on the use case. However, we recommend that developers consider a number of simple options to enhance user experience for the model. Our basic guidelines include:

- A non-expert summary of model and its use
- Interactive experimentation options
- Advanced experimentation or experimental design (aka scenario analysis) facilities
- A page describing the project and funding used to develop the model
- Model citation instructions
- Links to any external code repositories and model technical documentation

With a Streamlit implementation of a Python simulation model we propose that these can be organised as separate *pages* within the web app. The user can then navigate to the appropriate page to access the functionality. A developer will need to decide on a *landing page* for the app. In our example, we have opted for the non-expert summary. Writing a non-expert summary is a non-trivial task and out of scope of this article. It is worth noting that a health care simulation deployed via a web-app may be private to an organisation, organisational sub-group such as a department or team, or as in the case of streamlit.io public on the internet. Particularly in the latter case we recommend a high quality non-expert summary that provides a general overview of the model along with its use case.

7 DISCUSSION

We present a simple, practical framework for sharing free and open simulation of healthcare systems developed in Python.

7.1 Strengths and Contributions

Our FOSS approach benefits health services, such as the NHS in the UK, students of simulation and researchers other than the authors of the original simulation. Our framework has the following benefits for the simulation and health services communities:

Previewing models for potential users

The FOSS tools within our framework enable potential users of a model to test it out without going through any installation, source, or dependency management. This can enable fast feedback and revision of a model to fix mistakes or better meet user requirements. More technical users have direct access to executable code in this process.

Sharing runnable code with an organisation that cannot run python

NHS organisations have slowly been adopting Python, but when working with new organisations IT departments may be initially reluctant to allow model installation. Python and other free and open software are sometimes viewed with suspicion by IT departments and cannot be installed on a collaborators machine.

Making the results of a publication repeatable

In science, particularly within computational science, there's a concept called the *reproducibility crisis*. This is a problem with understanding how results of a particular study were generated, and an inability to (fully) reproduce (repeat) the results with the same code and data. In lay terms it means that most academic authors write a nicely crafted paper that doesn't manage to fully reveal how they got the figures they reported in their simulation study. Modern simulation projects are complicated and writing up a brief summary of methods may fail to enable reuse, verification, or peer learning. Our framework supports authors to provide an executable version of their model with their publications.

7.2 Limitations

Our framework aims to share Python models, although several of technologies used are language agnostic and overall it is easily adapted for *R* and to some extent Julia. Modifications for *R* would include switching from StreamLit to *R Shiny* (and Shinyapps.io). *R* users are less likely to be familiar with rigorous dependency management than Python and Julia developers; however, dependency management can be achieved via the *renv* library (or to a lesser extent via conda). Julia currently lacks a native StreamLit, equivalent, but our framework would allow sharing with more technical users. A limitation of all free remote hosting services is the amount of compute available. For expensive computer simulation, two potential modifications could be considered. The first is to pay to host a web app (or Binder server) that provides sufficient compute (e.g via Heroku). The second is to *containerise* the app and model via a technology such as *docker* and *dockerhub*. Model users will require docker installed on their local machine in order to run it. One drawback is that for commercial organisations this will involve a licence fee.

ACKNOWLEDGEMENTS

This article is independent research supported by the National Institute for Health and Care Research Applied Research Collaboration South West Peninsula. The views expressed in this publication are those of the author(s) and not necessarily those of the National Institute for Health and Care Research or the Department of Health and Social Care.

REFERENCES

Allen, M., A. Bhanji, J. Willemsen, S. Dudfield, S. Logan, and T. Monks. 2020, Aug. "A simulation modelling toolkit for organising outpatient dialysis services during the COVID-19 pandemic". *PLOS ONE* 15 (8): e0237628. Publisher: Public Library of Science.

Anagnostou, A., S. J. E. Taylor, D. Groen, D. Suleimenova, N. Anokye, R. Bruno, and R. Barbera. 2019, December. "Building Global Research Capacity in Public Health: The Case of a Science Gateway for Physical Activity Lifelong Modelling and Simulation". In *2019 Winter Simulation Conference (WSC)*, 1067–1078. ISSN: 1558-4305.

Bovim, T. R., A. N. Gullhav, H. Andersson, J. Dale, and K. Karlsen. 2021, December. "Simulating emergency patient flow during the COVID-19 pandemic". *Journal of Simulation* 0 (0): 1–15. Publisher: Taylor & Francis _eprint: https://doi.org/10.1080/17477778.2021.2015259.

Dagkakis, G., and C. Heavey. 2016. "A review of open source discrete event simulation software for operations research". *Journal of Simulation* 10 (3): 193–206.

Executable Books Community 2020, February. "Jupyter Book".

Fishwick, P. A. 1996, November. "Web-based simulation: some personal observations". In *Proceedings of the 28th conference on Winter simulation*, WSC '96, 772–779. USA: IEEE Computer Society.

Goldberg, A. P., and J. R. Karr. 2020. "DE-Sim: an object-oriented, discrete-event simulation tool for data-intensive modeling of complex systems in Python". *Journal of Open Source Software* 5 (55): 2685.

Hunter, J. D. 2007. "Matplotlib: A 2D graphics environment". *Computing in Science & Engineering* 9 (3): 90–95.

JupyterLab 2022, October. "JupyterLab". original-date: 2016-06-03T20:09:17Z.

McInnes, A. I., and B. R. Thorne. 2011. "ScipySim: Towards Distributed Heterogeneous System Simulation for the SciPy Platform".

McKinney, W. 2011. "pandas: a foundational Python library for data analysis and statistics". *Python for High Performance and Scientific Computing* 14.

Monks, T., C. S. Currie, B. S. Onggo, S. Robinson, M. Kunc, and S. J. Taylor. 2019. "Strengthening the reporting of empirical simulation studies: Introducing the STRESS guidelines". *Journal of Simulation* 13 (1): 55–67.

Monks, T., A. Harper, A. Anagnostou, and S. J. Taylor. 2022, Jul. "Open Science for Computer Simulation".

Nelson, B. 2013. *Foundations and methods of stochastic simulation: a first course*. London: Spinger.

NHS England 2022. "NHS England Open Source Programme". https://www.england.nhs.uk/digitaltechnology/open-source/.

Palmer, Geraint and Tian, Yawen 2021a, March. "Source code for Ciw hybrid simulations.".

Palmer, G. I., and Y. Tian. 2021b. "Implementing hybrid simulations that integrate DES+ SD in Python". *Journal of Simulation*:1–17.

Penn, M., T. Monks, A. Kazmierska, and M. Alkoheji. 2020, April. "Towards generic modelling of hospital wards: Reuse and redevelopment of simple models". *Journal of Simulation* 14 (2): 107–118. Publisher: Taylor & Francis _eprint: https://doi.org/10.1080/17477778.2019.1664264.

Penn, M. L. and Monks, T. 2018, October. "Generic Ward Configuration Simulation Model".

Pfeiffer, A., M. Hellerer, S. Hartweg, M. Otter, and M. Reiner. 2012. "PySimulator–A simulation and analysis environment in Python with plugin infrastructure".

streamlit.io 2022. "Streamlit. The fastest way to build and share data apps". https://streamlit.io/.

Team SimPy 2020. "SimPy 3.0.11". https://simpy.readthedocs.io/en/latest/index.html.

Tyler, J. M., B. J. Murch, C. Vasilakis, and R. M. Wood. 2022, June. "Improving uptake of simulation in healthcare: User-driven development of an open-source tool for modelling patient flow". *Journal of Simulation* 0 (0): 1–18. Publisher: Taylor & Francis _eprint: https://doi.org/10.1080/17477778.2022.2081521.

van der Ham, R. 2018. "Salabim: open source discrete event simulation and animation in python". In *Proceedings of the 2018 Winter Simulation Conference*, 4186–4187.

van der Walt, S., S. C. Colbert, and G. Varoquaux. 2011. "The NumPy Array: A Structure for Efficient Numerical Computation". *Computing in Science Engineering* 13 (2): 22–30.

AUTHOR BIOGRAPHIES

ALISON HARPER is an ESRC post-doctoral Research Fellow at the University of Exeter Medical School, in the Peninsula Collaboration for Health Operational Research and Development. Her email address is a.l.harper@exeter.ac.uk.

THOMAS MONKS is Associate Professor of Health Data Science at University of Exeter Medical School and Turing Fellow at the Alan Turing Institute. His email address is t.m.w.monks@exeter.ac.uk. His website is https://arc-swp.nihr.ac.uk/about-penarc/people/tom-monks/.

Proceedings of the Operational Research Society Simulation Workshop 2023 (SW23)
C. Currie and L. Rhodes-Leader, eds.
https://doi.org/10.36819/SW23.031

HYBRID MODELS WITH REAL-TIME DATA: CHARACTERISING REAL-TIME SIMULATION AND DIGITAL TWINS

Navonil Mustafee

Centre for Simulation, Analytics and Modelling
The Business School
University of Exeter
Exeter, EX4 4ST, UK
N.Mustafee@exeter.ac.uk

Alison Harper

PenCHORD, PenARC
University of Exeter Medical School
St. Luke's Campus, Heavitree Rd
Exeter EX1 1TE, UK
A.L.Harper@exeter.ac.uk

Joe Viana

Department of Accounting and Operations Management
BI Norwegian Business School
Nydalsveien 37, Oslo, 0484, NORWAY
Joe.Viana@bi.no

ABSTRACT

Real-time Simulation (RtS) and Digital Twins (DT) are terms generally associated with hybrid models that use real-time data to drive computational models. Additionally, in the case of DTs, real-time data is often used to create virtual replicas of the physical system as it progresses through real-time. There is an increasing volume of literature on RtS and DT; however, the field of OR/MS is yet to coalesce on accepted definitions and conceptualisations. This has arguably led to the cascading usage of these terms. The objective of the paper is threefold: (1) distinguish between RtS and DT, (2) present RtS-DT conceptualisation in four dimensions, and (3) present methodological and technical insights on developing RtS with limited data. We argue that the evolution of conventional simulation models to fully-fledged hybrid DTs may necessitate a focus on a transitional stage; namely, RtS models primarily driven using historical distributions with limited real-time data feeds.

Keywords: Real-time Simulation, Digital Twin, Hybrid Modelling, Conceptualisation

1 INTRODUCTION

A defining characteristic of the present Millennium is the exponential increase in data availability, made possible through advances in data acquisition technologies such as sensors, Global Positioning System (GPS)-enabled devices and Radio-frequency identification (RFID) chips and the underlying network and communication, storage, and computing infrastructures. Developing novel ways of making sense of this explosion of data, often referred to as data deluge (Bell et al., 2009), presents a contemporary challenge to researchers.

Modelling and simulation (M&S) techniques predates the era of Big Data (Taylor, 2019). The data requirements of such conventional simulation models were often limited to distributions computed from historical data and observational data, the latter necessary to model the system of interest using an overarching M&S methodology. With the increase in both the volume and velocity of data, the challenge to our community is to develop hybrid modelling approaches that combine the traditional M&S methods (and their reliance on historical distributions) with updated data streams. An example of such a hybrid approach is the application of M&S approaches with methods from Data Science, for example, machine learning with DES (Elbattah and Molloy, 2016), process mining with DES (Abohamad et al., 2017) and neural networks with SD (Abdelbari and Shafi, 2017). However, hybridisation could also be achieved by using values from historical distributions with real-time data and using it to drive computer simulations (Powell and Mustafee, 2017). In the M&S community, Real-time Simulation (RtS) and Digital Twins (DT) are generally recognised as computational models using real-time data feeds. However, the lack of common definitions and conceptualisation has, arguably, given rise to cascading usage of these terms. To address this gap, the objective of this paper is, ***first***, to distinguish between RtS and DT based on the literature; ***second***, to present a conceptualisation of RtS and DT based on the following four dimensions of a simulation study – *modelling objectives*, *data requirements*, *implementation,* and *experimentation*.

The paper also draws on the authors' experiences in developing a platform that publishes real-time data from several NHS Trusts in the South West of England – *NHSQuicker* (Mustafee and Powell, 2020) – and using data from NHSquicker in an Accident & Emergency (A&E) model developed using AnyLogic™. A conventional Discrete-event Simulation (DES) is used as the core computational model (Harper, 2021). The core model is being incrementally developed to integrate "limited" real-time data feeds available to the authors. The ***third*** objective of the paper is thus to present methodological and technical insights from our experience in developing RtS with limited real-time data.

For the remainder of the paper, the terms Operations Research/Management Science (OR/MS) DT and DT are used as synonyms for digital twins. Similarly, the terms "computational models", "conventional models" and OR/MS M&S are used interchangeably and refer to conventional simulation models. Finally, RtS refers to OR/MS simulations with limited real-time data feeds.

2 HYBRID MODELS WITH REAL-TIME DATA STREAMS

A simulation study comprises several well-defined stages. The model implementation stage concerns the development of a computer model using programming languages/libraries and software packages. However, prior to implementation, conceptualisations of the system must be developed and translated into modelling artefacts. Abstractions built on constructs specific to M&S techniques often lead to single-technique implementations. For example, an abstraction of a service delivery system viewed through the lens of queues and servers would often lead to a DES implementation; the same system could be modelled in agent-based simulation (ABS) if the conceptualisation draws on agent classes, inter-agent communication and the concept of emergence. Distinct from such single-technique DES and ABS implementations, *Hybrid Simulation* uses multiple simulation techniques in the context of a single simulation study. As reported by Brailsford et al. (2019), since 2000, there has been rapid growth in publications that have used combined approaches, such as the use of hybrid agent-based DES (Viana et al., 2020), DES with system dynamics (Xu et al., 2018) and mixed DES-ABS-SD modelling (Roemer and Strassburger, 2019).

The combined application of simulation with frameworks, methods, tools, and techniques that have been developed outside the field of M&S is referred to as Hybrid Modelling (Tolk et al., 2021). Unlike hybrid simulation, which mainly focuses on the model implementation stage, there are opportunities to leverage cross-disciplinary methods in conceptual modelling, verification and validation (V&V), scenario development, experimentation, and other stages of an M&S study (Powell and Mustafee, 2017). Mustafee et al. (2020) present several examples of hybrid models from the literature that have combined OR/MS simulations with approaches from applied computing, for example, distributed simulation, parallel computing, cloud computing, and real-time computing.

DT and RtS are hybrid models that use real-time data streams. They are considered hybrid models since their execution necessitates integration with backend data acquisition systems using distributed computing approaches such as socket programming and web application programming interfaces (APIs). Database-centric methods (including spreadsheets) could also be used wherein real-time data stored in the backend system is accessed using the *Open Database Connectivity (ODBC)* interfaces (Microsoft ODBC, 2022) to specific database management systems. Yet another approach is through the development of database triggers which access APIs defined in DT and RtS. A trigger is a code snippet stored in a database and executed following a defined database event, for example, an arrival of a patient in a hospital or a trigger executed at pre-defined intervals. The triggers are developed using procedural languages such as *PL/SQL* (Oracle) and *PL/pgSQL* (PostgresSQL). Flat-file-based systems (e.g., data from a data acquisition system (DAS) dumped in a shared directory) are yet another option, although this often requires parsing the data before being used by DT and RtS.

Our discussion has assumed that the front-end DASs were primarily developed to meet an organisation's data requirements and that they provide APIs, flat files, ODBC interfaces, etc., for external applications to access data. However, our experience working with multiple A&E patient flow systems has shown that this may not always be the case. A DAS could be a "closed" system in cases where primitive data structures are written in binary format using languages such as C/C++, and thus only accessible to program code (other programs cannot decipher records from these data files). On the opposite end of the spectrum, bespoke DAS and DT/RtS systems may be developed, which motivates the need for strong coupling in the design phases. In such cases, some of the complexities of real-time data communication are abstracted through higher-level libraries and API calls.

3 DEFINITIONS: DIGITAL TWINS AND REAL-TIME SIMULATION

VanDerHorn and Mahadevan (2021) reviewed 46 definitions of DTs and proposed the following generalised definition: *"(DT is) ... a virtual representation of a physical system (and its associated environment and processes) that is updated through the exchange of information between the physical and virtual systems"*. Arguably, the authors' definition rests on the characterisation of DT originally proposed by Michael Grieves (Grieves, 2014, as cited in VanDerHorn and Mahadevan, 2021), which necessitated the existence of a **physical system** in the real world, the **virtual representation** of the real system, and **communication channels** between the real and the virtual systems for information exchange and synchronisation. In the following paragraphs (including Tables 1 and 2), we critique DT and conventional M&S models, as applied to OR/MS, based on Grieves' characterisation of the definition of digital twins. Through this critique, we lay the foundation for the definition of real-time simulation, or RtS, which is introduced in Section 3.1.

Physical Reality: A fundamental difference between M&S and DT is whether the system being modelled exists in the real world. For OR/MS simulations, this could be a system that does not yet exist. Here the purpose of the simulation study is to experiment with system configurations before implementing the physical infrastructure. The work conducted by British Airways' Operations Research department is a good example. They developed generic simulation models to assess infrastructure requirements, such as the number of check-in counters for the new Heathrow Terminal 5 (Beck, 2011).

A physical system, a physical environment and physical processes are identified by VanDerHorn and Mahadevan (2021) as the three essential elements of the physical reality and which a DT must model. When a physical system does not exist, a DT implementation cannot be realised as there are no entities, no physical environment for the entities to reside and interact in, and no existing physical processes exist.

Virtual Representation: The five definitional elements associated with virtual representation are presented in Table 1 (column one). The definition presented by VanDerHorn and Mahadevan (2021) is a very broad and encompassing definition of DT. Thus, we assessed the definitions from the standpoint of OR/MS researchers. We consider both the traditional OR/MS simulations of physical systems that exist in reality (Table 1; column two) and DTs that include an OR/MS focus (Table 1; column three).

Table 1 *A comparison of conventional OR/MS simulation and OR/MS DT based on VanDerHorn and Mahadevan (2021) definition of Virtual Representation*

Definitional elements of virtual representation	Conventional OR/MS Simulation of a Physical System	OR/MS DT
Idealised representation of physical reality	For conventional OR/MS simulation, the conceptual modelling stage helps develop an abstraction of the system of interest based on factors such as the objectives of the simulation study.	The idealised representation of physical reality often has to consider the real-time data available to model the abstraction of the physical system in the DT model.
System states and parameters	A simulation model of a physical system often requires a modeller to observe the real-world system or develop familiarity through interaction with the problem stakeholders, reading literature, etc. The modeller then relies on this understanding	A key element of an OR/MS DT is to monitor the physical reality as it transitions through time (also referred to as the wall clock time). The terms system states and parameters are concepts that are

	and uses technique-specific constructs available to implement a computer model. The computer model will include system states which will transition through time and various parameters, all of which are defined based on the modeller's understanding of the system in question.	used in state space modelling; state estimation methodology is frequently used in DTs to change information between physical reality and virtual representation (VanDerHorn and Mahadevan, 2021).
Virtual system: The virtual system consists of data and models of the entities from the physical system at the chosen level of abstraction.	*Data:* For conventional models, the sources of data include primary and secondary data, observational data, estimates from the literature, hypothetical values, expert opinion, or a combination. *Models of Entity:* For detailed-level DES/ABS modelling, individual entities can be modelled as work units and agents. SD can provide a higher level of abstraction using stocks and flows. Hybrid simulation can provide different levels of abstraction.	*Data:* Generally acquired from the physical reality, e.g., manufacturing facility, using data acquisition systems. *Models of Entity:* Generally, represent the flow of real-world entities through physical reality. For example, in a manufacturing facility for mobile phones, every unit of the phone could be modelled as a virtual entity that interacts with entities that represent manufacturing sub-processes.
Virtual environment: Virtual representation of the physical environment at a chosen level of abstraction. The objective is for the virtual environment to mimic the physical system's interaction with the physical environment (if the latter affects the former).	Certain aspects of the physical environment (if they are deemed important) can be modelled through the conventional OR/MS models. For example, virtual representation could be spatial in nature. They may include the dimensions of a factory floor, machines, and conveyor belts. Commercial, off-the-shelf DES packages often provide options for 2D and 3D visual displays (Akpan and Brooks, 2012). The use of Virtual Reality (VR) in DES is also an active area of research (Turner et al., 2016). Spatial aspects can also be modelled in a conventional way. For example, the placement of inventory of semi-finished goods may affect the travel time associated with moving items from the store for further processing. Some packages like Simul8™ can include distance travelled which affects the overall processing time.	Sensors can monitor aspects of the physical environment like occupancy level and air quality. Sensor data can thus be used to create a virtual environment at the chosen level of abstraction. However, it is arguable that the inclusion of a virtual environment may only be necessary if its physical counterpart affects either the entities or the processes in the physical reality. For example, monitoring air quality in high-precision manufacturing facilities may be considered important. However, occupancy levels at the same facility may not have a bearing on the physical processes and may thus be excluded from the virtual environment.
Virtual processes: Virtual processes represent the process-specific interaction among the entities that together comprise the virtual system and/or virtual environment. The expression of virtual processes is often achieved through computational modelling of the underlying physical processes existing in physical reality.	As these are dynamic models, the time element associated with the transformation of physical entities (e.g., plastic blocks) through a physical process (e.g., a machine which melts plastic) is crucial. For example, in a DES model, the (virtual) process flows are modelled using networks of queues and servers; the servers can be initialised with distributions that represent the processing times of the physical entities. Thus, in an OR/MS M&S model, the virtual processes are implemented through a combination of model-specific values and the underlying M&S methodology and a simulation engine.	For OR/MS DTs, it is important to consider the granularity of virtual process representation. For a plastic injection moulding factory, a DT that models the physical transformation of raw material (e.g., polymer) from solid to a liquified state and then to the final product, may not be essential (although this may be possible through physics-based modelling).

<u>Communication Channels for Information Exchange:</u> The third component of the VanDerHorn and Manadevan (2021) definition of DT relates to the interconnection between physical reality and its virtual DT representation. The information exchange component is subdivided into three elements,

physical-to-virtual connections, *virtual-to-physical connections,* and *information fusion* (Table 2). Similar to our approach in Table 1, we assess the OR/MS conventional models and OR/MS DTs with these definitional elements. Our discussion on information fusion is particularly important for the RtS definition introduced in section 3.1.

Table 2 *A comparison of conventional OR/MS simulation and OR/MS DT based on VanDerHorn and Mahadevan (2021) definition of Information Exchange*

Def. elements of info exchange	Conventional OR/MS Simulation of a Physical System	OR/MS DT
Physical-to-virtual connection	Although data can be automatically collected from a physical system and stored in a database, distribution fitting and other forms of analysis will require the intervention of the modeller. As there is a manual element, the frequency of information exchange is minimal. Indeed, numerous OR/MS models are driven using distributions computed from very old data that were never updated.	Data acquisition is mostly automatic; updating virtual representation using measurements derived from the physical system does not usually require complex analysis; the frequency of data updates is in real-time. However, the experimental element of a DT will need to include distributions for faster than real-time simulations. Automation may enable the DT distributions to be recomputed at pre-defined intervals or when new data arrives.
Information fusion	In this paper, we define information fusion as combining historical data with real-time data. Going by this definition, since conventional models only rely on historical data and distributions, there is no scope for information fusion.	Synchronisation of the virtual representation of the physical systems (the digital replica) is usually achieved through real-time data. However, for the experimental element of the DT, the underlying computational model will need to use information fusion (i.e., using historical data with real-time data). This is further explained in the context of RtS in the later sections of the paper.
Virtual-to-physical connection	The objective of the virtual-to-physical connection for information exchange is to draw insights from the virtual system and make appropriate changes to the physical processes to achieve the desired state of the physical system (VanDerHorn and Mahadevan, 2021). A simulation is a decision support system; making changes to the physical system based on experiment results is the implementation stage of an M&S study. However, there is often a time lag due to the need for further considerations (including investments and stakeholder trust); the learnings from a study are often not implemented.	In an OR/MS DT, the virtual-to-physical connection often has mechanisms to control real-time feedback to the physical system based on either the visual replicas in DT (e.g., accumulation of stock taking place in the physical reality) or through faster than real-time experimentation. Interfacing of virtual simulation to physical systems has also been referred to as symbiotic simulation or on-line simulation (Aydt et al., 2009; Onggo et al., 2018). As DTs are primarily used for real-time decision-making, the time lag associated with making changes to the physical processes is minimal (indeed, it may also be automatic, e.g., through actuators which receive control feedback from DTs).

3.1 Defining Real-time Simulation (RtS)

Conventional OR/MS models, in addition to being able to represent "future" systems, are also widely implemented to capture "existing" physical realities. Examples include simulation models of hospitals and factories, distribution hubs and supply chain networks, airports and container ports. Like DTs, such models represent the physical system, the physical environment, and the processes of the physical reality (VanDerHorn and Mahadevan, 2021). However, unlike real-time DT execution, OR/MS simulations are implemented to execute faster than real-time (also referred to as simulated time).

For short-term decision-making, DTs may include OR/MS models for faster than real-time experimentation and optimisation. However, there also exists the opportunity to transition a conventional OR/MS simulation model of a physical system (which uses only historical data) to a DT (which uses real-time data), such that the former model is incrementally developed by replacing the distributions used to drive the model with real-time data feeds, as and when they become available. We refer to these transitional models of conventional OR/MS simulations as *real-time simulations* or *RtS*. To distinguish the conventional models from the hybrid RtS and DTs, we present a conceptualisation based on key aspects of a modelling study, which we refer to as its dimensions. This is presented next.

4 CONCEPTUALISING REAL-TIME SIMULATION AND DIGITAL TWINS

For this discussion, we define a conventional OR/MS simulation as using historical distributions to populate a core computational model. During the experimentation phase, such models are initialised with a warmup period, followed by the results collection until the simulation end time—results from experimentation inform medium to long-term decision-making. Figure 1 maps the conventional OR/MS simulation, depicted as a black circle, along the dimensions of *modelling objective*, *data needs*, *model implementation* and *experimentation*.

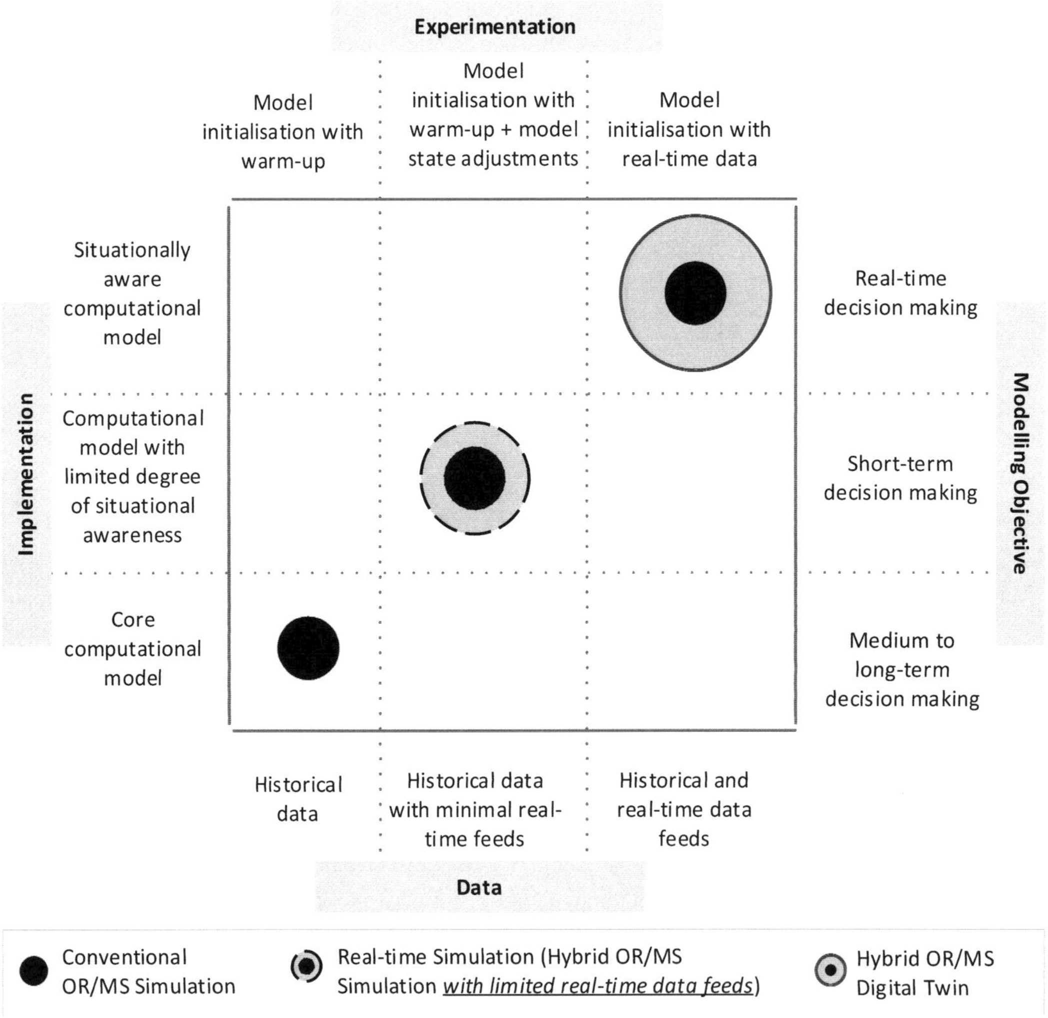

Figure 1 *Conceptualisation of RtS and DT using the dimensions of modelling objective, data requirements, model implementation and experimentation.*

A RtS, illustrated as a concentric circle with a dashed line, has, at its core, the core computational model (represented as the black circle), but with some of the historical distributions complemented with real-time data feeds. The grey outer circle represents real-time data; the dashed line illustrates that only limited real-time data is available to a RtS. Mapping of RtS in the four dimensions shows that such models are mostly used for short-term decision-making (*modelling objective axis*), it is dependent mostly on historical data but with some real-time data (*data axis*); the latter has implications in terms of model initialisation as it now includes the added element of making intrusive changes to the model state after warmup (*experimentation axis*). With the *implementation axis*, we note that RtS have limited situational awareness (this is not surprising considering these models have access to minimal real-time data), and the core remains the computational model— the conventional OR/MS model discussed in the earlier paragraph.

A DT is shown in Figure 1 as a concentric circle similar to RtS but with two differences. (1) Our definition of DT assumes integration with real-time data streams necessary for the virtual representation of entities, processes, and resources. As such, the dashed line used for RtS to illustrate information deficiency is replaced with a continuous red border. (2) Similar to the illustration for RtS, the grey outer circle represents real-time data; however, compared to the former, the latter has a larger diameter. This visually represents that the dependency of DT on real-time data is far more than compared to RtS. However, similar to RtS, the OR/MS DT also includes a computational model for experimentation (represented as a black circle); thus, DTs are equally dependent on distributions computed from historical data (refer to the *data axis* for the DT). Moving on to the *modelling objective axis*, a DT can be especially suited for real-time decision-making, although intuitively, it could also be viable to assist decision support in the short-term, for example, 2-4 hours. What is defined as "short-term" is subjective, and its viability will generally be determined based on the context of the application.

We define DT as a situationally aware computational model (*implementation axis*) that fulfils the following two objectives: (1) The virtual representation of the physical elements of a system as it evolves through real-time. (2) Faster-than-real-time simulation experimentation at specific pre-defined trigger points. (1) is likely to be a critical determinant for (2); for example, in a manufacturing facility, bottlenecks identified through the real-time assessment of key performance indicators (KPIs) could trigger simulation experiments to inform stakeholders of possible options, thereby assisting in real-time decision-making. Compared to DTs, we argue that the primary objective of RtS is (2). The *experimentation axis* notes that a DT will be initialised using real-time data streams, diminishing the need for warmup time and model state adjustments.

5 METHODOLOGICAL AND TECHNICAL INSIGHTS IN DEVELOPING REAL-TIME SIMULATION (RTS)

In Figure 1, the step-wise illustration of conventional simulation (bottom-left), RtS (centre) and DT (top-right) represents the evolution of conventional models to hybrid RtS and DT models. Transitioning from simulations using only historical distributions to fully-fledged DTs may be challenging since the gamut of real-time feeds needed for the DT realisation may not be available. Also, it may be technically challenging to integrate the data feeds with DT (refer to section 2). In such cases, a RtS can effectively use the real-time values available to the modeller by substituting them, in the underlying computational model, for values derived from historical distributions. Indeed, this was the approach taken by the authors while developing a real-time model of an A&E department in AnyLogic™. The A&E RtS uses secondary data from *Symphony* (hospital patient flow system) and real-time data from *NHSquicker* (https://nhsquicker.co.uk/). The remainder of this section provides insights on developing RtS based on the authors' experience. The discussion is structured around the four dimensions of modelling objectives, data needs, implementation, and experimentation.

5.1 Insights on Modelling Objectives for RtS

Organisational decisions are often challenged by shifting or competing goals and uncertain, dynamic environments. The decision-makers' situation awareness – an up-to-date state of knowledge about the current situation - can be influenced by the provision of appropriate real-time information. This can be achieved through descriptive information about the current state of the system, or through predictive information, such as short-term forecasts of system state metrics. Prescriptive information using RtS projects the development of a situation over a short time horizon and can support operational decisions informing system recovery through model experimentation. For the development of RtS models, focusing efforts on technical challenges is essential, however design principles are also required to support the needs of decision-makers who will interface with the model for decision-support (Harper, Mustafee & Pitt, 2022). Design decisions include, for example, intuitive output presentation and visualisations, and should be considered part of the set of modelling objectives from the outset of the RtS study. This generally requires a collaborative approach with the aim of satisfying the needs and requirements of users that cannot be anticipated at the study outset, and to adapt design as users, needs, and environments evolve during the development phases (Barricelli et al., 2019).

5.2 Insights on Data Requirements for RtS

One of the four main components of conceptual modelling is the model *input*; Robinson (2008) defines the inputs as those elements that can be altered to improve the problem situation. As an RtS is a transition model, a mid-point, in the shift from its current state as a conventional OR/MS implementation to the future state of a DT, it uses both historical data and real-time data feeds as inputs. An objective of input data analysis for RtS is, thus, to undertake a technical evaluation of the stakeholder organisation's data acquisition systems (DAS) for their potential to relay data automatically. From our experience, this necessitates broadening the participation from the stakeholder organisation to include database managers and technical team leads. For RtS it is assumed that not all data can be captured or sent in real-time. For these data points, the traditional mechanisms are to be used. Thus, the inputs in a hybrid RtS model will include values extracted from both real-time data and those computed from historical distributions.

5.3 Insights on RtS Implementation

RtS models will have a limited degree of situational awareness as they include minimal real-time data feeds. Even with data received from DAS in real-time (integrated into the RtS models as variable values), the stakeholders may determine the frequency of updates since the DAS will have other business functions to fulfil. Our implementation of A&E RtS uses feeds from patient flow systems (DAS in A&E departments) and includes a backend database trigger fired at a pre-determined frequency; the trigger executes a SQL (Structure Query Language) and sends the information through a Web API. RtS will also require a parsing function to process the incoming data stream and automation that implements a throttling behaviour in terms of model execution, i.e., the RtS is updated as new data comes in; the frequency of the update also determines the degree of situational awareness. Yet another element of implementation is the definition of trigger points which would start the automated execution of experiments. The trigger points are often based on KPIs. The threshold values of the KPIs will continually have to be checked by the RtS when new data is received.

5.4 Insights on RtS Experimentation

In conventional OR/MS modelling, multiple scenarios are developed for experimentation. The scenarios enable the stakeholders to test different strategies for (possible) implementation in the medium to long term. In the case of RtS, the objective is to assist the stakeholders with short-term decision-making. In RtS, experimentation can be executed automatically when the assessment of new data against pre-determined KPIs thresholds indicates a breach. In our experience, it is important to associate specific breaches with pre-developed RtS scenarios that must be *automatically loaded* and executed. There are also several challenges associated with model initialisation. For example, before experimentation, the current simulation time will need to reflect the time associated with the last tranche of data updates; appropriate adjustments to warmup time must be made before experimentation since the current time is continually progressing; real-time values for model components like queues and servers must be injected into the model, which will override the values assumed by RtS after the model warmup.

6 CONCLUSION

The availability of increasing volumes of data presents a challenge to modellers to maximise the value that could be potentially derived from this data. The use of real-time data streams with OR/MS computational models holds the promise of increasing situational awareness and assisting with near real-time decision-making. Implementing such hybrid models requires deploying knowledge from Computer Science/Applied Computing, Information Systems/Databases and OR/M&S. In the literature, the emergence of the Digital Twins (DT) concept has meant that multiple definitions are being used. In this paper, we approach DTs from the standpoint of researchers in OR/MS. We reflect on conventional OR/MS simulation models and their meaning in the new DT landscape. We argue that the computational model must exist in an OR/MS DT to enable faster than real-time experimentation, although we agree that a parallel objective of such DTs is to enable visualisation through virtual replicas of the physical system as it transitions through real-time.

How might the community transition from the conventional OR/MS approach that relies on historical distributions to a fully-fledged OR/MS DT with real-time data sources, the latter realising the dual objective of presenting virtual replicas and also enabling situationally aware real-time experimentation? From our experience in developing RtS, we identified that the transition from conventional models to pure OR/MS DTs might be assisted through the implementation of hybrid models that are driven using historical distribution *but also* include limited real-time data. We define these models as real-time simulations (RtS). Further, we have distinguished RtS from OR/MS DTs in our conceptualisation of the four modelling dimensions: modelling objectives, data requirements, model implementation and experimentation.

RtS implementation is technically challenging as it must fuse real-time data with values generated from distributions to represent, at the start of the experimentation (and after the model warmup), the best possible approximation of the physical system at the current wallclock time. After initialisation, the RtS will rely on distributions to populate the stochastic elements in a model. As more data becomes available, an RtS could potentially include an element of virtual representation, but distributions and model state adjustments will still be necessary for experimentation.

Future work will expand on the methodological and technical insights on RtS implementation we briefly discussed in section 5 of the paper. Articulation of design principles for RtS is an area of future work. Yet another stream of research is empirical work. In a subsequent publication, we will present our RtS of an Urgent Care system in the South West of England, which is integrated with *NHSquicker* (Mustafee and Powell, 2020) and which acts as a data acquisition system.

System failures associated with data acquisition systems would result in the non-availability of real-time data, impacting an RtS. Thus, an avenue for future research is to investigate novel algorithms that provide the best data estimates during such interruptions. Towards this, researchers could investigate using Machine-Learning and AI-based approaches to generate synthetic data for use in RtS and DTs. Similarly, Parallel and Distributed Simulation (PADS) techniques, such as optimistic synchronisation (Fujimoto, 2001), could enable the rollback of computations when real-time feeds are (eventually) restored, which is yet another opportunity for future research.

REFERENCES

Abdelbari, H. and Shafi, K. (2017). A computational intelligence-based method to 'learn' causal loop diagram-like structures from observed data. *System Dynamics Review* **33(1)**:3–33.

Abohamad, W., Ramy, A., and Arisha, A. (2017). A hybrid process-mining approach for simulation modelling. In *Proceedings of the 2017 Winter Simulation Conference*, pp. 1527-1538. Piscataway, NJ: IEEE.

Akpan, I. J. and Brooks, R. J. (2012). Users' perceptions of the relative costs and benefits of 2D and 3D visual displays in discrete-event simulation. *Simulation* ***88*(4)**: 464-480.

Aydt, H., Turner, S. J., Cai, W., and Low, M. Y. H. (2009). Research issues in symbiotic simulation. In *Proceedings of the 2009 Winter Simulation Conference*, pp. 1213-1222. Piscataway, NJ: IEEE.

Barricelli, B.R., Casiraghi, E., & Fogli, D. (2019). A survey on digital twin: Definitions, characteristics, applications and design implications. *IEEE Access*, **7**, 167653-71.

Beck, A. (2011). Case study: modelling passenger flows in Heathrow Terminal 5. *Journal of Simulation* **5(2)**: 69-76.

Bell, G., Hey, T., and Szalay, A. (2009). Beyond the data deluge. *Science*, **323(5919)**: 1297-1298.

Brailsford, S., Eldabi, T., Kunc, M., Mustafee, N., and Osorio, A. (2019). Hybrid simulation modelling in operational research: A state-of-the-art review. *European Journal of Operational Research* **278(3)**: 721-737.

Elbattah, M. and Molloy, O. (2016). Coupling simulation with machine learning: A hybrid approach for elderly discharge planning. In *Proceedings of the 2016 ACM SIGSIM Conference on Principles of Advanced Discrete Simulation*, pp. 47-56. ACM.

Fujimoto, R. M. (2001). Parallel and distributed simulation systems. In *Proceeding of the 2001 Winter Simulation Conference*, pp. 147-157. Piscataway, NJ: IEEE.

Grieves, M. (2014). Digital twin: manufacturing excellence through virtual factory replication. White paper 1, pp: 1-7. *Digital Twin Institute.*

Harper, A. (2021). *A hybrid modelling framework for real-time decision-support for urgent and emergency healthcare*. PhD Thesis, University of Exeter Business School, UK.

Harper, A., Mustafee, N., and Pitt, M. (2022). Increasing situation awareness in healthcare through real-time simulation. *Journal of the Operational Research Society*, 1-11. https://doi.org/10.1080/01605682.2022.2147030.

Microsoft ODBC. (2022). Microsoft Open Database Connectivity. https://learn.microsoft.com/en-us/sql/odbc/microsoft-open-database-connectivity-odbc (last accessed 1st Dec 2022).

Mustafee, N. and Powell, J. H. (2020). Providing real-time information for urgent care. *Impact* **2021(1)**: 25-29.

Mustafee, N., Harper, A., and Onggo, B. S. (2020). Hybrid modelling and simulation (M&S): driving innovation in the theory and practice of M&S. In *Proceedings of the 2020 Winter Simulation Conference*, pp. 3140-3151. Piscataway, NJ: IEEE.

Onggo, B. S., Mustafee, N., Smart, A., Juan, A. A., and Molloy, O. (2018). Symbiotic simulation system: Hybrid systems model meets big data analytics. In *Proceedings of the 2018 Winter Simulation Conference (WSC)*, pp. 1358-1369. Piscataway, NJ: IEEE.

Powell, J. H. and Mustafee, N. (2017). Widening requirements capture with soft methods: An investigation of hybrid M&S studies in healthcare. *Journal of the Operational Research Society* **68(10)**:1211-1222.

Robinson, S. (2008). Conceptual modelling for simulation Part I: definition and requirements. *Journal of the Operational Research Society*, **59(3)**: 278-290.

Roemer, A. C. and Strassburger, S. (2019). Hybrid system modeling approach for the depiction of the energy consumption in production simulations. In *Proceedings of the 2019 Winter Simulation Conference*, pp. 1366-1377. Piscataway, NJ: IEEE.

Taylor, S. J. E. (2019). Distributed simulation: state-of-the-art and potential for operational research. *European Journal of Operational Research*, **273(1)**: 1-19.

Tolk, A., Harper, A., and Mustafee, N. (2021). Hybrid models as transdisciplinary research enablers. *European Journal of Operational Research* **291(3)**: 1075-1090.

Turner, C. J., Hutabarat, W., Oyekan, J., and Tiwari, A. (2016). Discrete event simulation and virtual reality use in industry: new opportunities and future trends. *IEEE Transactions on Human-Machine Systems* **46(6)**: 882-894.

VanDerHorn, E. and Mahadevan, S. (2021). Digital Twin: Generalisation, characterisation and implementation. *Decision Support Systems* **145**: 113524.

Viana, J., Simonsen, T. B, Faraas, H. E., Schmidt, N., Dahl, F. A. and Flo, K. (2020). Capacity and patient flow planning in post-term pregnancy outpatient clinics: A computer simulation modelling study. *BMC Health Services Research* **20(1)**:1-15.

Xu, X., Wang, J., Li, C. Z., Huang, W., and Xia, N. (2018). Schedule risk analysis of infrastructure projects: A hybrid dynamic approach. *Automation in Construction* **95**: 20-34.

AUTHOR BIOGRAPHIES

NAVONIL MUSTAFEE is a Professor of Analytics and Operations Management at the University of Exeter Business School, UK. His research focuses on Modelling & Simulation (M&S) methodologies and Hybrid Modelling and their application in healthcare, supply chain management, circular economy and resilience and adaptation due to climate change. He is a Joint Editor-in-Chief of the Journal of Simulation (UK OR Society journal) and Vice-President of Publications at The Society of Modeling and Simulation International (SCS). His email address is n.mustafee@exeter.ac.uk.

ALISON HARPER is a research fellow at the University of Exeter Medical School, in the *Peninsula Collaboration for Health Operational Research and Development*. She gained her PhD from University of Exeter Business School. Her research interests are applied health and social care research using data science and quantitative methods to model and improve services. Her email address is a.l.harper@exeter.ac.uk.

JOE VIANA is a Researcher on the MIA – Measures for Improved Availability of medicines and Vaccines research project, at BI Norwegian Business School, Oslo. He holds a Ph.D in Operational Research from University of Southampton, UK. His interests are the application of hybrid simulation to improve the operation of health systems. His email address is joe.viana@bi.no.

Proceedings of the Operational Research Society Simulation Workshop 2023 (SW23)
C. Currie and L. Rhodes-Leader, eds.
https://doi.org/10.36819/SW23.032

SIMULATION FOR EVALUATING LONG-TERM MAINTENANCE PLANS FOR COMPLEX SYSTEMS

Dr. Luke A. Rhodes-Leader

Department of Management Science
Lancaster University
l.rhodes-leader@lancaster.ac.uk

Dr. David J. Worthington

Department of Management Science
Lancaster University
d.worthington@lancaster.ac.uk

Ian Griffiths

Decision Lab Ltd
V.301 Vox Studios, 1-45 Durham Street
London, SE11 5JH
ian.griffiths@decisionlab.co.uk

Panagiotis Samartzis

Decision Lab Ltd
V.301 Vox Studios, 1-45 Durham Street
London, SE11 5JH
panagiotis.samartzis@decisionlab.co.uk

ABSTRACT

The optimisation of maintenance plans for complex systems involving many components is not an easy problem. Analytical and mathematical models are possible, but often need to make significant assumptions and are unable to look at the distribution of costs and failures. This paper discusses a project in which a discrete-time simulation model was added onto an existing optimisation model in order to go beyond just estimating the mean performance and give a better picture of the risk and variability involved with potential maintenance plans.

1 INTRODUCTION

The scheduling of maintenance for complex systems of assets involving many components can be a very challenging problem. The aim is to produce a schedule of when and how to preventively maintain each asset that balances the cost of the scheduled maintenance with the expected cost due to failure within the system. The schedule may be intended for a long planning horizon, often up to 20 years. For large organisations, there can be thousands of assets, with maintenance costs measured in tens of millions of pounds. Furthermore, many assets are dependent structures, made up of heterogeneous components combined in a multi-layered hierarchy, each with their own specific maintenance needs. The structural combinations can be linked in series, parallel or k-out-of-N subsystems. There are often limited budgets on cost and/or manpower that constrain the amount of maintenance that can be performed in a given time window. Thus, the size and structure make maintenance schedule optimisation a complex problem, and the uncertainty in the usable life of each component introduces stochasticity as well.

One approach to maintenance optimisation is using analytical approaches based on stochastic processes (such as Markov chains) and renewal theory (Scarf 1997). These can be very powerful tools, allowing the probabilistic nature of the system to be modelled, but are only tractable for systems with a few components or many identical components (de Smidt-Destombes et al. 2007). This is certainly not the case in many real-life settings. For more realistic cases, the analytical approaches can only be used to motivate heuristic policies.

An alternative is to develop optimisation models, such as Mixed Integer Linear Programmes (MILPs) or Stochastic Programmes (Zhu et al. 2021), that utilise statistical models for the lifetime of components to estimate the risk of failure. Decision Lab have developed such a model, included in their Combined Health Asset Risk Model (CHARM). CHARM is based on an earlier model called CONCEPT, developed for the Canal and River Trust, which was a runner-up in the President's medal OR60 in 2018 (Griffiths and Wilson 2019). These models are much better suited to solving resource constrained problems than the stochastic models. However, in complex hierarchical structures, converting the probabilities of failure for each component into the probability of the asset failing is non-trivial, with approximations often required. And whilst the probabilities include some aspects of stochasticity, the output of the models cannot generate information beyond the mean costs over the time-horizon.

To help overcome the issues associated with stochastic modelling and optimisation models, simulation has been used to test the solutions proposed by those models, such as the work of Barata et al. (2002). Simulation can handle both the complexity and stochastic elements, and enable a full evaluation of a policy with fewer simplifying assumptions. Much of the literature focus on estimating the expected values of cost or lifetime. This is a risk neutral approach and does not take full advantage of simulation to empirically estimate the distribution of these quantities for more informed decision making.

In this paper, we discuss a project that aimed to add simulation to the CHARM tool built by Decision Lab. This multi-fidelity modelling approach enables the plans generated by a MILP to be tested more thoroughly, providing estimates for characteristics of the total maintenance cost distribution (such as the 95th percentile) and the availability of machines that function in parallel.

One of the key aims was to build a simulation model that could be generalised to the various contexts in which Decision Lab use this type of modelling without the need for significant modification. The model needed to be able to take the input data for CHARM and the optimised maintenance schedule and construct the hierarchical structure of the assets. For this reason, the simulation was built to be as generic as possible.

The rest of this paper is organised as follows. Section 2 gives a brief review of some of the literature on asset maintenance, particularly mentioning examples which feature simulation. The proposed simulation model is described in Section 3, with a discussion on methods for sensitivity analysis in Section 4. In Section 5 we demonstrate the model and its insights on a realistic randomised asset structure, before concluding in Section 6.

2 LITERATURE REVIEW

There is a large literature on maintenance optimisation, covering many methods from Markov decision processes (Olde Keizer et al. 2016) to stochastic programming approaches (Zhu et al. 2021). Here we focus on papers where simulation has played a significant role. For a recent and thorough review of the other areas of the maintenance optimisation literature, we direct readers to the work of de Jonge and Scarf (2020).

Simulation has been applied widely within the optimal maintenance literature. Its usage largely falls into two categories. The first is to evaluate the performance of maintenance policies derived from simplified analytical models. For example, de Smidt-Destombes et al. (2007) consider multiple large k-out-of-N homogeneous systems which all share spare parts and a repair shop. The solution methodology is based on approximating queueing behaviour, and applied to a problem with systems of up to $N = 3000$ components. A discrete-event simulation model is used to evaluate the accuracy of the approximate models. Wu et al. (2016) consider the multi-component case where preventive maintenance is only carried out when a component fails. They propose an importance measure to decide which components to maintain and evaluate this policy with a simulation model. These papers use simulation as an experimental paradigm, whereas in this paper seeks to use simulation as part of the decision making process.

The second use of simulation is as the primary modelling paradigm, and several frameworks have been suggested. Barata et al. (2002) considered condition-based maintenance with constant monitoring of components. They produced a discrete-time simulation model that modelled failure due to both deterioration and random shocks, and also allowed for random improvements from preventive maintenance. Zhou et al. (2015) consider a similar problem to Wu et al. (2016). However, they include planned preventive maintenance as well, and their analysis uses a simulation model to find the optimal policy. Both Barata et al. (2002) and Zhou et al. (2015) use simulation as part of a grid search optimisation. Huseby and Natvig (2013) utilise a Discrete Event Simulation (DES) model to estimate multiple importance measures for network flow systems. Hong et al. (2014) model the degradation of components for condition-based maintenance using Gamma processes. For multiple components that cannot be treated independently, Gamma processes can be analytically intractable, so the authors simulate these systems in discrete time to find optimal inspection periods, focusing on systems in which the degradation of components is dependent. Each of these models target a specific type of maintenance policy, whether condition based or time based. Our simulation is more flexible to evaluate alternative or mixtures of policies.

Chiacchio et al. (2020) developed a framework for simulating systems in which the physical and state environment can alter the lifetime distributions of the components, as well as complex structural dependence, such as spare parts that can also deteriorate when not being used. Their hybrid simulation model uses continuous-time simulation to model the physical system evolution alongside DES mechanisms for more basic failure and repair processes. This complexity leads to significant computational costs. The authors also acknowledge that in many situations, the more complicated behaviour is not known, but that this need not be included within their modelling framework. The simulation model described here uses a similar mechanism of combining DES and time-driven simulation, but not with the aim of modelling the physically processes directly. The time steps are larger influenced by the failure processes we seek to model and reducing computational cost. Another example of a hybrid simulation model for maintenance planning is Mulyana et al. (2020), who apply a combination of Monte Carlo simulation and Systems Dynamics simulation for periodic preventative maintenance for the packing department of a flour mill.

Beyond the maintenance planning discussed in this paper, simulation can also be used directly within the joint optimisation of maintenance and other considerations. Wakiru et al. (2019) use a DES model and simulation optimisation to find good policies involving maintenance and management of the spare-parts inventory. Zahedi-Hosseini et al. (2017) solve a similar problem, testing out various inventory control policies. In both cases, the optimisation is performed over a few variables using a commercial heuristic. Bouslah et al. (2018) focus on a two machine manufacturing line, considering maintenance, production and quality as an integrated problem, with dependent reliability and quality deterioration. The model combines continuous-time simulation and DES. The simulation optimisation uses a Response-Surface Methodology approach. Assid et al. (2015) also look at the joint preventive maintenance and production problem in manufacturing, but in this case examining a single machine that can create two products. The simulation is a hybrid between continuous-time simulation and DES. Similarly, the optimisation is formulated with policy parameters as the decision variables and is solved using a one-step Response Surface Methodology.

3 SIMULATION MODELLING STRUCTURE

Consider an asset constructed from a collection of components. Groups of components form subsystems, which can operate in series (one component failure breaks the subsystem), in parallel (all components must fail to break the subsystem) or in a k-out-of-N structure ($N-k+1$ component failures break the subsystem). Groups of subsystems form Level 1 items in similar structures. (More complex systems may have more levels.) Groups of Level 1 items make up the asset. This structure forms a hierarchical Reliability Block Diagram dictating the reliability behaviour of the asset.

As an example, consider an asset system called Emergency Vehicles. Figure 1 shows part of the reliability block diagram. This asset system is made up of Level 1 items such as Fire Engines and Pumping Systems. In turn, the Fire Engines and Pumping Systems (Level 1 items) are decomposed further into subsystems or components. For example, each Pumping System instance includes two Hose components, which are connected in parallel. In this case, if one Hose breaks down, the Pumping System can still function. Each Fire Engine contains one Camshaft and one Exhaust Manifold as its components and they are connected in series. If either the Camshaft or the Exhaust Manifold fails, the whole Fire Engine they belong to becomes unavailable.

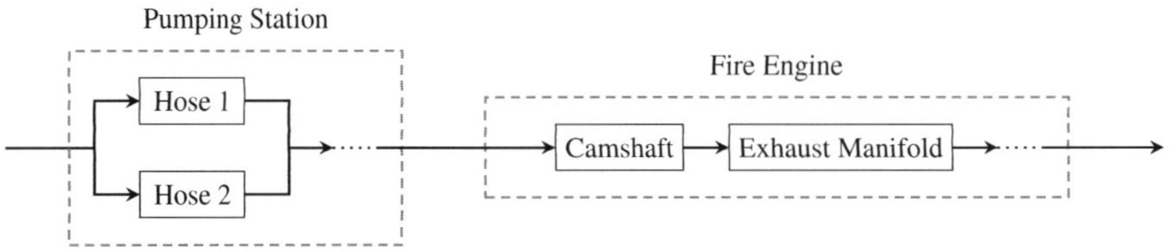

Figure 1: Part of the reliability block diagram of the Emergency Vehicles asset system.

The aim of the model is to simulate the evolution of the asset including its (planned and unplanned) maintenance over the time horizon. Overall, it is the reliability of the asset and the cost of the proposed maintenance interventions schedule that we wish to estimate.

3.1 Model Content

The model used here is a discrete-time simulation written in Python, and utilises the agent-based modelling package MESA (Kazil et al. 2020). We simulate at the component level over a fixed time horizon. Each component is modelled using the 'agent' class of MESA, though little interaction occurs between the components. At each time-step, the component's state (consisting of its age and/or health score, its status (functioning or failed), and accumulated planned and unplanned maintenance costs) is updated. Figure 2 shows the logical steps of each agent update. The information on which components have failed and the Reliability Block Diagram structure is then used to calculate whether the system has failed and any associated costs.

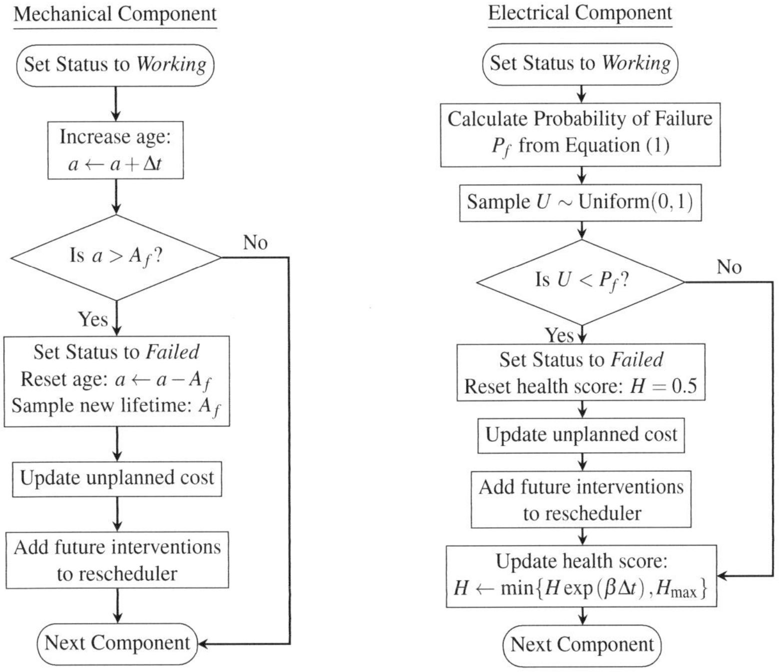

Figure 2: Update procedure for the two types of component. Here Δt is the time-step of the model.

The mechanism behind the failure of each component depends on its type. For some components, their lifetimes can be easily modelled using a time-to-failure distribution (such as a Weibull or Log-Normal distribution), in which case we partially utilise Discrete-Event Simulation ideas, mirroring the approach of Chiacchio et al. (2020). Given the current age of the component, A_0, we sample a failure age A_f from the lifetime distribution (conditional on being greater than A_0). At each time-step, the age is updated, and if it exceeds A_f, the component fails. For other components, where the probability of failure is often linked to a condition or health score, this approach is not possible. For example, the U.K. industry standard for electrical components used in power transmission is the CNAIM framework (Booth et al. 2017), which models the annual probability of failure as:

$$\Pr\{\text{Failure in next year}|H\} = K\left(1 + C\max\{H,4\} + \frac{C^2}{2!}(\max\{H,4\})^2 + \frac{C^3}{3!}(\max\{H,4\})^3\right) \quad (1)$$

where K and C are component-dependent constants and H is the current health score of the component expressed as a function of the component's age a:

$$H(a) = \min\{H_0\exp(\beta a), H_{\max}\}, \quad (2)$$

with $H_0 = 0.5$, $H_{\max} = 10$ and β a component dependent ageing rate. Thus, components following the CNAIM framework will fail in the current time-step with the probability given in equation (1). Using a discrete-time model enables both types of component to be modelled in the same simulation.

When a component fails it is replaced by a new component and, if necessary, a new failure age A_f is generated. Once all components have been updated, the simulation uses the Reliability Block Diagram to check the status of the asset. A significant assumption of the time-step approach is that all failures are assumed to be concurrent. This can be mitigated with smaller step-sizes, but as it is assumed maintenance must be completed within a time-step, there is a lower limit on the time-step.

The key decision input to the simulation model is the maintenance intervention schedule. This describes when each intervention will be performed, and which components are affected by the intervention. An intervention will reduce the age or health score of each of the components it impacts. We do not assume the intervention is perfect. After the intervention, a new failure-age is then generated (if required). As a simplification, we perform interventions before the ages and health condition of components are updated. This effectively assumes that all the interventions take place at the beginning of the time-step, rather than distributed through the time window.

An important element in the simulation model is that components do fail. In this case, the future planned interventions for that component must be rescheduled. In reality, a new maintenance plan would be developed each year using the optimisation model. As this can take hours to run, it is impractical to achieve within each replication of the simulation model. Thus, we allow a user-specified heuristic rule for rescheduling interventions. We describe the default heuristic here. Suppose that a component failed during a year. At the end of that year, any future planned interventions will be moved to a year after which all impacted components have exceeded their specified age or health score threshold.

3.2 Model Outputs

The simulation model outputs a data set stating the cumulative costs incurred over the planning horizon and whether each component, Level 1 item or the asset failed during each time-step. Thus, by performing many replications, the simulation can be used to estimate the time-dependent reliability of the asset, Level 1 items and components (as measured by the probability of failure), as well as quantifying the uncertainty of the future total cost.

The total cost of maintenance over the time horizon is the sum of the planned intervention costs, the costs due to component failures and repairs, the disruption cost of downtime should the whole asset fail and the cost of lost usable life due to early replacement. A key advantage of using the simulation model in addition to the optimisation model is that we can analyse various possible future costs rather than simply a single prediction, enabling percentiles and prediction intervals to be generated.

Whilst the reliability of the asset is important, we can also look at availability of Level 1 items. This is defined as the number of a particular type of Level 1 item operating in each time-step. If these items work in parallel, a few failures may not lead to asset level disruption, and thus the optimisation model may not penalise this. However, it may impact the levels of productivity that can be achieved. This information is much harder to quantify from the optimisation model output, but can be naturally quantified by running the simulation many times and looking at the detailed output.

4 SENSITIVITY ANALYSIS

One of the points made in the literature review by de Jonge and Scarf (2020) is that relatively little work in the field of maintenance optimisation accounts for uncertainty in the lifetime distributions of components. Getting reliable data on the lifetime distributions is often difficult. For this reason, the CHARM tool allows engineering judgement to be used when fitting distributions. The consequence of this is that uncertainty in the parameter estimates cannot be quantified, making it very important to study the sensitivity of the performance to errors in the lifetime distribution estimates, as well as other parameters such as the cost of failure and the downtime caused by a failure.

As a system can involve hundreds of components, a full sensitivity analysis could be very computationally expensive. For this reason, we took a factor screening approach, aiming to identify the most influential input parameters on the overall cost. The value of this is that if the proposed maintenance schedule is found to be particularly sensitive to certain parameters, the optimisation could be re-run using conservative estimates for only those particular parameters, rather than considering the worst case in all parameters (which could be very conservative).

We applied Improved Controlled Sequential Bifurcation (CSB-X), proposed by Wan et al. (2010). The key idea behind CSB-X is that, if the sign of the effect is known (if it increases or decreases the total cost), parameters can be grouped and then screened together. If the total effect of the group is unimportant, each parameter in the group can be declared unimportant. Otherwise, the group can be divided in two, and each tested again. This greatly reduces the amount of time required to perform the factor screening, and CSB-X does this in such as way as to control the misclassification error rate.

In our application, many of the input parameters have an intuitive sign. For example, increasing the cost of failure is likely to increase the overall cost, and increasing the scale parameter of a Weibull lifetime distribution will make failure before an intervention less likely, decreasing the total cost. For these parameters, CSB-X can be highly effective. The speed of CSB-X depends on how well the factors are ordered; if all the unimportant parameters are grouped together, it is much easier to screen them out. But if this was known beforehand, it would be unnecessary to perform the screening. To help, we first run the simulation 1000 times and split the total maintenance costs across the components. We then list parameters in order of decreasing cost contribution of the associated component. This heuristic appears to speed up the approach considerably.

The impact of increasing the shape parameter of lifetime distributions is much more difficult to determine. So CSB-X could not be used for this purpose. An alternative that does not require the sign of the effect to be known is the hybrid methodology proposed by Shen et al. (2010), which includes CSB-X. We did implement this method but found that it came with a considerable computational cost.

5 APPLICATION

The asset system at hand is a Waste Management (WM) system of an industrial complex, which follows the structure outlined in Section 3. The overall structure of the WM system is shown in Figure 3. The Level 0 WM system is broken down into four Level 1 asset systems, which are connected in series. On the top of the diagram, there is one Mobile Cleaning System (MCS), while instances of Portable Tank Systems (PTS) and Static Cleaning Systems (SCS) are in parallel with each other and instances of Super Portable Tank Systems (SPTS) are in series. The MCS item consists of 24 components, the PTS items consist of 27 components, whilst both SCS and SPTS items are made up of 19 components each. The individual Level 1 items are given a label, e.g. SCS 34 refers to the fourth SCS item. All Level 2 components of the four Level 1 asset systems discussed are in Series with each other. It should be noted that Figure 3 depicts assets of Levels 0 and 1, however the same idea generalises for deeper Levels. This WM asset system, being used in an industrial complex is quite a complicated hierarchical structure with 476 individual components being aggregated into their respective parents, something that outlines the computational cost of the problem at hand, as well as the scalability of the model created.

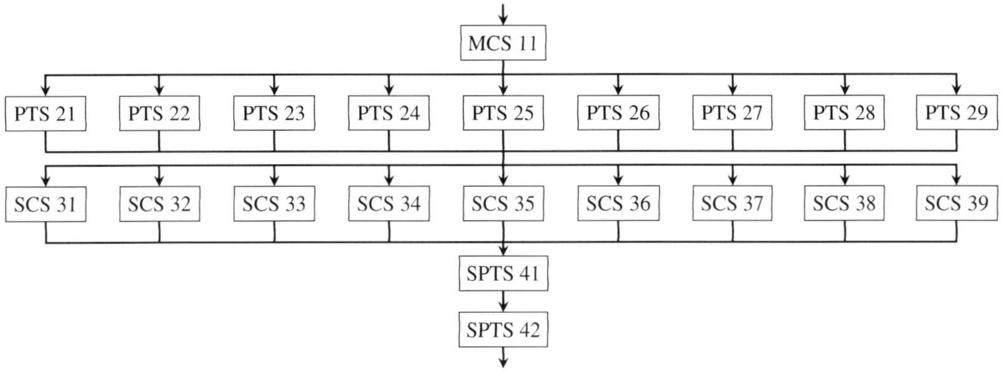

Figure 3: A Reliability Block Diagram of the Waste Management hierarchy.

The CHARM tool was used to model the time to failure for all components and then to optimise a maintenance schedule between 2021 and 2031, accounting for budget constraints for each year. The simulation used the same input data as the optimisation model as well as the resulting maintenance plan to estimate the cost distribution, the reliability of the system and the availability of Level 1 items. The results presented are based on 10,000 replications of the simulation with a time-step of 6 months. All cost units are scaled to be between 0 and 1.

The annual total and unplanned maintenance costs are shown in Figure 4. Whilst the highest annual costs are incurred upfront through a lot of planned maintenance, the variance in the costs increases considerably over time, caused largely by increases in the unplanned costs due to component failures. The maintenance plan reduces failure costs in the first half of the period, but is less effective later on.

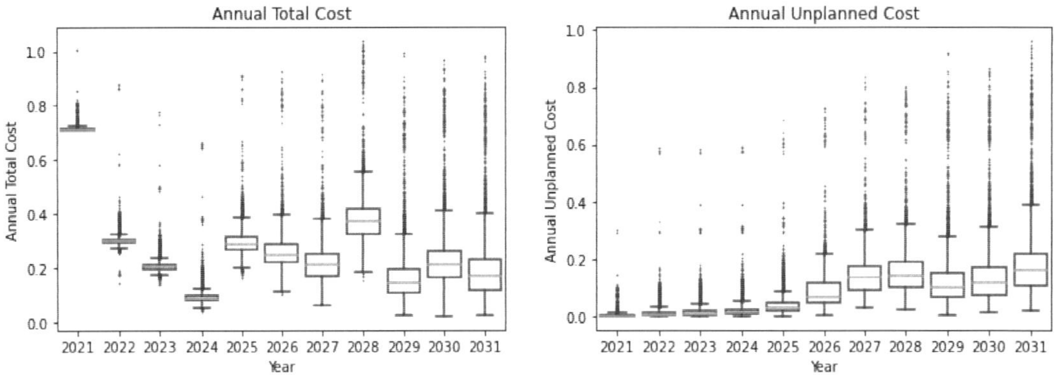

Figure 4: Box plots of the annual total and unplanned costs over the time horizon.

The stacked area plots in Figure 5 demonstrates how the annual costs breakdown into both cost types and Level 1 item groups. This can be done by different percentiles of the cost distribution, Figure 4 shows the 50th and 95th percentiles. Unsurprisingly, the planned costs remain fairly constant and it is the unplanned maintenance costs that dominate the latter years in the higher percentile case, with the PTS and SCS groups contributing significantly.

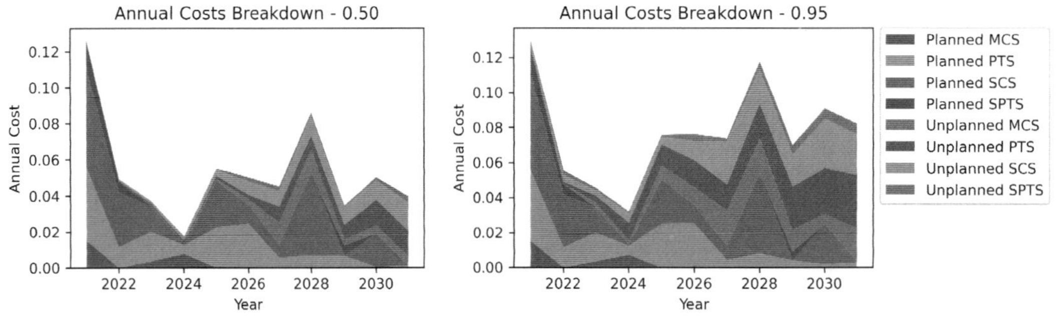

Figure 5: How the costs breakdown at different stages of the cost distribution.

The simulation is able to estimate the availability of groups of items over time. Whilst the PTS and SCS groups act in parallel (and are modelled as such in the optimisation model), lower availability leads to lower productivity. Hence, there are targets for the availability; seven PTS items and six SCS items are required to be available in each time period. Figure 6 indicates the probability of meeting these targets (green), having at least half of the target (amber) or falling below that (red). These plots highlighted a significant issue that is not obvious from the output of the optimisation model: that as time goes on the number of PTS items often fails to meet the target of 7. This indicates that PTS items need to be maintained more often, so that failures are reduced and we are able to meet the target.

Analysing the data further, Figure 7 demonstrates the cumulative probability of failure for seven items for PTS 21 (the other eight PTS items have similar plots). For several components, such as the panel hose and electrical supply, the cumulative probability of failure climbs very quickly between 2026 and 2030. This sort of result from the simulation model could then be used to feedback information into the optimisation model, as we could potentially change the maintenance thresholds of these components

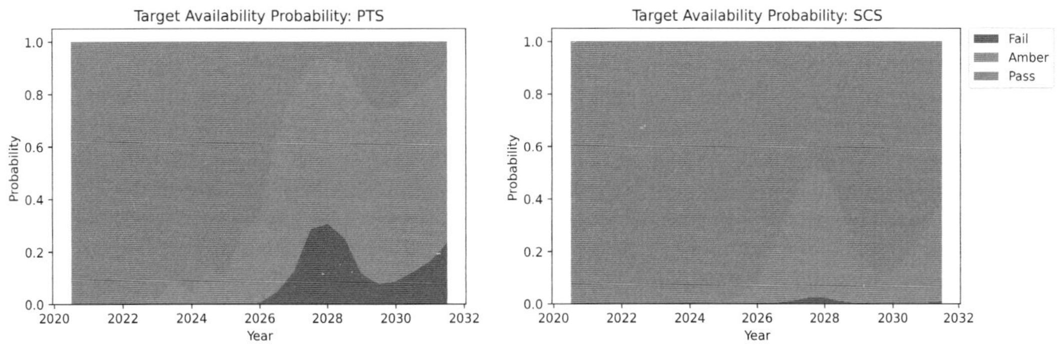

Figure 6: The probability of meeting the target availability for the PTS and SCS groups over the time horizon.

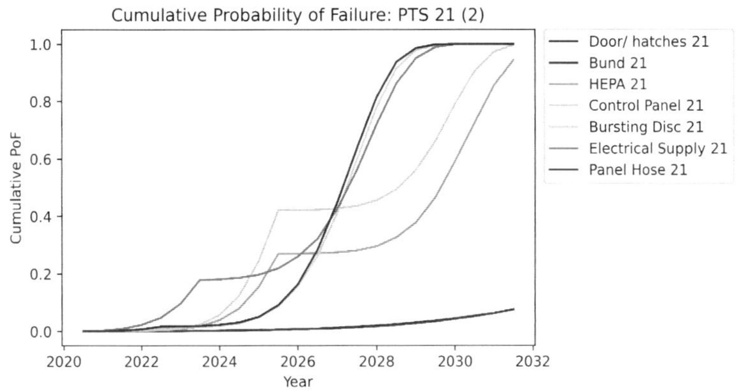

Figure 7: Cumulative probability of failure for the most vulnerable components of PTS 21.

to be more conservative. The likely consequence would be a greater prioritisation for these components in selecting maintenance and thus an improvement in the reliability of the Level 1 PTS group, which is going to drive down the unplanned costs of the asset type discussed earlier.

5.1 Application of Sensitivity Analysis

We applied the CSB-X screening procedure to the scale, cost and disrupted days parameters for each component type and the initial condition parameter of each individual component (constituting a total of 743 parameters). The aim here is to draw out parameters that have the largest influence on the expected total cost. This will allow us to see which parameters have the greatest requirement for additional data.

The thresholds for defining important and critical factors were 0.1% and 0.2% of the mean total cost, respectively. The experimental regions for scale parameters had problem specific definitions, whilst the others were varied by plus or minus 10%.

Despite the very large number of factors, we were able to identify the most influential input parameters within 30 minutes using 2 processors on a laptop and controlling the error rate at 5%. On average, the heuristic mentioned in Section 4 reduced the computational cost by over 30% (based on 50 replications of each procedure). This demonstrates the utility of CSB-X for sensitivity analysis when there are many input parameters. Ten input factors were identified, relating to eight component types. The most influential parameters were the cost and scale parameters for the bund component of the MCS and the flexible pipework of the SCS.

6 CONCLUSION

This paper has presented a discrete-time simulation model for improving the understanding of the stochastic nature of future maintenance costs and reliability of complex systems of assets under a particular optimised maintenance schedule. The simulation model is flexible and able to adapt to many systems based on the input data provided. The discrete-time nature gives the model the ability to model failure processes that are more easily modelled in either continuous or discrete time.

The results from our application demonstrate the advantages of using a simulation model alongside an optimisation model. As well as giving a clearer picture of the cost distribution and risk associated with any proposed maintenance plan, the simulation is able to highlight problematic behaviour of the optimised maintenance schedule, particularly the fact that several items that function in parallel can fail at the same time.

In further work, we anticipate that this framework will be able to form a feedback loop with the optimisation to improve solutions. The optimisation model is based on certain age or health thresholds that determine a desired window for maintenance. If the simulation model highlights undesirable behaviour, such as low availability of Level 1 items reducing productivity in certain periods, the detailed simulation output could help identify which of these thresholds should be made more conservative (i.e. shifting the window earlier) to improve the solutions the optimisation model generates.

We could also develop a more direct usage of simulation within the optimisation procedure. The applications of simulation optimisation within the maintenance optimisation literature are limited to either exhaustive grid searches (Barata et al. 2002) or low dimensional search problems (Zahedi-Hosseini et al. 2017). As large complex systems with budget constraints create high dimensional combinatorial optimisation problems, many of the cutting-edge simulation optimisation algorithms may not perform well. This could further motivate a multi-fidelity modelling approach to the optimisation.

ACKNOWLEDGMENTS

This project was funded through an EPSRC Impact Acceleration Account project, with financial contributions from Decision Lab Ltd.

REFERENCES

Assid, M., A. Gharbi, and A. Hajji. 2015. "Joint production, setup and preventive maintenance policies of unreliable two-product manufacturing systems". *International Journal of Production Research* 53 (15): 4668–4683.

Barata, J., C. Guedes Soares, M. Marseguerra, and E. Zio. 2002. "Simulation modelling of repairable multi-component deteriorating systems for 'on condition' maintenance optimisation". *Reliability Engineering and System Safety* 76 (3): 255–264.

Booth, J., B. Wells, D. Seeds, M. Black, G. Howarth, M. Nicholson, G. Boyd, P. Sherwood, J. Hurley, R. Sharma, L. Johnston, J. Smart, I. Butler, R. Friel, R. Wakelen, P. Mann, and D. Tighe. 2017. "DNO Common Network Asset Indices Methodology Version 1.1". Technical Report January, Ofgem. https://www.ofgem.gov.uk/system/files/docs/2017/05/dno_common_network_asset_indices_methodology_v1.1.pdf.

Bouslah, B., A. Gharbi, and R. Pellerin. 2018. "Joint production, quality and maintenance control of a two-machine line subject to operation-dependent and quality-dependent failures". *International Journal of Production Economics* 195:210–226.

Chiacchio, F., A. Iacono, L. Compagno, and D. D'Urso. 2020. "A general framework for dependability modelling coupling discrete-event and time-driven simulation". *Reliability Engineering and System Safety* 199:106904.

de Jonge, B., and P. A. Scarf. 2020. "A review on maintenance optimization". *European Journal of Operational Research* 285 (3): 805–824.

de Smidt-Destombes, K. S., M. C. van der Heijden, and A. van Harten. 2007. "Availability of k-out-of-N systems under block replacement sharing limited spares and repair capacity". *International Journal of Production Economics* 107 (2): 404–421.

Griffiths, I., and S. Wilson. 2019. "Improving the waterways of England and Wales". *IMPACT* 9 (Spring 2019): 5–9.

Hong, H. P., W. Zhou, S. Zhang, and W. Ye. 2014. "Optimal condition-based maintenance decisions for systems with dependent stochastic degradation of components". *Reliability Engineering and System Safety* 121:276–288.

Huseby, A. B., and B. Natvig. 2013. "Discrete event simulation methods applied to advanced importance measures of repairable components in multistate network flow systems". *Reliability Engineering and System Safety* 119:186–198.

Kazil, J., D. Masad, and A. Crooks. 2020. "Utilizing python for agent-based modeling: The Mesa framework". In *Social, Cultural, and Behavioral Modeling*, edited by R. Thomson, H. Bisgin, C. Dancy, A. Hyder, and M. Hussain, 308–317. Cham: Springer International Publishing.

Mulyana, I. J., I. Gunawan, Y. V. Angelia, and D. Trihastuti. 2020. "A hybrid simulation study to determine an optimal maintenance strategy". *Jurnal Optimasi Sistem Industri* 19 (2): 91–100.

Olde Keizer, M. C. A., R. H. Teunter, and J. Veldman. 2016. "Clustering condition-based maintenance for systems with redundancy and economic dependencies". *European Journal of Operational Research* 251 (2): 531–540.

Scarf, P. A. 1997. "On the application of mathematical models in maintenance". *European Journal of Operational Research* 99 (3): 493–506.

Shen, H., H. Wan, and S. M. Sanchez. 2010. "A hybrid method for simulation factor screening". *Naval Research Logistics* 57:45–57.

Wakiru, J. M., L. Pintelon, P. N. Muchiri, and P. K. Chemweno. 2019. "A simulation-based optimization approach evaluating maintenance and spare parts demand interaction effects". *International Journal of Production Economics* 208:329–342.

Wan, H., B. E. Ankenman, and B. L. Nelson. 2010. "Improving the efficiency and efficacy of controlled sequential bifurcation for simulation factor screening". *INFORMS Journal on Computing* 22 (3): 482–492.

Wu, S., Y. Chen, Q. Wu, and Z. Wang. 2016. "Linking component importance to optimisation of preventive maintenance policy". *Reliability Engineering and System Safety* 146:26–32.

Zahedi-Hosseini, F., P. Scarf, and A. Syntetos. 2017. "Joint optimisation of inspection maintenance and spare parts provisioning: a comparative study of inventory policies using simulation and survey data". *Reliability Engineering and System Safety* 168:306–316.

Zhou, X., K. Huang, L. Xi, and J. Lee. 2015. "Preventive maintenance modeling for multi-component systems with considering stochastic failures and disassembly sequence". *Reliability Engineering and System Safety* 142:231–237.

Zhu, Z., Y. Xiang, and B. Zeng. 2021. "Multicomponent maintenance optimization: A stochastic programming approach". *INFORMS Journal on Computing* 33 (3): 898–914.

AUTHOR BIOGRAPHIES

LUKE A. RHODES-LEADER completed his PhD in Statistics and Operational Research at Lancaster University, where he is now a Lecturer in Management Science. His research interests are in analysis of simulation models, and their use alongside other modelling paradigms.

DAVID J. WORTHINGTON is a senior lecturer in Operational Research in the Department of Management Science in Lancaster University Management School. He researches the modelling and management of time-dependent queueing systems, and applications of Management Science in health-care. These research interests often coincide.

IAN GRIFFITHS is the Chief Technology Officer of Decision Lab where he oversees technical delivery within the business and its strategic development. He has a career in science and technology spanning 25 years and multiple sectors, including defence and security, water and infrastructure.

PANAGIOTIS SAMARTZIS is a Senior Consultant in Decision Lab, having completed his MSc studies in Operational Research at Lancaster University. He is an expert in Optimisation, Mathematical Modelling and Deep Learning, which he practices in a variety of sectors such as water, aerospace and asset management.

POSTER ABSTRACTS

URBAN DISTRIBUTION IN VIENNA WITH HUBS USING AGENT-BASED SIMULATION

Aitor Ballano, Anas Al-Rahamneh, Adrian Serrano-Hernandez, and Javier Faulin

Institute of Smart Cities. Public University of Navarre

With the rise of e-commerce and door-to-door sales, last-mile deliveries are gaining more and more importance. As a result, last-mile distribution has become one of the most sensitive logistics processes due to its uniqueness, difficulties in meeting schedules, and high costs. Therefore, this work explores the use of urban consolidation centers to ease these last-mile difficulties. For that purpose, a hub in the city center of Vienna has been selected to deliver up to 150 clients disseminated by the city. This suitability is assessed by defining convenient simulation settings in order to replicate parcel demands in the city. Experiments are based in different hub-based fleets (traditional internal combustion vehicles or electric cargo bikes), demand patterns, and delivery frequency strategies by means of a biased randomization vehicle routing optimization heuristic. Results quantify the effects of having an urban consolidation center and highlight the use of electric cargo bikes for the last-mile distribution.

MODELLING THE RECOVERY OF AN ELECTIVE CARE ORTHOPAEDIC PATHWAY IMPACTED BY COVID-19

Matthew Howells, Paul Harper, Daniel Gartner and Geraint Palmer

Cardiff University

Following the disruption caused by COVID-19, the NHS is struggling with a backlog of patients and growing elective waiting lists. We present work in collaboration with Orthopaedic managers and surgeons working in the Cardiff and Vale University Health Board. Orthopaedics is especially under pressure with high demand for elective surgery driven by both the backlog and an ageing population. We present the results of our systematic review into operational research methods applied to orthopaedic settings and treatments. We further have developed a discrete-event simulation that models patient flows through a holistic orthopaedic surgical pathway, enabling hospital staff to make more informed decisions on demand and capacity planning. Furthermore, we discuss plans to combine our discrete-event simulation with optimisation approaches to evaluate robustness of scheduling algorithms for alleviating backlogs, and to incorporate into the model patient deterioration whilst waiting for surgery.

MODELLING PATIENT FLOW ALONG CANCER PATHWAYS USING DISCRETE EVENT SIMULATION

Amalia Gjerloev, Sonya Crowe, Christina Pagel, Yogini Jani, and Luca Grieco

University College of London

With the Covid-19 pandemic, healthcare systems have seen a huge influx of patients, a strain on hospital resources and increased pressure to operate efficiently in order to save patient lives. Cancer services have been particularly impacted and face long waiting times following a dramatic fall in referrals at the start of the pandemic. In order to improve resource allocation, scheduling and understanding of the pandemic's impact on patient care, we developed a flexible model of the cancer pathway. The complexity of cancer pathways and lack of available data demand an adaptive model. We present a framework for analysing cancer pathways that employs Discrete Event Simulation and heavily involves clinical and operational managers. A case study is presented, in which we use our simulation framework to inform operational decisions for lung cancer services at a London-based hospital. Our model has been parameterised and validated by clinicians.

CONTINUOUS IMPROVEMENTS TO AN ANALYTICAL DELIVERY FRAMEWORK AT SELLAFIELD LTD

Paul Hinsley

Sellafield Ltd

Sellafield Advanced Business Analytics (SABA) is a decision support capability within Sellafield Ltd that provides a portfolio of analytical tools and solutions across the business. The foundation of our multidisciplinary team is in operational research with particular focus on the development and analysis of simulation models to assess and improve plant performance. To ensure continuous improvement in our process and quality of analytical findings, and to accommodate our broadening capability, we recently reviewed our project delivery framework and have begun to implement changes - this includes greater focus on defining the business problem during initial engagement with customers to ensure that a fit-for-purpose analytical solution is proposed. This poster presentation provides an overview of the changes we have been making to our framework to improve the quality of our capability.

ROBUST MODELS FOR SIMULATION ANALYTICS

Drupad Parmar

Lancaster University

Simulation models are increasingly being integrated with machine learning methods. The application of these methods to dynamic sample path or system state data is often referred to as simulation analytics. This produces models that are then able to facilitate real-time prediction and assist with system control. If the input distributions that drive the simulation are sampled randomly, then the machine learning model may be trained over a narrow input space. We investigate the use of a designed experiment for input distribution sampling, to ensure the data on which the machine learning model is trained covers a larger proportion of the input space. The motivation behind this is to produce a robust machine learning model, which performs well over a greater amount of the input space.

USE OF HYBRID SIMULATION (HS) WITH ARTIFICIAL INTELLIGENCE (AI) TO CAPTURE DYNAMISM IN DIGITAL TWINS (DT) AND INTELLIGENT PRODUCTION ENVIRONMENTS (I4)

Lambros Viennas

University of Surrey

Modern manufacturing and other economic and social systems are complex structures. Research and literature show that the different types of complexity identified in these systems are strongly related and affect the dynamic capabilities of the system. This study systematically reviews the literature to investigate the utilisation of Hybrid Modelling & Simulation (M&S) in conjunction with the implementation of Digital Twins, AI/ML, and other enabling technologies in Intelligent Production Systems. The study's objective is to identify cases where the modelling practice using Hybrid M&S and Digital Twin in coexistence acknowledges and includes aspects of organisational dynamism in the implemented solution (conceptual or empirical), contributing to enhancing the dynamic capabilities of the organisation. The study concludes with realisations made and suggestions for further research.

INVESTIGATING THE SEASONAL EFFECTS ON PATIENT ATTRIBUTES IN A&E TO IMPROVE THE ACCURACY OF A SYMBIOTIC SIMULATION MODEL

Alex Heib, Christine Currie, Stephan Onggo, Honora Smith and James Kerr

University of Southampton, Hampshire Hospitals NHS Foundation Trust

The objective of symbiotic simulation is to aid short-term decision making by initialising the simulation to represent the current state of the system, and simulating only into the near future. Usually, weak seasonal effects can be safely ignored in a simulation, as the goal is to evaluate steady-state performance.

However, these seasonal effects should be included in a symbiotic simulation, because we want to accurately model how the system will develop over a short time period. In this work, we use patient data from an A&E department to investigate how seasonality can affect the probability distributions of attributes associated with patients arriving in the department. Generalised linear models that included crossed factors were applied to data, and the model that minimised the Bayesian Information Criterion was selected. We found that the time of day strongly affects the distributions of patient attributes, with the time of year having a lesser effect.

SHAPELETS FOR SIMULATION: DYNAMIC TRAJECTORY ANALYSIS

Graham Laidler

Lancaster University

The dynamic sample path of discrete-event simulation contains a wealth of insight into the working behaviour and characteristics of a system. However, traditional output analysis often overlooks this information, with dynamic behaviour being averaged across both time and replications. Here, we look at the individual trajectories of typical simulation outputs such as the throughput or the number in system. Our aim is to uncover insights and draw comparisons among competing system designs based on their dynamic performance. For this, we consider the use of shapelets, which represent locally characteristic patterns, and have emerged as a novel method for non-parametric time series classification. We demonstrate their application to continuous-time simulation trajectories, and their ability to discover and discriminate the dynamic behaviours of different systems.

HYBRID MODELLING AND SIMULATION FRAMEWORK FOR MODEL COMBINATION IN HEALTH AND SOCIAL CARE

Eyup Kar, Masoud Fakhimi and Christopher Turner

University of Surrey

The use of Hybrid Simulation (HS) for modelling of healthcare systems increased as problems have become more complex and multidimensional, with a particular focus on healthcare systems. Such complexities make it challenging for single simulation models to provide the right support for decision-making. This research categorises the Modelling and Simulation (M&S) techniques employed and analyses the application type, software packages, trends, opportunities, and challenges of HS in healthcare. Current limitations of the literature and opportunities for future research are discussed. Findings show that combining discrete and continuous M&S methods is the most common HS in healthcare. However, the popularity of combining other Operational Research and Data Science techniques with M&S is on the rise. This study also discusses model combinations for each existing HS design and proposes a framework to show information transformation standards between the models.

THE DEVELOPMENT OF A HYBRID SIMULATION TO ENHANCE PORT OF DOVER'S COMPETITIVENESS

Siti Fariya, Kathy Kotiadis, Jesse O'Hanley, Paola Scaparra, Timothy Van Vugt, and Christian Pryce

Kent Business School, Port of Dover

The Port of Dover is a critical piece of UK infrastructure and one of the world's busiest ports, handling £144bn worth of trade annually, equivalent to 33% of EU-UK trade. However, the Port faces various challenges, including increased border processing times causing in-port congestion which can overflow onto the strategic road network. This causes disruption to both port-bound traffic and the local community using the surrounding road network, leading to significantly longer journey times. To address this issue and improve port operations, detailed traffic modelling is required. Our research aims are to develop a range of discrete-event and hybrid simulation models to more accurately forecast traffic congestion and inform investments to manage traffic flows. Here, we focus on our initial model of inbound exit from the port. This study aims to evaluate the potential impact of proposed operational and infrastructure changes on traffic flows, congestion, travel time, and fuel consumption.

EVERYTHING YOU ALWAYS WANTED TO KNOW ABOUT DR SIEBERS (ACADEMIC DETAILS ONLY :) - SECOND EDITION

Peer-Olaf Siebers

University of Nottingham

After more than 10 years of publishing the first edition of my research agenda I felt that it's time for an update. My current research can be subsumed under the umbrella of "collaboratively creating artificial labs for better understanding current and future human and mixed human/robot societies". I am a strong advocate of agent-based simulation, but open to hybrid solutions. My poster provides references to my previous and ongoing work, identifies current gaps, and offers ideas for future collaborations.

My research aligns towards standardisation of methods and towards considering future scenarios of human/robot interactions in an operational and service oriented context. Wherever possible, I aim to introduce techniques from computer science, and in particular software engineering, to come up with a more structured and transparent approach to simulation modelling. I also like to embed simulation into other analysis tools to enable these to consider uncertainties in a transparent way.

TO WHAT EXTENT CAN SIMULATION OPTIMISATION BE USED IN WILDLIFE CONSERVATION?

Shengjie Zhou, David Worthington, Luke Rhodes-Leader and Richard Williams

Lancaster University

Reserve design is an important problem in wildlife conservation. The underlying aim of reserve design is to guarantee species' survival whilst minimising conservation costs. In the case of the Grey Wolf, their survival is based on stochastic dispersal, breeding, and death processes. And the conservation cost is related to the number of areas allocated for wolves to live in. To estimate the survival probability, we have built a model that simulates the whole population based on individual wolves in discrete time.

The overall aim of the research is to investigate the extent to which Simulation Optimisation can be used to tackle the wolf conservation problem. The results of experiments using the Chance Constrained Selection of the Best (CCSB) algorithm will be presented and discussed on the poster.

SUBSET SELECTION FOR NOISY BLACKBOX OPTIMIZATION, USING GAUSSIAN PROCESSES

Sasan Amini and Inneke Van Nieuwenhuyse

Data Science Institute, Hasselt University

At the end of any simulation optimization procedure, the algorithm needs to identify the optimal solution(s). The current literature on subset selection/ranking and selection assumes that the analyst only has information on the solutions that were actually simulated during the search process; yet, in recent years, metamodel-based simulation optimization approaches (in particular, Bayesian optimization using Gaussian processes) have become increasingly popular. These metamodels provide predictions for the outcomes across the whole search space; strikingly, though, current algorithms tend to simply identify the solution with the best prediction as the final optimum, neglecting the (intrinsic and extrinsic) uncertainty inherent in the model. Our research highlights the need for an identification approach that accounts for this uncertainty. We propose a subset selection approach that uses the Gaussian Process information in view of returning all solutions that are estimated to be non-inferior in the (discretized) search space.